BIOLOGICAL ASYMMETRY AND HANDEDNESS

The Ciba Foundation is an international scientific and educational charity. It was established in 1947 by the Swiss chemical and pharmaceutical company of CIBA Limited — now CIBA-GEIGY Limited. The Foundation operates independently in London under English trust law.

The Ciba Foundation exists to promote international cooperation in biological, medical and chemical research. It organizes about eight international multidisciplinary symposia each year on topics that seem ready for discussion by a small group of research workers. The papers and discussions are published in the Ciba Foundation symposium series. The Foundation also holds many shorter meetings (not published), organized by the Foundation itself or by outside scientific organizations. The staff always welcome suggestions for future meetings.

The Foundation's house at 41 Portland Place, London W1N 4BN, provides facilities for meetings of all kinds. Its Media Resource Service supplies information to journalists on all scientific and technological topics. The library, open five days a week to any graduate in science or medicine, also provides information on scientific meetings throughout the world and answers general enquiries on biomedical and chemical subjects. Scientists from any part of the world may stay in the house during working visits to London.

Ciba Foundation Symposium 162

BIOLOGICAL ASYMMETRY AND HANDEDNESS

A Wiley-Interscience Publication

1991

JOHN WILEY & SONS

Chichester · New York · Brisbane · Toronto · Singapore

©Ciba Foundation 1991

Published in 1991 by John Wiley & Sons Ltd.
Baffins Lane, Chichester
West Sussex PO19 1UD, England

Other Wiley Editorial Offices

John Wiley & Sons, Inc., 605 Third Avenue,
New York, NY 10158-0012, USA

Jacaranda Wiley Ltd, G.P.O. Box 859, Brisbane,
Queensland 4001, Australia

John Wiley & Sons (Canada) Ltd, 5353 Dundas Road West, Fourth Floor,
Etobicoke, Ontario M9B 6H8, Canada

John Wiley & Sons (SEA) Pte Ltd, 37 Jalan Pemimpin 05-04,
Block B, Union Industrial Building, Singapore 2057

Suggested series entry for library catalogues:
Ciba Foundation Symposia

Ciba Foundation Symposium 162
ix + 327 pages, 81 figures, 14 tables

Library of Congress Cataloging-in-Publication Data
Biological asymmetry and handedness.
      p.   cm.—(Ciba Foundation symposium; 162)
    Editors, Gregory R. Bock and Joan Marsh
    Papers presented at the Symposium on Biological Handedness and
Symmetry, held at the Ciba Foundation, London, 20–22 Feb. 1991.
    "A Wiley–Interscience publication."
    Includes bibliographical references and indexes.
    ISBN 0 471 92961 1
    1. Stereochemistry—Congresses.   2. Left- and right-handedness—
Congresses.   3. Laterality—Congresses.   4. Morphology (Animals)—
Congresses.   5. Embryology—Congresses.   I. Bock, Gregory.
II. Marsh, Joan.   III. Symposium on Biological Handedness and
Symmetry (1991: Ciba Foundation)   IV. Series.
QP517.S83B55 1992
591.4—dc20   91-22440
                                          CIP

British Library Cataloguing in Publication Data
A catalogue record for this book is
available from the British Library

  ISBN 0 471 92961 1

Phototypeset by Dobbie Typesetting Limited, Tavistock, Devon.
Printed and bound in Great Britain by Biddles Ltd., Guildford.

Contents

Participants

M. Annett Department of Psychology, University of Leicester, University Road, Leicester LE1 7RH, UK

H. C. Berg Department of Cellular & Developmental Biology, Harvard University, Biological Laboratories, 16 Divinity Avenue, Cambridge, MA 02138, USA

N. A. Brown MRC Experimental Embryology & Teratology Unit, St George's Hospital Medical School, Cranmer Terrace, London SW17 0RE, UK

M. Brueckner Division of Pediatric Cardiology, Yale University School of Medicine, 333 Cedar Street, New Haven, CT 06510, USA

J. Burn Division of Human Genetics, University of Newcastle upon Tyne, 19/20 Claremont Place, Newcastle upon Tyne NE2 4AA, UK

C. H. Chothia Cambridge Centre for Protein Engineering and MRC Laboratory of Molecular Biology, Hills Road, Cambridge CB2 2QH, UK

R. L. Collins Jackson Laboratory, Bar Harbor, ME 04609, USA

M. C. Corballis Department of Psychology, University of Auckland, Private Bag, Auckland, New Zealand

T. J. Crow Division of Psychiatry, MRC Clinical Research Centre, Northwick Park, Harrow, Middlesex HA1 3UJ, UK

R. Dohmen Department of Experimental Zoology, State University of Utrecht, PO Box 80058, Padualaan 8, NL-3584 Utrecht, The Netherlands

J. Frankel Department of Biology, University of Iowa, Iowa City, IA 52242, USA

M. Fujinaga   (*Ciba Foundation Bursar*) Department of Anesthesia, Stanford University School of Medicine & Palo Alto VA Medical Center, 3801 Miranda Avenue 112A, Palo Alto, CA 94304, USA

A. M. Galaburda   Department of Neurology, Harvard Medical School, Beth Israel Hospital, 330 Brookline Avenue, Boston, MA 02215, USA

J. W. Galloway   Cancer Research Campaign, 2 Carlton House Terrace, London SW1Y 5AR, UK

R. P. S. Jefferies   Department of Palaeontology, The Natural History Museum, Cromwell Road, London SW7 5BD, UK

D. Kondepudi   Department of Chemistry, Wake Forest University, Box 7486, Winston-Salem, NC 27109, USA

P. A. Lawrence   MRC Laboratory of Molecular Biology, Hills Road, Cambridge CB2 2QH, UK

W. M. Layton   Department of Anatomy, Dartmouth Medical School, Hanover, NH 03756, USA

J. Lewis   ICRF Developmental Biology Unit, Department of Zoology, University of Oxford, South Parks Road, Oxford OX1 3PS, UK

S. F. Mason   Department of Chemistry, King's College London, The Strand, London WC2 2LS, UK

I. C. McManus   Department of Psychology, University College London, Gower Street, London WC1E 6BT and Department of Psychiatry, St Mary's Hospital Medical School, Imperial College of Science, Technology and Medicine, Norfolk Place, London W2 1PG, UK

M. J. Morgan   Laboratory for Neuroscience, Department of Pharmacology, University of Edinburgh, 1 George Square, Edinburgh EH8 9JZ, UK

M. H. Peters   Department of Psychology, University of Guelph, Guelph, Ontario, Canada N1G 2W1

C. D. Stern   Department of Human Anatomy, Oxford University, South Parks Road, Oxford OX1 3QX, UK

J. J. Thwaites   Department of Engineering, Trumpington Street, Cambridge CB2 1PZ, UK

K. Weber Max-Planck Institute for Biophysical Chemistry, PO Box 2841, Am Fassberg, D-3400 Göttingen, Germany

L. Wolpert (*Chairman*) Department of Anatomy & Developmental Biology, University College & Middlesex School of Medicine, Windeyer Building, Cleveland Street, London W1P 6BD, UK

W. B. Wood Department of Molecular, Cellular & Developmental Biology, University of Colorado, Box 347, Boulder, CO 80309, USA

H. J. Yost Department of Cell Biology and Neuroanatomy, University of Minnesota, 4–135 Jackson Hall, 321 Church Street S.E., Minneapolis, MN 55455, USA

Introduction

Lewis Wolpert

Department of Anatomy & Developmental Biology, University College & Middlesex School of Medicine, Windeyer Building, Cleveland Street, London W1P 6DB, UK

This is a very exciting meeting: so far as I know it is unique in its attempt to discuss left–right asymmetry, from molecules to brains. It is a brave attempt and I have no idea how it is going to go. In developmental biology left/right-handedness is rather neglected. The psychologists have been much more to the fore in thinking about this problem. Nigel Brown and I a little while ago wrote a review and sent it to what we thought was a leading journal in the field, Developmental Biology: not only did they loathe the paper but they made it quite clear that they couldn't possibly see why anybody should want a review of this subject since no one was interested in it. I am delighted to say that I think they are entirely wrong.

I have a personal interest in this question: nobody who is left handed can not be interested in left–right asymmetry. However, my theory that people study left–right asymmetry because they are left handed is disproved by the observation that there are only two left-handed people out of 29 participants at this conference.

Perhaps the issue is the following. If life had used D-amino acids instead of L-amino acids, would we all have had our hearts on the other side? And would the laterality of the brain have been the other way around? I believe the answer is both yes and no. The 'yes' answer comes from Nigel Brown's and my belief that left–right asymmetry has its origin in a handed or asymmetrical molecule. Obviously, if you change the handedness of this molecule, you will change the position of left–right asymmetrical structures in the body. The 'no' response is that it needn't have been that way—the fact that the heart is on the left side is purely fortuitous. Once you have established asymmetry in development you can do what you like with it; evolution could have selected the heart to be on the left or the right. In other words, evolution could have been even handed as to where structures are put once a basic asymmetry has been established. It will be interesting to see if we can come up with a better answer during the meeting. If we can, I think we will have made quite a lot of progress.

There are a lot of problems that we have to consider, not least is why there is a handed symmetry at all. If a vertebrate has to fit the gut and other organs into the body cavity, it is perfectly reasonable to push them all to one side.

It is much less clear to me why you should do it always to the one side. What difference does it make to an animal that the heart is always put on the left side rather than the right side? I hope someone will provide an answer to that particular question.

We know that asymmetry is very primitive. Joe Frankel's ciliates' ancestors were probably asymmetrical too, so I would say that asymmetry in cell structure is a very primitive character. At the multicellular level, I simply don't know.

We want to understand the generation of left–right asymmetry. One problem is the question of linkage: what is the relationship between one asymmetry and another, for example handedness and arm-folding? Even in development, handedness of asymmetrical structures is not necessarily tightly linked; that's a tremendous puzzle to me. If you have a mechanism for putting things on one side and not the other, surely you do it as a unitary system, but I am not sure that's actually the case.

Origins of the handedness of biological molecules

Stephen F. Mason

Department of Chemistry, King's College London, The Strand, London WC2R 2LS, UK

Abstract. Pasteur (1860) showed that many organic molecules form enantiomeric pairs with non-superposable mirror-image shapes, characterized by their oppositely signed optical rotation but otherwise apparently identical. Equal numbers of left-handed and right-handed molecules resulted from laboratory synthesis, whereas biosynthetic processes afforded only one of the two enantiomers, leading Pasteur to conclude that biosynthesis involves a chiral force. Fischer demonstrated (1890–1919) that functional biomolecules are composed specifically of the D-sugars and the L-amino acids and that the laboratory synthetic reactions of such molecules propagate with chiral stereoselectivity. Given a primordial enantiomer, bio-molecular homochirality follows without the intervention of a chiral natural force, except prebiotically. Chiral forces known at the time were found to be even handed on a time and space average, exemplifying parity conservation (1927). The weak nuclear force, shown to violate parity (1956), was unified with electro-magnetism in the electroweak force (1970). *Ab initio* estimations including the chiral electroweak force indicate that the L-amino acids and the D-sugars are more stable than the corresponding enantiomers. The small energy difference between these enantiomeric pairs, with Darwinian reaction kinetics in a flow reactor, account for the choice of biomolecular handedness made when life began.

1991 Biological asymmetry and handedness. Wiley, Chichester (Ciba Foundation Symposium 162) p 3–15

Molecular dissymmetry

Pasteur (1860) reviewed the three methods he had discovered for separating the two enantiomeric molecules of a racemic mixture. The first was manual sorting of the crystals separating out from a solution of a racemate, sodium ammonium *para*tartrate, into two enantiomorphous sets: small facets distinguished the crystals of one set from those of the other, relating them as non-superposable mirror-image forms. A solution of crystals of one set gave a specific optical rotation of polarized light equal in magnitude but opposite in sign to one of crystals from the enantiomorphous set, accounting for the optical inactivity of the racemate solution. At the time, the main guide to molecular form was the concept that a crystal and its constituent molecular building blocks must be 'images of each other' morphologically. From this, Pasteur inferred that the molecular shapes of dextrorotatory (+)-tartaric acid and laevorotatory (−)-tartaric acid have non-superposable mirror-image forms. Pasteur termed such forms 'dissymmetric': the equivalent term 'chiral' (handed) was introduced later by Kelvin.

The second method for the optical resolution of a racemate came with Pasteur's discovery of diastereomers: these compounds contain two or more different chiral units and, unlike enantiomers, have inequivalent chemical properties. An optically active alkaloid base, such as quinine, with (+)- and (−)-tartaric acid forms salts that have different solubilities, thus allowing a ready separation of the two enantiomers from racemic *para*tartaric acid.

The third method had particular significance for Pasteur. He found that the mould *Penicillium glaucum*, grown on racemic *para*tartrate, preferentially uses (+)-tartrate as a carbon source, leaving the (−)-isomer, which Pasteur isolated as the ammonium salt. In his 1860 lectures Pasteur emphasized that the production of the optically active molecules then known was confined to the biosynthetic activity of living organisms, the laboratory syntheses of the time affording only optically inactive products. Accordingly, it appeared to Pasteur that in living organisms, 'dissymmetric forces exist at the moment of the elaboration of natural organic products; forces which are absent or ineffectual in the reactions of our laboratories'. Pasteur later described some of the inconclusive experiments he had carried out in an attempt to characterize the chiral natural forces. These forces might have been magnetic, for Faraday had shown that a magnetic field induces optical activity in glass and other isotropic media, or they might have arisen from the radiation of the Sun and the rotatory motions of the Earth. Pasteur pointed out that the solar system as a whole is dissymmetrical, for it is not superposable on its mirror image.

Stereoselective chiral synthesis

Pasteur's hopes for chirally selective synthesis in the laboratory became realizable when the purely morphological concept of molecular dissymmetry was given a three-dimensional structural basis by the tetrahedral model for the orientation of the four valencies of the carbon atom (Le Bel 1874, van't Hoff 1874). The new stereochemistry provided detailed expectations in the laboratory investigation of organic natural products. In particular, the predictions of the number and the type of stereoisomers resulting from a chain of n bonded chiral carbon atoms, A–$[CXY]_n$–B, proposed by van't Hoff (1874), were tested and subsequently used as a guide by Emil Fischer in his studies of the sugar series, 1884–1908 (Freudenberg 1966).

Fischer found, as van't Hoff had foreseen, that there are 2^n stereoisomers, all optically active, if A and B are inequivalent groups, as in the aldose sugar series, $HOCH_2$–$[CHOH]_n$–CHO. The 2^n stereoisomers are made up of 2^{n-1}

chemically distinct diastereomers, each diastereomer consisting of a pair of mirror-image enantiomeric molecules, which were generally supposed to be chemically equivalent in reactions with achiral reagents.

Fischer (1894) and his colleagues evolved reaction sequences for ascent and descent of the sugar series, for a unit increase or decrease in the number n of chiral carbon atoms. In this way Fischer distinguished two enantiomeric series of sugars, the D-series related back chemically to the parental $(n = 1)$ triose, D-(+)-glyceraldehyde, and the L-series similarly related to the mirror-image triose. Subsequently, from his studies of proteins and synthetic polypeptides, Fischer characterized chemically two enantiomeric series of amino acids. He found that the homochiral biochemistry of living organisms is dominated by the D-series of sugars and the L-series of amino acids.

In his studies of the sugars, Fischer discovered that, contrary to the common understanding of the time, the synthetic reactions of an enantiomer with achiral reagents are chirally selective. In the ascent of the sugar series from the aldopentose, arabinose $(n = 3)$, to the two related hexose diastereomers, mannose and glucose $(n = 4)$, he showed that mannose is the major product and glucose the minor one, formed in such a small yield that it had remained undetected in previous researches on the sugars. Only achiral reagents, such as hydrogen cyanide, were involved in the ascent of the sugar series. The synthetic reactions of a chiral molecule with chiral reactants were even more stereoselective, and they became stereospecific in reactions mediated by enzyme catalysts.

These observations led Fischer to the view that 'the difference frequently assumed in the past to exist between the chemical activity of living cells and of chemical reagents, in regard to molecular asymmetry, is non-existent . . . once a molecule is asymmetric, its extension proceeds also in an asymmetric sense'. Starting with a single enantiomer, synthetic reactions lead inevitably to a dominant diastereomeric product favoured by the steric congruence of the reaction intermediates. The propagation of the best stereochemical fit in a synthetic reaction, Fischer's 'key and lock' hypothesis, offered 'a simple solution to the enigma of natural asymmetric synthesis', obviating any need for the chiral force of Nature active in biosynthesis conjectured by Pasteur. Fischer's conclusion left unsolved the problem of the origin of the primordial enantiomer from which stereoselective synthesis began, as he himself appreciated (Freudenberg 1966).

Chiral force fields

Pierre Curie (1894), in an analysis of the symmetry relations between a cause and its effect in physical phenomena, showed that each of the natural forces Pasteur considered dissymmetrical was, in fact, symmetrical to mirror-plane reflection. But a collinear combination of two complementary kinds of force, one rotatory (axial) and the other linear (polar), provides two possible chiral force fields, the antiparallel combination being the mirror-image enantiomorph of the corresponding parallel combination. Rotatory and linear motion combined give a helical motion, right-handed if the axial and the polar vectors are parallel, left-handed if they are antiparallel. Similarly, a magnetic (axial) field and an electric (polar) field lose their individual mirror-plane symmetry in a collinear union. If the two components oscillate at a common frequency in an electro-magnetic field, the two enantiomorphous combinations are represented by left- and right-circularly polarized light.

With his proposed structural explanation for optical isomerism, Le Bel (1874) suggested that circularly polarized light might favour the formation of one of the two possible enantiomers in a photochemical reaction. The surmise, supported by Curie, was provided with a physical basis by Cotton (1895), who discovered that an optical isomer in solution differentially absorbs left- and right-circularly polarized light (circular dichroism): for the enantiomeric molecule the differential absorption at the same wavelength is equal in magnitude but opposite in sign, analogous to the optical rotation (circular birefringence) in the longer wavelength region of optical transparency. Thus, a racemic mixture of photolabile enantiomers irradiated with circularly polarized light is expected to become optically enriched in the enantiomer with the smaller absorption coefficient for the particular hand of circularly polarized light employed (left or right).

Cotton searched for the expected effect, but without success, and the first authentic photoresolutions of racemic mixtures by irradiation with circularly polarized light were achieved by Kuhn (1930). Meanwhile, the surmise of Le Bel, reinforced by the studies of Curie and Cotton, led to a minor but enduring tradition that invoked circularly polarized sunlight as the original physical cause of an enantiomeric enrichment among the prebiotic biomolecules, which subsequently became elaborated to chiral homogeneity through competitive stereoselective reactions during the course of biochemical evolution.

Kuhn (1930) indicated a restriction on the photochemical theory of the prebiotic origin of biomolecular handedness. He showed that only mono-chromatic circularly polarized radiation, tuned to a particular light-absorption band of the racemate, discriminates between the two enantiomers, pro-ducing a photochemical change larger for one isomer than the other. Circu-larly polarized broad-band 'white' radiation, like sunlight, lacks discrimination, because the differential photoreaction at one wavelength is compensated by a converse differentiation at another: over the electromagnetic spec-trum as a whole the chiral photodiscrimination between two enantiomers sums to zero.

Other restrictions on the theory of a photochemical origin of biomolecular handedness followed. Solar radiation has, in fact, a minor circular polarization of up to 0.5% at twilight, owing to scattering by dust particles. But the overall diurnal effect is very small, since the right-circularly polarized component in excess at sunrise is almost wholly compensated by the left-circularly polarized component in excess at dusk. Similarly, the excess of right-circular light from the north pole of the Sun virtually cancels out the excess of left-circular light from the south pole in the total flux of solar radiation reaching the Earth (Mason 1988).

Parity and its non-conservation

During the 1920s it appeared that, on a time and space average, the classical chiral force fields characterized by Curie were even handed; singular circumstances—a particular time or place, or adventitiously monochromatic circularly polarized irradiation—had to be invoked to account for the origin of molecular handedness through their agency. The apparent even-handedness of all the forces of Nature was elevated to the principle of the conservation of parity in 1927 with the postulate that all physical causes and the laws linking them to the effects produced are invariant to spatial inversion through a coordinate origin (the parity operation) or, what is equivalent, they are unchanged by mirror-plane reflection.

The natural forces assumed to conserve parity included the newly discovered strong and weak nuclear forces responsible for α- and β-radioactivity, respectively. It was found observationally that the strong force conforms to the parity conservation principle but, from 1928, a few experiments involving the weak force gave puzzling results. Ultimately, the accumulation of anomalies led Lee & Yang (1956) to conclude that parity is not conserved in the weak nuclear interaction. They designed tests for parity violation and its occurrence was soon confirmed. The experiments establishing parity violation in the weak interaction showed that the fundamental particles have an intrinsic handedness or helicity. The electrons emitted in radioactive β-decay are inherently left handed, with the spin axis preferentially orientated antiparallel to the linear momentum direction, whereas the corresponding antiparticles, β-positrons, are right handed, with a preferred parallel alignment of the spin axis and the momentum direction. Although parity itself is violated in the weak interaction, the combination (CP) of parity (P) and charge conjugation (C)—the conversion of a particle into the corresponding oppositely charged antiparticle—is conserved in good approximation (strictly, time-reversal T must be included to give the more complete principle of CPT conservation).

The aspects of parity violation in the weak nuclear force that were initially studied involved charge-changes and particles moving with relativistic velocities, as in radioactive β-decay. Subsequent investigation of the neutral features led, around 1970, to the unification of the weak force with electromagnetism in the electroweak interaction, which gives rise to chirality effects in systems moving at non-relativistic velocities. In particular, the theory of the electroweak force predicts new chiral properties for atoms and molecules in their electronic ground state or other stationary states. According to the principle of CP conservation, the negatively charged electron and the positively charged positron are CP mirror-image forms; the hydrogen atom, composed of an electron and a proton, has a CP enantiomer, made up of a positron and an antiproton. Viewed as chiral entities in the CP mirror, all atoms are expected to be optically active, giving an optical rotation proportional to the sixth power of the atomic number. This idea was pursued throughout the 1970s, and progressively increased accuracy in the measurement of optical rotation culminated in the early 1980s with the recording of an optical activity for heavy metal atoms in the gas phase (bismuth, lead and thallium) in agreement with the sign and magnitude calculated (Emmons et al 1983).

Again, according to the principle of CP conservation, a chiral molecule composed of normal atoms, such as L-alanine, has a true CP mirror-image analogue made up of anti-atoms in a counterworld of antimatter, but the terrestrial enantiomer with the standard particle composition (D-alanine) has properties dependent upon the electroweak interaction which are inequivalent because of the violation of the simple parity equivalence. The inclusion of the electroweak interaction in *ab initio* quantum mechanical calculations of the binding energy in the electronic ground state show that the L-α-amino acids, and the L-polypeptides in either the α-helix or the β-sheet regular conformation, are slightly more stable than their respective terrestrial D-enantiomers. Similar calculations indicate that the parent triose of the D-sugar series, D-(+)-glyceraldehyde, and the salient furanose, D-ribose, are energetically stabilized relative to the corresponding L-enantiomers (Mason 1988).

The electroweak energy difference between the enantiomers of an amino acid or a sugar is small: it amounts to approximately 10^{-14} J mole^{-1}, corresponding to an excess of some 10^6 molecules of the stabilized enantiomer per gram mole of racemate (6×10^{23} molecules) in thermodynamic equilibrium at the surface temperature of the Earth. Such an enantiomeric excess at equilibrium is not measurable as yet, but it has implications for dynamic reaction sequences remote from equilibrium.

Frank (1953) proposed a general mechanism for the evolution of biomolecular homochirality from a racemic basis by means of a flow reactor, in which each enantiomer acts as 'a catalyst for its own production and an anticatalyst for the production of its optical antimer'. Darwin's 'warm little pond', equipped with an input stream for substrates and an output stream for products, models such a flow reactor. The system is fed by an input of achiral molecules A, which react reversibly with each enantiomer, L or D, with common rate constants k_1 and k_{-1} to duplicate the enantiomer molecule:

$$A + L(D) \underset{k_1}{\overset{k_1}{\rightleftharpoons}} 2L(2D) \qquad (1)$$

In addition, the two enantiomers react together irreversibly, eliminating one another as the inactive side product, P, which constitutes, together with the excess enantiomers, the output of the flow reactor system:

$$L + D \xrightarrow{k_2} P \tag{2}$$

The system is stable, giving a racemic output with some inactive product P, as long as the input solution of achiral substrate remains dilute. At a larger input concentration the system becomes metastable, since the overall increase in the molecular populations results in greater competition between the enantiomers for the achiral substrate needed for self-propagation. The mutual elimination of the enantiomers (equation 2) now assumes more significance than their autocatalytic self-duplication (equation 1), and the system becomes hypersensitive to small perturbations that trigger the switching of the reaction sequences to production of one or the other single enantiomer.

Before the discovery of parity non-conservation, the direction of the switching appeared to be wholly a matter of chance, but afterwards it was demonstrated that the small electroweak energy difference (or concentration difference) between biomolecular enantiomers sufficed to determine which of the two single-enantiomer reaction sequences was adopted in the flow reactor at the metastable stage (Kondepudi & Nelson 1985). With typical rate constants (equations 1,2) and an allowance for thermal and photochemical racemization, the electroweak-stabilized enantiomeric series is selected with 98% probability if the passage through the metastable stage occupies some 10^4 years in a flow reactor corresponding to a lake one kilometre in diameter and four metres deep.

The electroweak interaction differs from the classical chiral fields in its universality: the former has operated at all times and places, whereas the latter require special conditions for their chiral discrimination. The discovery that the electroweak force promotes the particular enantiomeric series selected during the course of chemical evolution, the D-sugars and the L-amino acids, means that biomolecular handedness on the Earth connects to more general chiral inequivalencies: thus, particles overwhelmingly predominate over antiparticles in the universe, owing to the violation of the approximate principle of CP conservation, within the exact principle of CPT conservation (Mason 1991).

Acknowledgement

The author thanks Dr. G. E. Tranter for collaborative studies of the electroweak differential stabilization of biomolecular enantiomers.

References

Cotton A 1895 Absorption inégale des rayons circulaires droit et gauche dans certaines corps actifs. C R Acad Sci Paris 120:989–991

Curie P 1894 Sur la symétrie dans les phénomènes physiques. J Physique (3) 111:393–416

Emmons TP, Reeves JM, Fortson EN 1983 Parity-non-conserving optical rotation in atomic lead. Phys Rev Lett 51:2089–2092

Fischer E 1894 Sythesen in der Zuckergruppe II. Ber Dtsch Chem Ges 27:3189–3232

Frank FC 1953 On spontaneous asymmetric synthesis. Biochim Biophys Acta 11:459–463

Freudenberg K 1966 Emil Fischer and his contribution to carbohydrate chemistry. Adv Carbohydr Chem 21:1–38

Kondepudi DK, Nelson GW 1985 Weak neutral currents and the origin of molecular chirality. Nature (Lond) 314:438–441

Kuhn W 1930 The physical significance of optical rotatory power. Trans Faraday Soc 26:293–308

Le Bel JA 1874 Sur les relations qui existent entre les formules atomiques des corps organiques, et le pouvoir rotatoire de leurs dissolutions. Bull Soc Chim Fr (2) 22:337–347

Lee TD, Yang CN 1956 Question of parity conservation in the weak interaction. Phys Rev 104:254–258

Mason SF 1988 Biomolecular homochirality. Chem Soc Rev 17:347–359

Mason SF 1991 Chemical evolution: origins of the elements, molecules and living systems. Clarendon Press, Oxford, p 281–284

Pasteur L 1860 Researches on molecular asymmetry. (Alembic Club Reprint 14, Edinburgh, 1948)

van't Hoff JH 1874 Sur les formules de structure dans l'espace. Arch Neerl Sci Exactes Nat 9:445–454

DISCUSSION

Kondepudi: The electroweak energy difference between the enantiomers of an amino acid corresponds to an excess of 10^6 molecules of the stable form in a total of 10^{23}. You may wonder how such a small difference, one part in 10^{17}, can lead to differential production of L- and D-amino acids. How could the small difference have any noticeable effect when there are so many random fluctuations in factors, such as circularly polarized light, that could influence which enantiomer is formed?

The main point concerns an evolving system that comes to a critical point at which it has to choose between two alternatives. At that point it is extraordinarily sensitive. Even if a systematic bias is much smaller than the root mean square value of the fluctuations, the system is able to respond to that bias. The system does what electrical engineers call signal averaging. If there is a constant signal embedded in large random noise (the root mean square magnitude of which is much larger than the magnitude of the bias), by sampling the signal plus noise for a long enough time the system is able to pick out the small signal. For a random noise, the total output grows at a rate proportional to the square root of the time for which it is gathered; for a systematic signal the output grows at a rate proportional to the time. After enough time, the total output of the systematic signal will exceed that of the random noise.

If we assume that during molecular evolution there came a point at which either L-amino acids or D-amino acids had to dominate (we don't know why, we just have to assume that), then we find that evolutionary times of tens of thousands of years are long enough for the system to signal average the small bias, pick up this bias and respond to it (Kondepudi & Nelson 1985).

Morgan: In the competitive model, the D- and the L-morphs compete and destroy each other; is it envisaged that the D-forms compete with each other to the same extent or is the competition only intermorphic?

Mason: In the Darwinian reaction kinetics scheme of Frank, the competition is wholly interspecific. Molecules of the D-isomer annihilate only their L-enantiomers; they do not 'fight' other one another.

Morgan: Is there any justification for that?

Mason: At the molecular level there is none. In chemical reactions, molecules identical in all known respects—handedness, isotopic composition and so on—do compete with one another for substrate, but they do not 'annihilate' each other. In condensation reactions, identical molecules do eliminate one another, in the sense that they form dimers and polymers, but they are not 'annihilated' in the product formed.

Wolpert: Are you saying there is no basis for assuming competition?

Mason: No; but intraspecific competition in chemistry generally means only that the unsuccessful molecule lives to react another day, unlike the unsuccessful organism of intraspecific competition in the biological world.

Kondepudi: This question is something I have been pondering for many years. I was trained as a physicist, so I went to the Chemistry Department and asked if they could show me a system in which there was autocatalysis of L and D forms, and in which the two forms competed and destroyed each other. They did not know of any such system.

I have recently found that it can be accomplished very easily, in the crystallization of sodium chlorate, however, not with amino acids. The sodium chlorate molecule itself is not handed but the crystal has a definite chirality. That is similar to a non-chiral molecule changing to a chiral molecule. During crystallization there is a process of secondary nucleation. I found that if you crystallize sodium chlorate in a petri dish as usual, you find equal numbers of each form. But, if you stir the system constantly during crystallization, you find almost 100% of either L or D (Kondepudi et al 1990). The mechanism has to have two components, autocatalysis *and* competition. Otherwise, one enantiomer would not be able to dominate.

Autocatalysis arises from secondary nucleation. If you stir the solution, somehow one crystal is able to produce many others. This is an established fact, but the mechanism is not well understood. It is not simply breaking the crystal into lots of little pieces. The competition comes because the first crystal is formed and it multiplies into many other similar crystals; this soaks up the solute so fast that the concentration drops rapidly and no other crystals can nucleate. The original crystal completely dominates the system. Competition need not be direct. In this case, I start with a system that has no bias and finish with 99.8% of the crystals having the same handedness. Which handedness will dominate is completely random.

Mason: One of my students, John Gould, obtained similar results with sodium uranyl acetate, which also forms handed crystals but is achiral in solution. He carried out a crystallization just before going on holiday; he obtained a large number of very small crystals that he stored in a covered beaker for later attention. On his return, he found a single large crystal in the beaker, and of course it was uniquely handed. Here, one crystal had grown by consuming smaller ones, through the processes of redissolving and recrystallization. Small crystals of both hands had been consumed, illustrating both interspecific and intraspecific competition.

Lewis: Are you saying that given what we know, you can deduce that evolution *had* to go in the direction of the particular handedness it has, or that there is a conceivable mechanism that *might* have had that consequence?

Mason: The latter. Solutions to scientific problems are never absolutely certain, only probable to a greater or lesser degree. Dr Kondepudi estimates a probability of 98% for the evolution of biomolecular homochirality by the comprehensive kinetic mechanism he has developed, based on the small electroweak energy difference between enantiomers, under the conditions he specifies.

Another feature of the acceptability of solutions to scientific problems is the degree to which they unify individual fields that were formerly unconnected. Thus, the dominance of L-amino acids and D-sugars in the biochemical world is now linked to the overwhelming predominance of particles over antiparticles in the universe by the electroweak interaction; they respectively exemplify the violation of parity (P) conservation and CP conservation within the overall principle of CPT conservation.

Brown: Is one of the consequences of this hypothesis that wherever we are going to find organic molecules in the universe they are going to be our kind of organic molecules?

Mason: Yes.

Brown: How disappointing.

Lewis: But is that proven or a speculative hypothesis?

Mason: It is well supported.

Wolpert: Then how do right-handed amino acids arise, for example?

Mason: Many bacteria synthesize D-amino acids for incorporation into their cell walls, but D-amino acids don't occur in biochemically functional proteins. D-amino oxidases are widespread in higher organisms, presumably to remove the D-amino acids produced by racemization or synthesized by bacteria. Penicillin is a relatively benign antibiotic because it interferes with only the incorporation of D-amino acids into the bacterial cell wall.

Chothia: As I understand this, originally, for 10^{17} D-amino acids there were $10^{17} + 1$ L-amino acids. Professor Mason is arguing that the additional one eventually led to domination of the L-isomer.

Wolpert: The whole point about Darwinian evolution is that when one form dominates, the other simply disappears. In this case, I don't understand why the D-amino acids have persisted.

Mason: D-amino acids are generated continuously by slow racemization from L-amino acids. The half-life of this racemization is about one million years at Earth surface temperatures; there are wide variations, depending on environmental conditions (dry solid or solution, and thence pH, ionic strength, catalytic co-solutes) and their chemical state (free or combined as di- or polypeptides). L-isoleucine contains two chiral centres, the α- and the β-carbon atoms, the latter being very stable to epimerization. Normal racemization at the α-carbon atom gives D-allo-isoleucine, which is not generally found in the normal metabolism of living organisms and can readily be detected by analytical chemistry. Even the small amount of D-allo-isoleucine formed during a human lifespan in teeth, where there is virtually no turnover of the amino acids in the proteins once the teeth are fully formed, is reliably detected. This can be used to show that putative centenarians are often somewhat younger than they claim. The established rates of racemization of the L-amino acids are widely used to date recent fossil bones, up to some 200 000 years old (Sykes 1988).

Berg: But if that racemization can happen in 80 years, why doesn't it randomize the amplification mechanism discussed earlier?

Kondepudi: The dominance is maintained by the constant production of L-amino acids. They are removed by racemization but there is a constant input of L-amino acids from other compounds. The model takes racemization into consideration. For most amino acids the half-life of racemization is hundreds or thousands of years. Dentine is a rare example.

Crow: Isn't this also used as an archaeological dating system?

Mason: Yes. It has been used to show that there were people in North America at least 40 000 years ago.

Galaburda: What establishes dominance in one direction in the first place? Is it dust in the atmosphere?

Mason: The scattering of solar radiation by dust particles in the atmosphere gives sunlight a circular polarization, particularly at twilight when the rays of the Sun traverse a maximum pathlength through the atmosphere. But the right circular component of solar radiation in excess at sunrise virtually equals the excess of the left circular component at sunset, and the net effect over the course of a day is vanishingly small. One can envisage a pool of racemic amino acids on an east-facing slope that catches sunlight only at dawn. In prebiotic times, right circularly polarized ultraviolet radiation, which is now filtered out by the ozone layer and the oxygen in the atmosphere, could have produced a differential photolysis of the two enantiomers, leaving the L-amino acids in excess.

Analogous views of the origin of biomolecular handedness held between 1900 and 1980 that were derived from the characterization of classical chiral force fields by Pierre Curie were similarly dependent on special conditions—particular times and places, dawn or dusk, the northern or southern hemisphere, and so on. The electroweak interaction is a universal force: it has operated at all times and places; no special conditions are required to produce its determinate effects.

Wood: Don't we have to worry that the origin of life may not have been an average process but may have happened in a unique situation?

Mason: One can choose: particular times and particular places, or a universal force that's been in operation since the universe began.

Frankel: Presumably, there is some deterministic process that resulted in a predominance of the L-amino acids; these then became incorporated into the most important molecules in organisms. A secondary racemization provided D-amino acids, which appear in certain relatively unusual molecules such as the cell walls of bacteria. If we imagine some other planet on which life is originating, what would prevent the secondary racemization leading by chance not to some peculiar, relatively minor molecule but to the informational equivalent of a nucleic acid?

Mason: The world of 'chance' is open ended; in it, 'anything goes'. The concept of 'chance' as a kind of even-handed nescience becomes useful principally when it is tamed statistically into a normal-error or other type of distribution with determinate large-scale expectations, as in the Maxwell–Boltzmann distribution of energy over an assembly of gas molecules, or in the population genetics of a species in a given ecosystem.

McManus: Given that there are D-amino acids on our Earth and that most organisms use L-amino acids for building their bodies, surely there is a point at which it becomes selectively advantageous for an organism to build its body with the D-amino acids which are around because then it couldn't be eaten by the other organisms?

Mason: That may have been the original selective advantage of the microbes that incorporated D-amino acids into cell walls. But then fungi evolved penicillin and other antibiotics which countered that advantage. The D-amino acids do not occur in any functional metabolic proteins, such as ribosomes or enzymes. They are found only as a structural component of bacterial cell walls, and in other products of secondary metabolism, such as antibiotics.

Thwaites: But in terms of biomass that is a very large amount. Are you implying that the relative amounts of D- and L-amino acids in the bacterial cell wall has changed over a long period of time?

Mason: The ratio of D- to L-amino acids in the bacterial cell wall probably has changed over time. Where it can be measured with any confidence, i.e. over the last 200 000 years or so, the ratio records only the slow racemization towards an equimolar equilibrium. In the early microfossils, which date back about three billion years, the residual organic material is in the form of a complex carbonaceous polymer, kerogen, and little evidence remains of the type of organic molecules from which this derived.

Galloway: It's not true that there are no functional D-amino acids. There have been reports of D-amino acids in neuropeptides, for example dermorphin. Some dipeptides may also include D-amino acids: these would not be coded for as parts of proteins but would be created by specialized enzyme activity.

Also, antibiotics are often composed of alternating sequences of L- and D-amino acids, for example gramicidin (produced by *Bacillus brevis*).

Mason: The neuropeptides and the antibiotics are not produced universally by all living organisms: they are products of specialized secondary metabolism, like the alkaloids and terpenes. Some species of the Camphor laurel produce D-camphor; other species produce L-camphor.

During the 1930s there was a vigorous debate on the significance of D-amino acids found in protein hydrolysates from human cancer cells. By 1940 these were shown to be artifacts of the isolation procedure that resulted in some racemization (see Gause 1941).

Corballis: Does the CPT conservation imply that the ultimate asymmetry is time? If time went the other way, would I scratch my head with the other hand? Time is the most obviously asymmetrical parameter.

Mason: Perhaps when the Big Bounce occurs and the universe begins to contract, you will use the other hand.

References

Gause GF 1941 Optical activity and living matter. Biodynamica, Normandy, Missouri p 129–132

Kondepudi DK, Nelson GW 1985 Weak neutral currents and the origin of biomolecular chirality. Nature (Lond) 314:438–441

Kondepudi DK, Singh N, Kaufman R 1990 Chiral symmetry breaking in sodium chlorate crystallization. Science (Wash DC) 250:975–976

Sykes GA 1988 Amino acids on ice. Chem Br 24:235–244

Macromolecular asymmetry

J. W. Galloway

Cancer Research Campaign, 2 Carlton House Terrace, London SW1Y 5AR, UK

Abstract. The helix is the most common and the most readily recognized example of an enantiomorphic structure. Helical proteins and DNA are good examples of structures where a clear explanation can be provided as to why they adopt one hand or the other. Proteins and DNA are composed of chiral building blocks, amino acids and nucleotides, respectively. Only the L-amino acids occur in proteins; this uniformity of handedness is a prerequisite for helix formation and thus, one could argue, for the development of higher life forms. Helical proteins form higher order helical strutures, from collagen and viral capsids to cotton fibres.

1991 Biological asymmetry and handedness. Wiley, Chichester (Ciba Foundation Symposium 162) p 16–35

I have never been able to decide whether the problem of left–right asymmetry is very deep or quite trivial—and in a sense that is what keeps me interested in it.

What is undoubtedly true is that although scientists may find the question interesting, artists do not. For example, Fig. 1 shows an engraving of Fragonard's *The Swing*. It is a mirror reflection of the original painting that is part of London's Wallace collection. The reason for its being reversed is easy to find—the engraver made a plate identical to the original painting, which then printed a reversed image. However much artists care about colour, mass, form and realism or drama, they do not seem to care which way the pictures actually point, i.e. their right–left asymmetry.

My earliest interest in the problem I can trace to reading (*Alice*) *Through the Looking Glass*. A prominent feature of the book is its numerous examples of helices. The picture by Tenniel shown in Fig. 2 contains toves, which are badger-like animals with corkscrew noses and tails. These are, as far as I know, the only pictures of toves and the helices are left handed.

The helix is biology's favourite shape. Because of its elementary geometry and distinctive appearance it is also the clearest instance of an enantiomorphic object—a helix and its mirror image are identical in all respects except their screw sense. This is a distinction that can be ignored from the points of view of pure geometry and pure group theory but any helical structure is available as either hand.

The potential for having a mirror-reversed 'twin' is not peculiar to helices with their screw symmetry; it is characteristic of any object that possesses no inverse symmetry elements. The simplicity of the helix makes it most suitable for demonstrating the principles underlying the existence (or not) of mutual mirror images. Conklin (1903) wrote in his paper, '*The cause of inverse symmetry*' that 'inversion of symmetry' (i.e. production of a mirror image) in animals, 'with its profound implications for embryology, is clearly seen in gastropods' (which are roughly helical) 'though doubtless taking place in other animals where it is obscured'.

The agreeable economy of the helix makes it the preferred solution for innumerable problems of growth, form and function in living things. Because it is so common, found at every anatomical level across about nine orders of magnitude (Table 1), it is possible to disentangle to some extent the mechanical or structural design principles behind the helix from the large number of ways the design can be realized. One reason for the popularity of the helix can be found in Needham's (1936) rather apt description of biology as 'largely the study of fibres'. Add to this the idea articulated by Crane (1950) that 'any structure which is straight or rodlike' (a category that includes fibres when the length greatly exceeds the diameter) 'is probably a structure having a repetition along a screw axis', i.e. a helix, and the central role played by the helix in biology is clear. More recently, Wainright et al (1976) have pointed out that the hollow cylinder provides the most common design for body walls on every scale. The cylinder lends itself to a helical mode of construction whether using discrete subunits, as in the cylindrical viruses, or helical winding, as in animal bodies and plant cell walls (Fig. 3).

Biological macromolecules: helices of one hand only?

The emergence of the helix as the structural paradigm of molecular biology can be traced to Linus Pauling: 'rolling paper scrolls on a sick bed in Oxford in 1948 before the helix was built as a model structure by Branson and Corey in Pasadena in 1950' (Hodgkin & Riley 1968). Before then the helix was not taken very seriously; afterwards it became the most common and important structure for those interested in big biological molecules and recognized as 'the classic element of protein structure' (Richardson 1981). In 1953, biology's most famous helix appeared in print for the first time with the publication of Watson and Crick's paper on the structure of DNA.

Questions that arose immediately and have continued to be asked up to the present day are what handedness these molecular helices would adopt—and why. Early structural work on fibrous proteins and nucleic acids relied on fibre diffraction, which could not readily distinguish helical hand. Electron microscope studies of helical viruses, like tobacco mosaic virus, and helical organelles, such as microtubules and bacterial flagella, suffered from a similar deficiency although for a rather different reason. The electron microscope could

FIG. 1. '*The Swing*' by Fragonard. The picture on the left is an engraving (courtesy of The British Museum). The original is shown on the right (courtesy of The Wallace Collection, London).

" *Twas brillig, and the slithy toves . . . .* "

FIG. 2.   'Twas brillig and the slithy toves. . . . Two toves can be seen in this picture which Sir John Tenniel painted for *Through the Looking Glass.* They have helical noses and tails. Both toves appear to be left handed. No other illustrations being extant, it is not known whether there exist right-handed as well as left-handed toves (courtesy, MacMillan Press).

TABLE 1 The range of structures that occur in helical forms

Type of structure		Example	Diameter of helix nm
Molecules	Fibre	α-Helix	1.0
		Collagen triple helix	1.5
		DNA double helix	2.0–2.5
Viruses	Hollow tubular assemblies of (usually) globular protein molecules	Bacterial filamentous virus	6.0
		Tobacco mosaic virus	20.0
		Turnip yellow virus polyhead	80.0
Organelles		Microtubules	25.0
Cells	Hollow tubes made from winding fibres	*Aquaspirillum* (a bacterium)	500–5000
		Cotton fibre	3×10^4–10^5
Animal skin		Worm cuticle, e.g. of *Wolstorffii parachordes*	7.5×10^5
		Eel skin	10^7–10^8
		Squid mantle	10^8
		Whale skin	10^9

Helical structures arise at every anatomical level across nine orders of magnitude. Note the range of diameters within each class of structure: for molecules the range is covered by a factor of about two; for molecular tubes by roughly an order of magnitude; for cells by roughly two orders of magnitude; for animal skin by three or four orders of magnitude.

'see' the back and front of the helix at the same time but could not distinguish which was which. Finch solved this particular problem for tobacco mosaic virus in 1972.

Although the helix was recognized to be crucial in both proteins and nucleic acids at about the same time, studies on the detailed structures of the two molecules and hence the real appreciation of molecular helices, including their sense of twist, had rather different histories. To begin with it was felt that the α-helix might appear in both right- and left-handed forms, it apparently having no preferred sense of twist. Linus Pauling's book *The Nature of the Chemical Bond* (1960) showed both left- and right-handed α-helices (Fig. 4). Cohen (1955), discussing optical rotation by globular proteins, suggested that optically opposing, i.e. left- and right-handed, chain configurations produce 'cancelling' effects to explain what was felt to be their low rotating power. However, Elliott & Malcolm (1956) later concluded on the basis of optical rotation that the α-helix was probably right handed.

Once high resolution information for a large number of globular proteins started to be obtained in the early 1960s, two things became clear. First, these molecules contained a lot of helix, up to 60% or even 70%. This was mostly

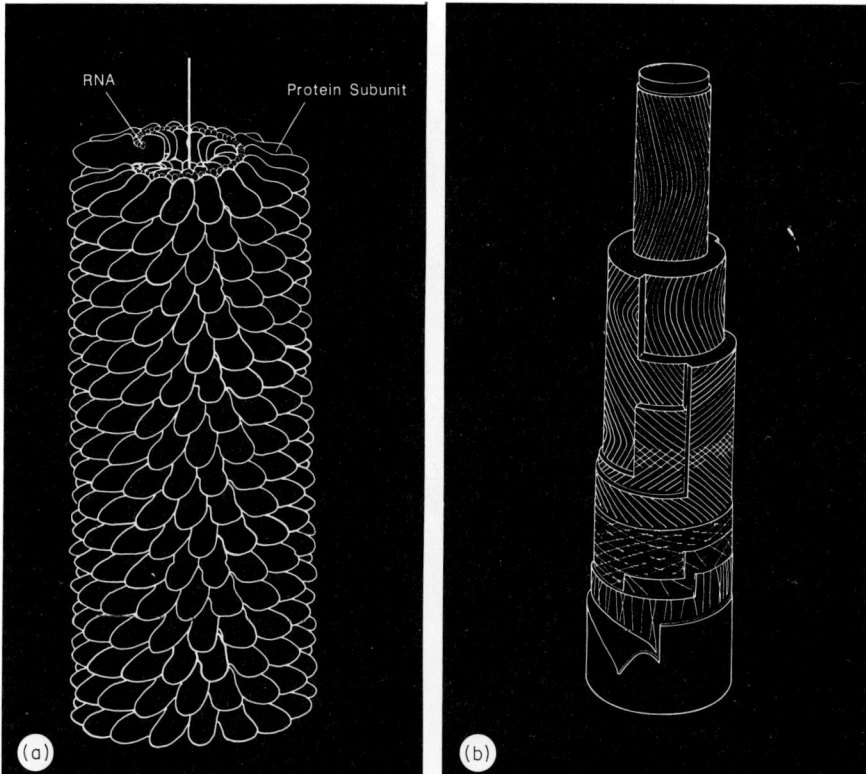

FIG. 3. Two kinds of helical structure. (a) Tobacco mosaic virus: a hollow tube constructed of globular protein molecules arranged on a continuous helical path—the genetic or ontogenic spiral. The helix is right handed and the viral RNA follows this helical path. About 16½ protein subunits constitute the pitch. From Butler (1984). (b) Cotton fibre cell: a complex multi-layered cylindrical structure made by winding polysaccharide fibres. The fibres follow both right- and left-handed paths. They often change hand abruptly. After Waterkeyn (1985).

α-helix but other helices were also found—the π-helix, the 3_{10}- and the α_{II}-helix. These were all realizations of a formula suggested by Bragg and co-workers in 1950, which gives the number of atoms in a hydrogen-bonded ring as $R = 3n + 4$, where n takes the values $1,2,3 \ldots$ Second, the helix is always right handed; the single exception so far is a solitary turn of left-handed helix in the enzyme thermolysin. All structural features of proteins where there is a choice of handedness tend to favour one hand over the other.

Another large family of helical proteins consists of the collagens, the chief proteins of animal skeletal and connective tissues, although many collagens are now known which play no obvious structural role. Collagens are triple helices based on repeating triplets of amino acids, Gly X Y. X is frequently proline

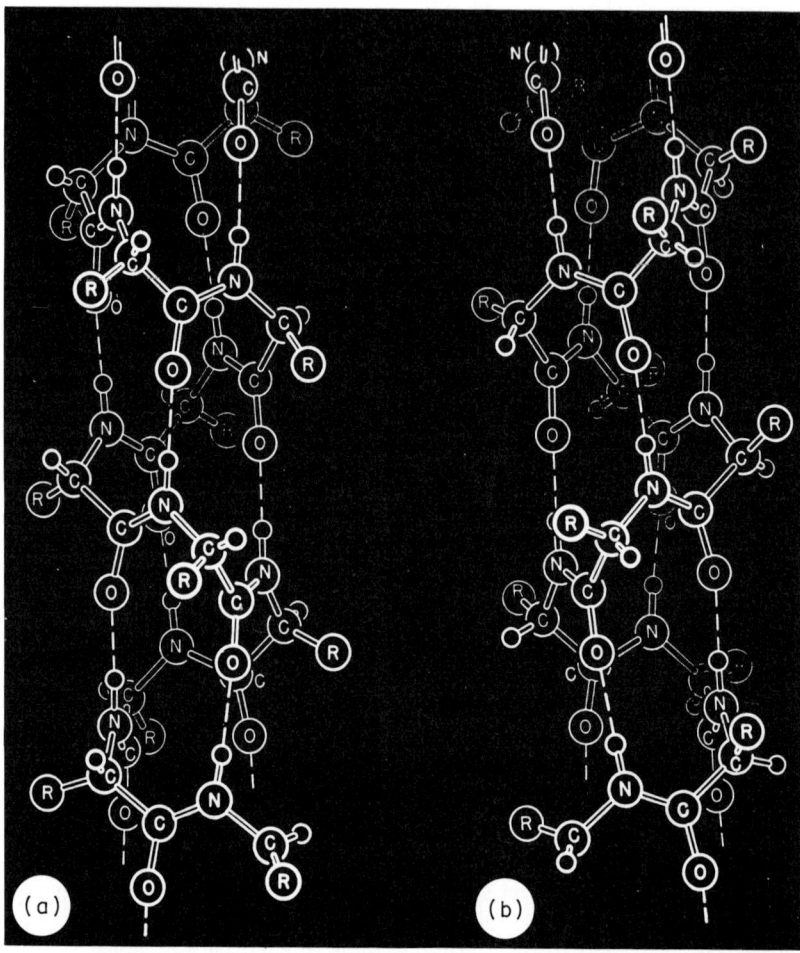

FIG. 4. (a) Hypothetical left-handed and (b) actual right-handed α-helices. These are not strict mutual mirror images, since both are constructed of L-amino acids; true enantiomorphism would require the D-form in one helix, the L-form in the other. From Pauling (1960).

and Y is often hydroxyproline which is rarely found outside the collagens. Each amino acid chain, which in the structural collagens of the vertebrates contains more than a thousand amino acid residues, is twisted into a left-handed helix with three residues in its pitch. Three of these helices wind around one another plectonemically (such that they cannot be separated without being unwound) to give a structure for which a close analogy is rope—usually a right-handed, three-stranded fibre.

The story for the nucleic acids is rather different. Crick and Watson found by trial (and error, presumably) that their double helical model for DNA could be built only in a right-handed sense. Steric effects involving van de Waals contacts between the nucleotides apparently ruled out the possibility that the helix was left handed. It became clear later that in detail their model did not correspond particularly well to any of the observed conformations of DNA. These fell into two families, A and B, that differed fundamentally in both the stacking of the base pairs and the nature of the double helical backbone. Fuller et al (1965), reporting part of a detailed analysis of DNA structure, pointed out that the Crick and Watson model resembled the B form rather than the A, and that although left-handed helices were very unlikely for the A form it was not easy to rule out the possibility that the B form was left handed. In the end they fell back on the argument that since A and B forms were mutually interchangeable and it was very unlikely that such a transition involved reversing screw sense, the B form must also be right handed.

Considerable conformational freedom is enjoyed by individual nucleotides and this leads to a great range of structural polymorphisms. Puckering of the furanose ring is now known to account for the differences between the A and B forms of DNA. C and D forms were both identified but turned out to be best thought of as members of the B family and right handed. The possibility of left-handed helices did not disappear. A proposal that polyd(IC) was left handed (Mitsui et al 1970), the first concrete proposal for a left-handed DNA, was discredited by Arnott et al (1974). S and Z DNAs were discovered in the late 1970s. In these, the structural unit was not a single nucleotide but a pair, one puckered in the form found in the A form of DNA and the other as in the B form, hence these DNAs are designated as A + B. The important point for this review is that S and Z polynucleotides are left-handed helices. These are, however, very different types of structure from those of the A and B families; they are in no sense left-handed versions of otherwise right-handed helices (Fig. 5). A useful article discussing the three families is the one by Dickerson et al (1982).

Thus, we have two major families of protein helices, that dominated by the α-helix and the collagens, and three major families of nucleic acid helices, A, B and A + B. All the members of a particular family (irrespective of amino acid or nucleotide sequence) possess the same helical handedness. For the α-helix and its close relations and for the A and B families of DNA, the building block is a monomer—amino acid or nucleotide; for A + B DNA, it is a pair of nucleotides C-G; for collagen it is a triplet of amino acids, Gly Pro Y.

Physical origins of molecular helices

It would be satisfying to account for this great uniformity of behaviour in some way. I want to suggest using the architectural analogy of the spiral, i.e. helical,

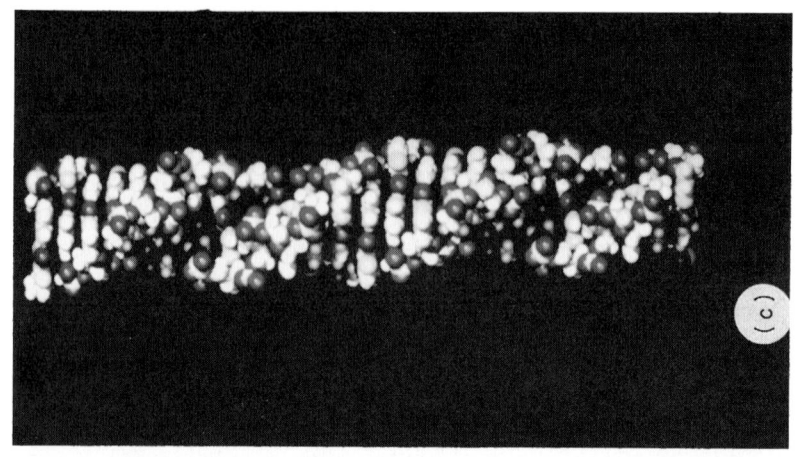

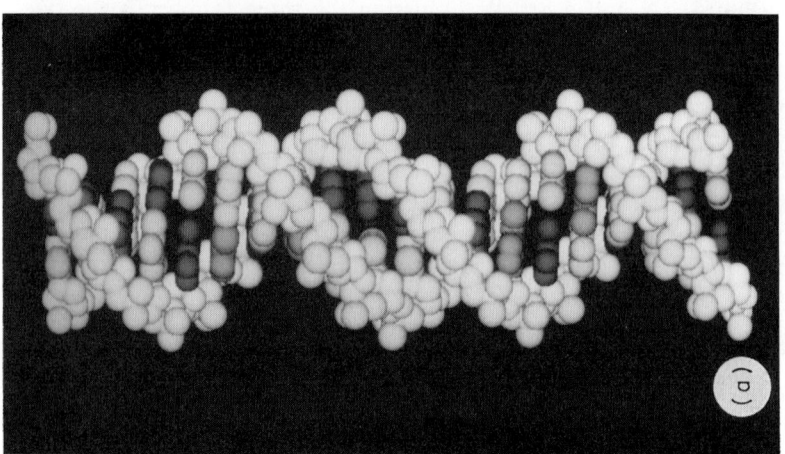

FIG. 5. Different forms of DNA. (a) A and (b) B are right-handed double helices (duplexes); (c) Z is left handed. Computer graphics by kind permission of Dr Rod Hubbard, York University, UK.

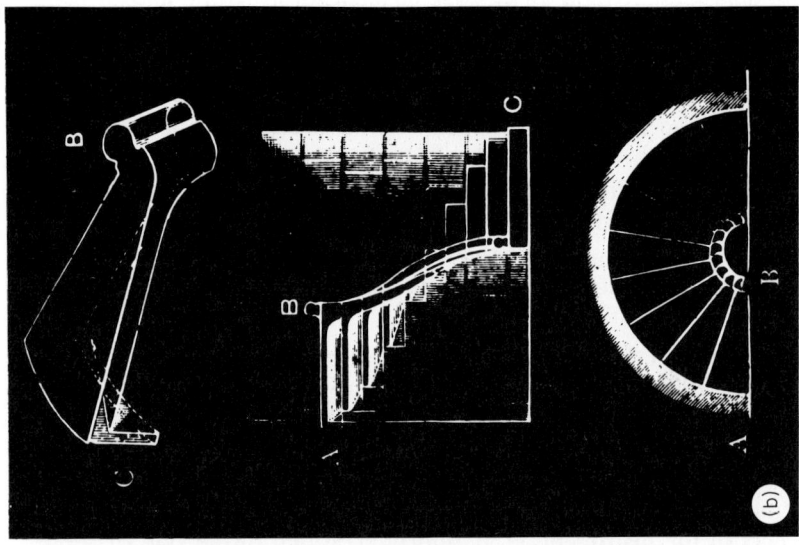

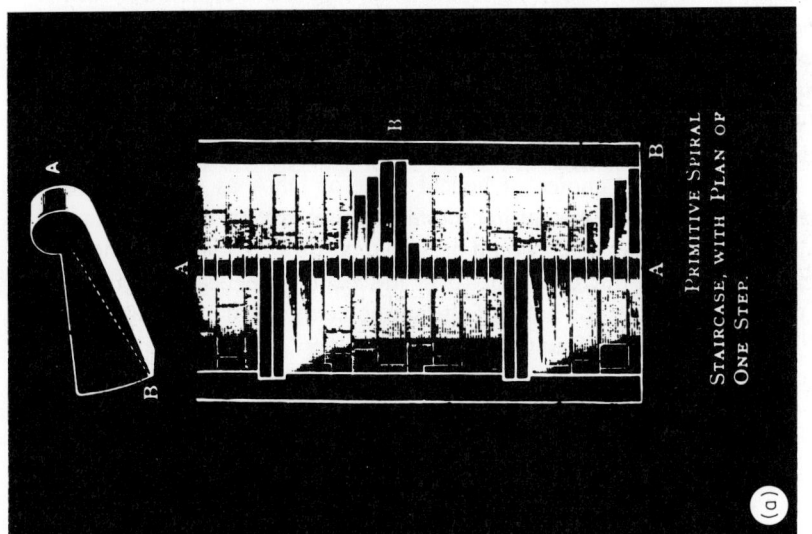

FIG. 6. Construction of spiral staircases; after Viollet-le-Duc. (a) Steps contain a mirror plane parallel to their upper and lower surfaces. Left-handed and right-handed staircases can be built from this unit. (b) Steps possess no symmetry elements, they are therefore enantiomorphic and can be assembled to produce only a right-handed staircase. Cook (1914).

staircase. Two drawings are reproduced (Fig. 6) from Cook's *The Curves of Life* (1914). He, in turn, had borrowed them from the works of Viollet-le-Duc, the great 19th century French restorer of mediaeval architecture.

Both these staircases are built up by stacking identical structural units which, together with a section of the central newel post, form the successive steps. In the second (b), the units are more elaborate than in the first and include a section of handrail. In the first staircase (a) the step has a mirror plane parallel to its top and bottom surfaces—its only symmetry element. The 'step' in the second staircase (b) has no symmetry elements at all. Both staircases are shown as right-handed, but by simply turning each step over before adding the next the first staircase could be converted to a left-handed one. This is not possible for the second staircase, where the shape of the step fixes the handedness of the helix quite unambiguously. A staircase of opposing handedness could be constructed only from mirror-image steps.

The importance of this analogy is that amino acids, with the exception of glycine, and nucleotides are themselves chiral, i.e. enantiomorphic, molecules. They exist potentially as mirror-image pairs, but only the L-forms of amino acids and nucleotides are found in proteins and DNA, respectively. Right-handed (D) amino acids are found in the peptidoglycans of bacterial outer coats, and in some antibiotics. In the eukaryotes, a single instance of a D-amino acid, D-alanine, occurs in the neuropeptide, dermorphin, found in the skin of some amphibians.

Of the 500 or so naturally occurring amino acids only about eight are D-forms. The 20 that make up proteins are all L-forms, except glycine which is achiral. As late as 1951, Bijvoet et al suggested that anomalous X-ray scattering would discriminate between L- and D-forms, and allow the hand to be determined absolutely. They showed that the convention guessed by Fischer half a century before for sugar molecules was the correct one, a point made by Crick & Watson (1954) in support of their choice of helical sense of DNA as right-handed.

Because only L-amino acids occur in proteins, Nature favours the right-handed α-helix. Chains of D-amino acids would presumably form left-handed α-helices. (Polyglycine crystallizes but does not form independent helical structures in solution.) So not only does the hand of the amino acids dictate the hand of the helix they form, but more profoundly, single-handedness appears to be necessary for helices to form at all. No helical protein molecules would mean no collagen—and therefore no skeleton and no muscle and hence no movement. In fact, no helices would mean no higher life forms at all. Thus, it is not unreasonable to suppose that the existence of life has depended on evolution being able to choose and then use just one of the two available handed forms of amino acids and nucleotides. Autocatalysis has been suggested as the way an early imbalance could be amplified in favour of one hand of amino acid or nucleotide. Joyce et al (1984), for instance, have recently studied this possibility in the formation of oligonucleotides. But what created the original imbalance that marginally favoured left-handed amino acids over right-handed ones? Mason (1985, this volume) argues that it originates in the electroweak force, the result of the non-conservation of mirror-image symmetry in fundamental particles and their interactions.

That the handedness of the amino acids determines the handedness of the α-helix suggests that the latter may, in turn, determine the handedness of helical assemblies of α-helical molecules. Both myosin and tropomyosin, proteins from which the filaments of striated muscle are created, are α-helical, as is the capsid of the filamentous bacterial virus Pf1. Crick (1953) made the first suggestion of this sort, pointing out that two α-helices would tend to 'cross' at about 20°, creating a two-stranded, twine-like, left-handed helical coiled coil. This double helix was later found to be very common and became firmly entrenched in the minds of molecular biologists (see Cohen & Parry 1986). Crossing in relation to packing within globular proteins was analysed by Chothia et al (1981), among others, who showed that helices also tended to cross at ~52°. This underlies Michel's (1982) observation that the folded α-helical molecules of rhodopsin crystallize into helical stacks with seven molecules in the helical pitch ($7 \times 52 \simeq 360$). The idea that right-handed α-helix forms left-handed helical aggregates and left-handed α-helix forms right-handed aggregates was tested by Tachibana & Kambara (1967). They synthesized poly-γ-benzyl D- and L-glutamate, PBDG and PBLG, respectively, and concluded that PBDG produced aggregates with a right-handed twist and PBLG produced their mirror images. This was consistent with D-amino acids producing left-handed α-helix which in turn gave a right-handed aggregate, and vice versa. However, less artificial circumstances produce structures less neatly explained.

Marvin and his co-workers (e.g. Marvin & Wachtel 1975) attempted to build the structure of the bacterial filamentous virus Pf1, whose capsid (coat) is constructed from a large number of α-helical subunits. They were unable to determine the hand of the viral helix experimentally, but fixed it as left-handed on the basis of the presumed interactions between the component α-helices. Later, using improved methods of structure determination, they were astonished to discover that the virus was right handed and that neighbouring α-helices 'crossed in an unusual negative sense' (Bryan et al 1983). A second counter example (see McGavin 1976) is provided by the collagenous cerato-trichia (fin rays) of dog fish and sharks, which always dry into right-handed helices.

The idea that considerations of subunit symmetry alone dictate the handed-ness of a helical structure is obviously appealing. It is, of course, only an approximation to the truth—which is that helical handedness is fixed physico-chemically through detailed patterns of atomic contacts. This is revealed in the filamentous viruses and in the single turn of left-handed helix in thermolysin. Bradbury et al (1960) have shown that the α-helix of poly-β-benzyl-L-aspartate (which is not found in proteins) is left handed, not right.

A challenge to the staircase analogy is presented by synthetic polypeptides with alternating L- and D-amino acids. Alternation is also a feature of the antibiotic Gramicidin A isolated from *Bacillus brevis*, which acts by forming channels through biological membranes. Koeppe et al (1978) and Heitz et al (1975) concluded that these alternating polypeptides have helical conformations. Heitz et al suggest that the conformation is very similar to an α-helix, but they do not say whether it is a left- or right-handed helix and one cannot be favoured over the other on energetic grounds. Do both hands exist? Is there here an example of true helical enantiomorphism? I doubt it. Despite the claims of theorists, the odd idiosyncratic exception and some explanatory difficulties, the message from molecular biology is clearly that ambidextrousness is not a feature of biology at the molecular level.

Acknowledgement

This article is based very closely on a review, 'Reflections on the ambivalent helix', published in Experientia (1989) 45:859–872 and reproduced with permission of Birkhäuser Verlag AG.

References

Arnott S, Chandrasekaran R, Hukins DWL, Smith PJC, Watts L 1974 Structural details of a double helix observed for DNAs containing alternating purine and pyrimidine sequences. J Mol Biol 88:523–533

Bijvoet JM, Peerderman AF, van Bommel AJ 1951 Determination of the absolute configuration of optically active compounds by means of X-rays. Nature (Lond) 168:271–272

Bradbury EM, Downie AR, Elliot A, Hanby WE 1960 The stability and screw sense of the α-helix in poly-β-benzyl-L-aspartate. Proc R Soc A Math Phys Sci 259:100–128

Bragg Sir Lawrence, Kendrew JC, Perutz MF 1950 Polypeptide chain configurations in crystalline proteins. Proc R Soc A Math Phys Sci 203:321–357

Bryan RK, Barsal M, Folkhard W, Nave C, Marvin DA 1983 Maximum entropy calculation of the electron density at 4 Å resolution of Pf1 filamentous bacteriophage. Biophysics 80:4728–4731

Chothia C, Levitt M, Richardson D 1981 Helix to helix packing in proteins. J Mol Biol 145:215–250

Cohen C 1955 Optical rotation and polypeptide chain configuration in proteins. Nature (Lond) 175:129–130

Cohen C, Parry DAD 1986 α-helical coiled coils—a widespread motif in proteins. Trends Biochem Sci 6:245–248

Conklin EG 1903 The cause of inverse symmetry. Anat Anz 23:577–588

Cook TA 1914 The curves of life. Constable & Co, London; republished in 1979 by Dover, New York

Crane HR 1950 Principles and problems of biological growth. Scientific Monthly 6:376–389

Crick FHC 1953 The packing of α-helices: simple coiled coils. Acta Crystallogr 6:689–697

Crick FHC, Watson JD 1954 The complementary structure of deoxyribonucleic acid. Proc R Soc A Math Phys Sci 233:89–96

Dickerson RE, Drew HR, Conner BN, Wing RM, Fratini AV, Kopka ML 1982 The anatomy of A-, B- and Z-DNA. Science (Wash DC) 216:475–485

Elliott A, Malcolm BR 1956 Absolute configuration and optical rotation of folded (α) polypeptides. Nature (Lond) 178:912

Finch JT 1972 The hand of the helix of tobacco mosaic virus. J Mol Biol 66:219–294

Fuller W, Wilkins MHF, Wilson HR, Hamilton LD 1965 The molecular configuration of deoxyribonucleic acid. J Mol Biol 12:60–80

Heitz F, Lotz B, Spach G 1975 α_{DL} and π_{DL} helices of alternating poly-γ-benzyl-D-L-glutamate. J Mol Biol 92:1–13

Hodgkin DC, Riley DP 1968 Some ancient history of protein X-ray analysis. In: Rick A, Davidson N (eds) Structural chemistry and molecular biology. DW Freeman, San Francisco

Joyce GF, Visser GM, van Boeckel CAA, van Boom JH, Argel LE, van Westrenen J 1984 Chiral selection in poly (C)-directed synthesis of oligo (C). Nature (Lond) 310:602–604

Koeppe RF, Hodgson KO, Stryer L 1978 Helical channels in crystals of gramicidin A and of a cesium-gramicidin A complex: an X-ray diffraction study. J Mol Biol 121:41–54

Marvin DA, Wachtel EJ 1975 Structure and assembly of filamentous bacterial viruses. Nature (Lond) 253:19–23

Mason S 1985 Biomolecular handedness. Chem Br 21:538–545

Mason SF 1991 Origins of the handedness of biological molecules. In: Biological asymmetry and handedness. Wiley, Chichester (Ciba Found Symp 162) p 3–15

McGavin S 1976 The handedness and chirality of biological structure at the molecular and at higher levels of structural organisation. Biosystems 8:147–152

Michel H 1982 Characterisation and crystal packing of three dimensional bacteriorhodopsin crystals. EMBO (Eur Mol Biol Organ) J 1:1267–1271

Mitsui Y, Langridge R, Shortle BE, Cantor CR, Grant RC, Kodama M, Wells RD 1970 Physical and enzymatic studies in Poly d (I-C). Poly d (I-C), an unusual double helical DNA. Nature (Lond) 228:1116–1169

Needham J 1936 Order and life. Yale University Press, New Haven, USA

Pauling L 1960 The nature of the chemical bond. 3rd edition. Cornell and Oxford University Press

Richardson JS 1981 The anatomy and taxonomy of protein structure. Adv Protein Chem 34:167–339

Tachibana T, Kambara H 1967 Enantiomorphism in the super helices of poly-γ-benzyl-glutamate. Kolloid Zeitschrift Zeitschrift Polymere 219:40–42

Wainwright SA, Vosburgh F, Hebrank JH 1976 Shark skin: function in locomotion. Science (Wash DC) 202:747–749

Watson JD, Crick FHC 1953 Molecular structure of nucleic acid. A structure for deoxyribose nucleic acid. Nature (Lond) 171:737–738

DISCUSSION

Frankel: You mentioned that helical handedness is extremely important in the formation of a fibre or in the formation of a hollow tube, and that the sense of handedness of the tube can be determined by that of its monomers in some way. Previously, you wrote an interesting note about systems involved in the winding of filaments in a bacterium (Galloway 1990). There, handedness was indeterminate: it could be right handed, left handed or neither, depending on environmental conditions. That structure is built up of monomers. In terms of your talk, do those monomers have to have a symmetry element because they can assume either right- or left-handed forms?

Galloway: I think not. Sometimes the helical asymmetry would be determined from the monomer hand through one helical structure building up another helical structure in a helical hierarchy. That's not always going to be the case and quite clearly it is not the case in these bacterial mutants. There are other conditions involved.

An obvious example is temperature. I am interested in foraminiferans. Some of these have helical shells or tests and it is well known, particularly in the oil industry, that whether they have left- or right-handed shells depends on water temperature. Oil prospectors actually decide where to drill on the basis of whether they find left- or right-handed fossil foraminiferans.

Frankel: What fascinated me was thàt a continuous process could lead to a conversion between a left- and right-hand form. That is possible if the symmetry is helical. If the symmetry were based on a closed ring, in which left and right were independent of anterior and posterior, one would have an entirely different topology in which that kind of conversion would be impossible.

Thwaites: The bacterium is a mutant of *Bacillus subtilis* which grows in long filaments. People have thought that rod-shaped bacteria grow just by extension. From experiments that Neil Mendelson and I have done, it is clear that they actually grow with a slight twist. We cannot demonstrate this for a single cell, but it can be seen in single filaments and measured in the multifilament structures that are formed under certain conditions. Whether the twist is right handed or left handed depends not only on the kind of mutant, but also on the ionic environment, what it feeds on, the temperature and so on. I can understand biologists wanting to relate this to molecular helices, but such a relation is unlikely because of the lack of regular structure in the cell wall at the molecular level.

The bacterium is about one micron in diameter. The peptidoglycan polymer which is the structural element of the cell wall is only about one hundred repeats long. Each backbone chain can stretch only a tiny way around the wall. The mechanical properties of the wall have been measured using aggregates of filaments (Thwaites & Mendelson 1989) and, as far as I can see, the peptido-glycan behaves like any other amorphous polymer. It doesn't seem to have any kind of regular structure.

Galaburda: Discontinuity in a helical structure commonly occurs when you make a loop in a telephone cord. It seems to me that it may not be accidental that anomalous, wrong-handed molecules are present much more commonly in structural peptidoglycans than in functional proteins. That could be a good mechanism for breaking a regularity in a three-dimensional structure and producing more variability in three-dimensional morphology. If a cell was building a membrane in a particular way, then introduced a D-amino acid, it could produce more variability in the types of three-dimensional structure that it could build. This is very important for people such as myself who are interested in variability in three-dimensional structures of large neuronal assemblies.

Galloway: I am not aware that D-amino acids do occur structurally except in bacterial coats.

This break in continuity is very important. But linking occurs without the need to switch hands of amino acids; the L-amino acid L-proline does that. It will kink or bend a straight piece of helix. Other L-amino acids are also called helix breakers; they will stop a regular structure at the point they occur or cause it to go in a different direction.

There is a protein called macrophage scavenger protein which is largely a triple helix, but it looks as if the helical hand is reversed half way along. At one end it consists of three α-helices twisted into a left-handed helix; at the other end it is like a collagen triple helix, being right handed. Another example is a liposome that has oppositely coiled helices at either end, reversing in the middle.

Thwaites: This sort of reversal in the telephone cord represents a property well known in the textile industry. Such reversals occur in many artificially textured fibres; they arise there, and possibly elsewhere, in one of two ways. If you take a long fibre that has the same curvature all along its length, i.e. it would form a closed coil if you let it, hold it straight and then reduce the tension while preventing relative rotation of the ends, it forms helices of left- and right-hand with a coupling piece. I don't know whether that is how the liposome is formed, but the two-helix form is in a lower energetic state than the straight or simply curved form. Similar structures can be obtained by twisting a fibre into a helix, heat-setting it in that form and de-twisting it by rotating one end. This is how it happens in the telephone cord. Depending on the set twist, helices of different pitch and different size can be obtained.

Galloway: There are two kinds of analogies you can use for fibrous molecules and other fibrous structures. I chose the spiral staircase because I was interested in self-assembly. Rope, or the twisted telephone cord, is the other analogy. They are not really related. The latter is a continuous fibre; the staircase analogy is for a discontinuous fibrous structure.

Wolpert: Can you explain Z DNA? You have given us the idea that the difference in handedness can reflect the building blocks used in self-assembly, but there is no change in the base sequence of the DNA.

Galloway: The difference between Z DNA and other forms of DNA is that in the former there is a peculiar regularity in the sequence. You always get one type of base followed by the other in a strict repeating pattern. That structural regularity seems to allow a structure of different hand. In both A and B forms of DNA the nucleotide sequences are essentially random, they don't have that extra regularity imposed on them. Z DNA is not a left-handed form of what would otherwise be a right-handed DNA structure.

The basic α-helices in proteins and DNA are, for me, the two elemental structural forms in molecular biology. Individual amino acids or individual bases are stacked together randomly, more or less. The only protein helix that is fundamentally different from the α-helix is the collagen triple helix; to get that, you have to have a very contrived amino acid sequence.

Wolpert: Why is it contrived? You are almost implying that forming a right-handed helix is somehow the natural way to do things. I don't see why it should be.

Chothia: It's contrived in terms of the amino acids you put in the sequence. An α-helix can have any amino acid anywhere within it. To form a collagen-type helix, you need a special sequence because the side chains interact with the main chains. The collagen triple helix does not occur just because you have asymmetrical centres in its constituent amino acids; there are also groups on specific side chains that occur in a specific order and interact in such a way as to determine the triple helix structure.

Galloway: The same is true of right- and left-handed DNA; to get left-handed DNA, you need greater order than in the sequences that form 'normal' DNA.

Galaburda: Are there specific base sequences that make the DNA even more right-handed than the DNA formed by a random mixture of bases?

Galloway: I don't know the details. The helical parameters for different polynucleotides are different, so there may be. If there is not a strictly random distribution of bases, the helical parameters will be altered in some way. But helical DNA structures do fall into obvious structural families.

Mason: DNA molecules containing alternating purine and pyrimidine bases tend to form a left-handed double-stranded helix, particularly those rich in guanine and cytosine.

Weber: A large number of protein sequences can give rise to α-helices. A limited number of these sequences have a certain pattern superimposed by the distribution of amino acid side chains. Such α-helices form coiled coils. These sequences have to have hydrophobic residues in the A and D positions, but can tolerate many changes in other positions. These will give the various coiled coils individual surfaces.

In muscle, there are α-helices in myosin, in tropomyosin and in the inter-mediate filaments; these are all coiled coils built of interacting α-helices. Helices have been known for a long time and they have become very popular. Recently, however, we have learned that β-structure is also a prominent feature in the myofibril. The superstructure running from the Z line to the M line is composed of titin molecules which have the length of a half-sarcomere. Titin and several myosin-associated proteins are built by repeats of β-structure. By structural criteria, all these molecules belong to the immunoglobulin G superfamily.

Wolpert: Structures themselves, like microtubules, do they ever go the other way? Are there any mutants that reverse them?

Weber: There are some observations that microtubules may have a different number of protofilaments in some species and in some situations.

Wolpert: But never with a different handedness?

Weber: I don't think so.

Burn: Is there any evidence of a connection between the handedness of helices at the molecular level, for example in collagen, and of the helices in the whole organism? Do you ever see contrast at the molecular level in helices in organisms that are going the opposite way to the norm?

Galloway: If you accept the symmetry argument that the chirality of the structural monomers (amino acids or sugars) determines the hand of the helical structures they form, it is natural to suppose that the hand of those helices will force the hand of higher order helical structures made from them. People have thought that for a very long time, but is it true? The answer, of course, is yes and no.

The first person to look at this question seriously was Francis Crick. He was interested in the idea that two α-helices could twist together to form a twine-like protein. He conjectured that two right-handed protein α-helices would form a left-handed double helix. In rope, the helical hand switches at each level in the hierarchy. If you form a three-stranded rope from three separate identical fibres, the rope has a sense of twist which is opposite to the sense of twist in the fibres. The reason is that as the fibres try to untwist they force themselves into each other. It looks as though that is also true of proteins. In proteins where α-helices are arranged in coiled coils, the amino acid sequence in those particular proteins is again contrived by evolution to get that effect: it cannot be done with a random amino acid sequence. Every three and a half residues there is an amino acid with a large hydrophobic side chain which interlocks with another on the inside of the coiled coil to make the whole structure stick together.

The idea of symmetry forcing the hand, then the amino acid sequence being contrived to support that structure, became ingrained in the thinking of molecular biologists and persisted for some time. That is the Yes part of the answer.

The No answer comes from the filamentous bacterial virus Pf1, in which right-handed helical protein subunits stack together to form a right-handed virus. The symmetry argument is very good, but it is not perfect. Structural hand at a higher level is determined by the interactions between amino acid side chains. In some cases, the symmetry may override the effects of the side chains, but it doesn't always.

That gets you from big molecules to the next level of structure. I don't know whether you could then carry the argument further and say something about the helical hand of, for example, spiral surface windings on neurons.

Lewis: Another example that handedness of the molecule doesn't determine the handedness of the higher order structure is bone, where the successive layers of collagen in a Haversian system have opposite helical hand.

Brown: When extrapolating from molecular symmetry to higher orders, it's easy to see how you might do that when the higher orders have spiral handedness but not when they have bilateral handedness as in the mammalian body plan. We should distinguish between those two situations.

Galaburda: There are epigenetic factors that could help relate the helix to bilateral asymmetry. They are based on the fact that motility is a very important part of embryogenesis. Cells are born in germinal centres and must migrate to their end organ. The direction of that migration may be biased by the helices of the motors that drive them. We have some indirect evidence that migration occurs asymmetrically in the developing brain (Rosen et al 1991).

Galloway: I don't want to give the impression that the only way in which structure at one level can influence structure at another level is by simple interaction. There may be much more interesting ways in which a helix might influence structure at a higher level.

Wood: Some biological structure is going to depend on interactions between sheets of helices. In nematodes, the cuticle has at least two layers of helical fibres arranged with a particular pitch. A mutation in a particular collagen gene locus can produce animals where these sheets are shifted relative to each other (Kramer et al 1988). The animal itself may have a left-handed helical twist, for example, and move in an abnormal way; these are called 'left-rollers'. Another allele at the same locus can produce a right-roller phenotype, where the helices are shifted a little differently (Kusch & Edgar 1986). These are single amino acid substitutions that presumably result in different interactions of the sheets with each other to affect the entire morphology of the animal.

Galloway: An area which I didn't touch on at all is this idea of helicoidal structures, where flat sheets interact face-to-face, and you may get stacks of many hundreds, if not thousands, with a gradual turn. Insect cuticle is a good example of that. Plant cell walls also often have a mechanism of that kind to change from one structural layer to another. These are helicoidal; they are flat sheets arranged with a definite screw sense.

Burn: My recollection of mediaeval castles is that they have left-handed spiral staircases so that the defenders coming down could fight with their right hand whereas the attackers coming up were forced to fight with their left hand. Your cathedral had a right-handed staircase (Fig. 6b). Was that because they wanted to be non-militaristic in their architecture or because they wanted to give the advantage to the person trying to fight his way up to heaven or was your restorer working from an engraving?

Galloway: There is no evidence that one hand of spiral staircase predominated over the other. A large survey by myself and the Scottish Office of National Monuments showed that there is no particular handedness!

McManus: You showed an engraving of a painting by Fragonard (Fig. 1) and said that artists don't care about right–left asymmetry. That is not always the case. Van Gogh wrote to his brother and complained that his painting 'The Potato Eaters' had been reversed when it was printed. The painting shows five right-handers eating potatoes; the print shows the statistically improbable occurrence of five left-handers! We showed people a series of slides of paintings, half the right way round, half the wrong way round, and asked which was the original (Blount et al 1975). People could do it, but when we analysed the results they could only tell which was the original if they had seen the picture before.

Peters: From an engraving, you can tell the handedness of the engraver, because right-handers cross-hatch in one direction and left-handers cross-hatch the opposite way.

Wolpert: From molecules to brains in one easy session!

References

Blount P, Holmes J, Rodger J, Coltheart M, McManus IC 1975 On the ability to discriminate original from mirror-image reproductions of works of art. Perception 4:385–389

Galloway J 1990 Putting a twist in the tale. Nature (Lond) 343:513–514

Kramer JM, Johnson JS, Edgar RS, Booth C, Roberts S 1988 The *sqt-1* gene of *C. elegans* encodes a collagen critical for organismal morphogenesis. Cell 55:555–565

Kusch M, Edgar RS 1986 Genetic studies of unusual loci that affect body shape of the nematode *C. elegans* and may code for cuticle structural proteins. Genetics 113:621–639

Rosen GD, Sherman GF, Galaburda AM 1991 Ontogenesis of neocortical asymmetry: a [^{3}H]thymidine study. Neuroscience 41:779–790

Thwaites JJ, Mendelson NH 1989 Mechanical properties of peptidoglycan as determined from bacterial thread. Int J Biol Macromol 11:201–206

Asymmetry in protein structures

Cyrus Chothia

Cambridge Centre for Protein Engineering and MRC Laboratory of Molecular Biology, Hills Road, Cambridge CB2 2QH, UK

Abstract. The asymmetry of L-amino acids determines the asymmetrical features of α-helices and β-sheets. These in turn determine two principal aspects of the three-dimensional structure of proteins: the preferred ways in which α-helices and β-sheets pack together, and certain topological features of the paths followed by polypeptide chains through structures. Though the asymmetrical nature of amino acids plays the central role in determining the asymmetrical aspects of protein structures, it has little or no influence on the next level of biological structures— assemblies of protein molecules.

1991 Biological asymmetry and handedness. Wiley, Chichester (Ciba Foundation Symposium 162) p 36–57

In the 1950s the existence of α-helices and β-sheets in proteins was widely accepted but there was no direct knowledge of their three-dimensional structure. The models proposed for protein structures tended to be simple and symmetrical (see, for example, the review by Hodgkin 1950). The complex and asymmetrical nature of the first protein for which the structure was determined caused a considerable shock. Kendrew (1961) described myoglobin as '. . . almost nothing but a complicated set of rods (of polypeptide) sometimes going straight for a distance then turning a corner and going off in a new direction . . . much more complicated and irregular than most of the early theories of the structure of proteins had suggested'. Since then, many more protein structures have become known and the degree of complexity has only increased.

Although proteins have complex structures, simple general principles that govern these structures have been discovered (for reviews see Richardson 1981, Chothia 1984, Finkelstein & Ptitsyn 1987, Chothia & Finkelstein 1990). The relation of these principles to the actual structures is similar to the relation between the rules of grammar and sentences. The principles define preferred conformations, packings and topologies: different combinations of these form different structures. In the same way, the rules of grammar, working with only a limited vocabulary, can generate a large number of complex sentences.

In this paper, I will mainly discuss the principles that relate the asymmetrical nature of the building blocks of proteins, the amino acid residues, to the major asymmetrical features of their three-dimensional structure. I shall suggest, however, that the importance of asymmetry of amino acids in biological structure is limited. Though it plays a major role in determining the internal structure of protein molecules, it plays little or no role in the higher levels that involve assemblies of protein molecules.

α-helices, β-sheets and the structure of proteins

Proteins are polymers made from amino acid residues. Most proteins contain between 50 and 1000 residues linked together in a chain which is folded so it runs back and forth to form a globular unit. There are 20 different types of residues and the structure and function of a protein is determined by their sequence in the chain. However, it is usual for each run of chain to either coil into a (α-)helix or form part of a (β-)sheet.

Mason (this volume) has discussed the asymmetrical nature of the amino acids and Galloway (this volume) has described how L-amino acids produce the right-handed α-helix. The β-sheet is formed by two or more stretches of protein chain running parallel or anti-parallel to each other. β-sheets are not flat but, when viewed along their strands, twist in a right-handed direction (Fig. 1). The twist is right handed because of the energetics of chains made from L-amino acids (Chothia 1973). Chains made from D-amino acids would produce sheets with a left-handed twist.

Two principal aspects of the three-dimensional structure of proteins are (1) the manner in which α-helices and β-sheets pack together, and (2) the topology of the pathways followed by the polypeptide chains through the structure.

The packing of α-helices and β-sheets

The general shape of α-helices and β-sheets governs the ways in which they pack together to form compact structures (Chothia et al 1977). Commonly, β-sheets form layer structures with helices or other β-sheets packed on their faces. They can also fold upon themselves to form barrel-like structures. The cylindrical shape of α-helices allows them either to stack around a central core or to form layer structures (Levitt & Chothia 1976, Richardson 1981, Finkelstein & Ptitsyn 1987, Murzin & Finkelstein 1988).

Protein interiors are close packed. This means that the exact manner in which α-helices and β-sheets pack together depends upon the exact shape of their surfaces. Because the different residues have different shapes, α-helices and β-sheets with different amino acid sequences will have different surfaces. In spite of this, the observed packing geometries do show clear preferences and regularities. These can be explained by simple models that embody just the average features of the surfaces of α-helices and β-sheets. Most observed packings can be seen as variations on the simple model packings in which

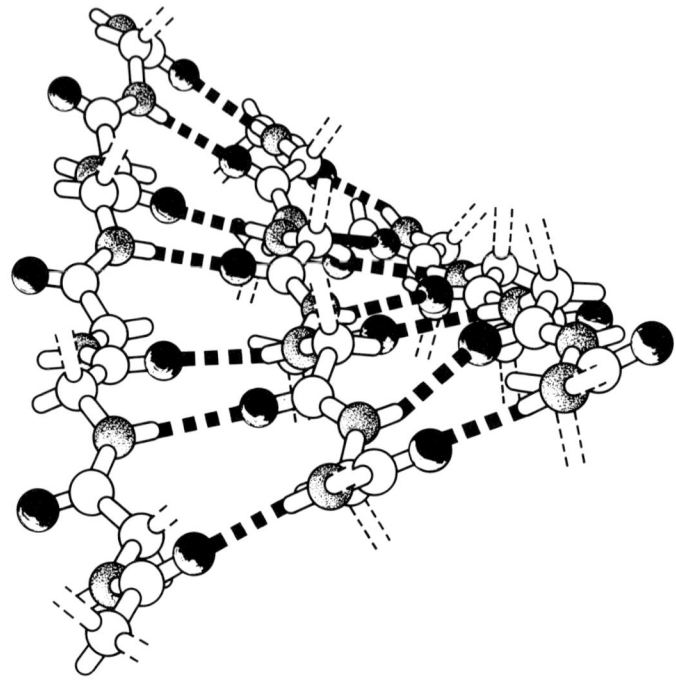

FIG. 1. The twist of β-sheets. β-sheets are formed by the parallel or antiparallel association of polypeptide chains. Usually, β-sheets are not flat but twisted as shown here. The twist is right handed: as you move down the sheet in this figure you rotate in a right-handed direction. From Chothia (1973) with permission.

the geometry is modified only a little by the residues that actually form the surfaces.

The packing of α-helices

The relative orientation of two α-helices packed face-to-face can be described by the angle between the helix axes when these are projected onto their plane of contact. Though the values for this angle in observed helix packings cover the whole range of possible relative orientations, about half pack with their axes inclined at approximately $-50°$ and about a quarter pack with their axes inclined at approximately $+20°$ (Fig. 2). The preference for these orientations arises from the arrangements of side chains on helix surfaces (Crick 1953, Chothia et al 1981).

   Residues on the surface of an α-helix tend to form ridges separated by grooves. Usually, ridges are formed by residues separated by three other residues in the sequence; less commonly, they are formed by residues separated by two others. The packing together of the ridges and grooves on the surface of two ideal α-helices will incline their axes at an angle of $-50°$ or $+20°$. These angles arise from the way the ridges are tilted relative to the helix axis, as shown in Fig. 2. D-amino acids would produce left-handed helices, in which the ridges would tilt in the opposite direction and so their preferred packing angles would be $+50°$ and $-20°$.

The packing of β-sheets

The intrinsic flexibility of β-sheets allows them not only to twist but also to coil and bend (local coiling). This means they can pack in several different ways (Chothia & Finkelstein 1990). Here I will discuss one example: the packing of independent, twisted β-sheets.

The right-handed twist of β-sheets means that their surfaces are also twisted. Two ideal twisted β-sheets close pack if the rows of side chains that form the contacts at the interface are aligned. Although these side chains are aligned, the twist of the sheets results in the main chain directions of the two β-packed sheets being at an angle of about $-30°$ (Fig. 3). In real structures, the variations in the residues that form the interface modulate the exact geometry (Chothia & Janin 1981, Cohen et al 1981). As before, if they were made from D-amino acids, the sheets would twist in the opposite direction and pack together at an angle of $+30°$.

The packing of α-helices on β-sheets

This packing involves two surfaces with similar asymmetries and their association produces a symmetrical arrangement. As mentioned above, β-sheets in proteins have a right-handed twist when viewed along the strands (Fig. 1). In an α-helix the two adjacent rows of residues, i, $i+4$, $i+8$, . . . and $i+1$, $i+5$, $i+9$, . . . form on one side of a helix a surface with a right-handed twist (Fig. 4). Ideal α-helices pack onto ideal β-sheets with their axes parallel to the strands, because in this orientation the two rows of helix residues form a surface complementary to that of the β-sheet (Fig. 4). The majority of the observed packing has geometries close to that given by this model; very occasionally, large departures from this geometry are produced by unusual sets of side chains at the interface between the β-sheets (Janin & Chothia 1980, Cohen et al 1982).

Although D-amino acids would give helices and sheets with twists opposite to those produced by L-amino acids, these would still be complementary and pack with the helix axis parallel to the sheet strands.

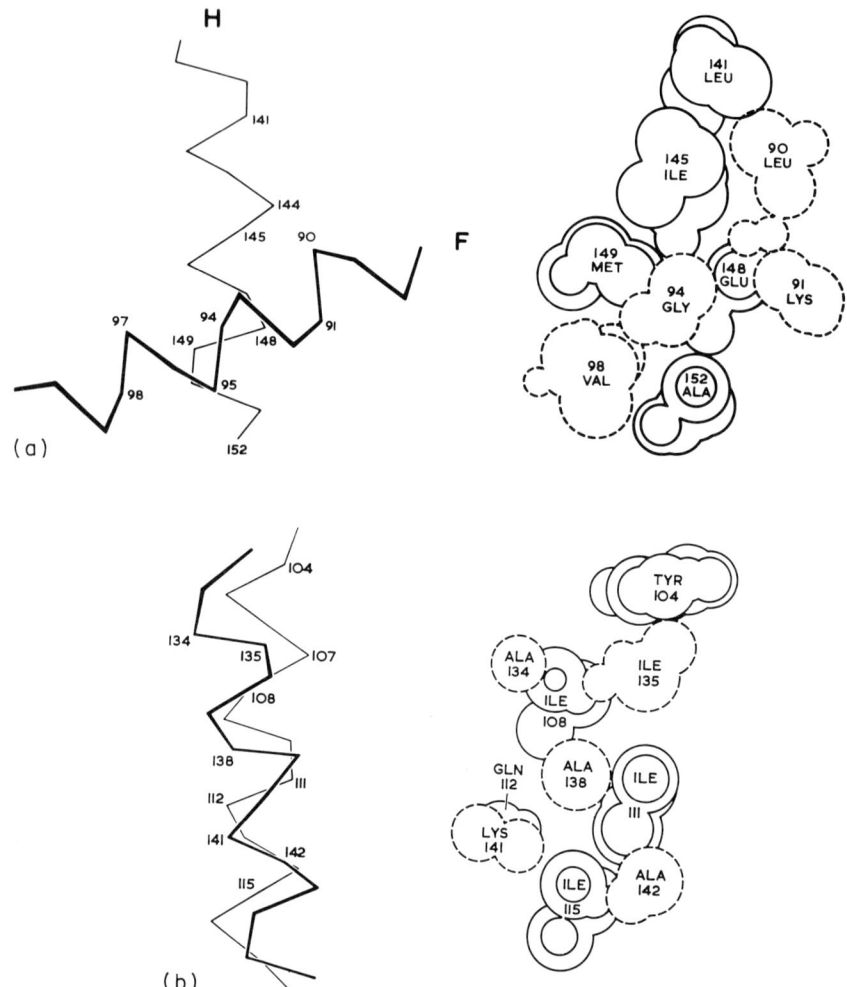

FIG. 2.   The packing of α-helices. The two preferred orientations of packed helices are those in which their axes are inclined at − 50° and + 20°. Here I show one example of each type taken from observed protein structures. On the left is a simplified representation of the packed helices; each residue is represented by one atom. On the right I show how the helices pack together. Broken lines represent residues in the helix above the plane of the page; unbroken lines represent residues in the helix below the plane of the page. The two different orientations are given by the packing of different rows of residues. (Top) The − 50° orientation is given by the packing of ridges formed by residues separated by three in the sequence: residues 90–94–98 of one helix pack between two ridges, 141–145–149 and 148–152, in the other helix. (Bottom) The + 20° orientation is given by one helix using ridges formed by residues separated by two in the sequence, e.g. 135–138–141, packing into a ridge formed by residues separated by three in the sequence, 104–108–112 and 111–115.

STRAIGHT   TWISTED
CHAINS   CHAINS

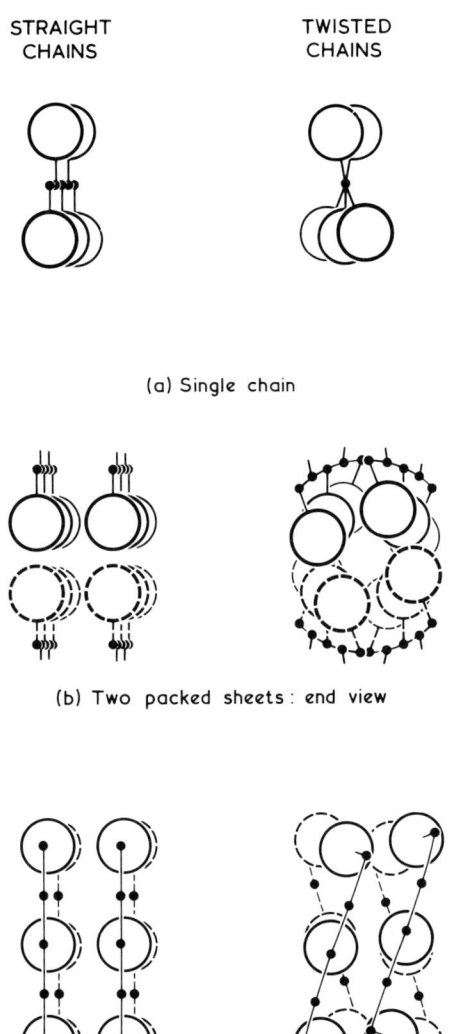

(a) Single chain

(b) Two packed sheets : end view

(c) Two packed sheets : top view

FIG. 3. The packing of β-sheets. Each residue in a protein chain is represented by a small closed circle (main chain atoms) and a large open circle (side chain atoms). On the left are structures formed by straight strands; on the right are structures formed by twisted strands. (a) single strands; (b) two two-stranded sheets packed face-to-face and viewed along the strands; (c) the same structures viewed perpendicular to the interface between the sheets. The twist of the sheets rotates the main chain direction of the sheet nearest the viewer by an angle of − 30° relative to the far sheet.

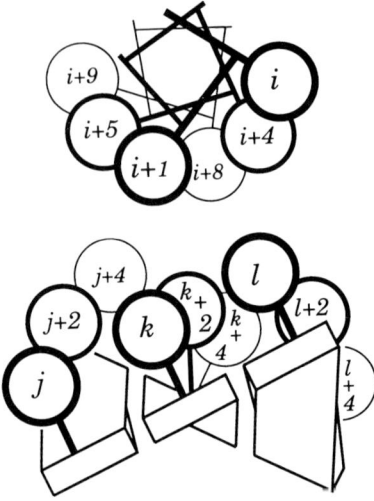

FIG. 4. The packing of α-helices and β-sheets. Schematic pictures of an ideal three-stranded β-sheet and an α-helix in which the side chains on one face of each are shown as large open circles. On the α-helix, the surface formed by helix residues i, i + 1, i + 4, i + 5, i + 8 and i + 9 has a right-handed twist. Because of the twist of the β-sheet, the side chains j, j + 4, j + 6, k, k + 2, k + 4, l, l + 2 and l + 4 also form a surface that has a right-handed twist. Therefore, these ideal secondary structures close pack when the helix axis is parallel to the strand direction, because in this orientation the two surfaces are complementary. See Fig. 7 for an actual example of the packing of α-helices and β-sheets in proteins.

Chain topology

Inspection of protein structures shows that the pathways followed by polypeptide chains are subject to limitations and preferences. These can be summarized as follows:

(1) Pieces of secondary structure that are adjacent in the primary sequence are often in contact in three dimensions and usually pack in an antiparallel, rather than parallel, manner (Fig. 5a).

(2) The connections in β–X–β units (where the βs are parallel strands in the same sheet, though not necessarily adjacent, and X is an α-helix, a strand in a different sheet, or an extended piece of polypeptide) are right handed (Fig. 5b).

(3) The connections between secondary structures neither cross each other nor make knots in the chain.

Combinations of these basic rules give further topology rules.

Initially, it was suggested that the three basic empirical rules might arise from structural features of the folded state or structural and kinetic features of intermediates in the folding pathway. However, it has recently been shown that a completely general explanation is provided by the statistical properties of polypeptide chains (Finkelstein & Ptitsyn 1987).

The energy required to bend a polymer chain is proportional to its stiffness (Landau & Lifshitz 1959, Flory 1968). The free energy required to bend a chain is given by

$$\Delta G = \frac{RTa(\Delta\Theta)^2}{2L}$$

where R is the gas constant, T the temperature, a the persistence length (a measure of the stiffness of a polymer and about 17 Å for proteins), L the chain length, equal to 3.5m Å for m residues, and $\Delta\Theta$ is the bending angle. At room temperature

$$\Delta G = \frac{6.3\,(\Delta\Theta)^2}{m}$$

This expression shows that the rigidity of polypeptide chains imposes restrictions on the paths they can follow (Finkelstein & Ptitsyn 1987). For a loop of 10 residues the bending free energy is about 6.2 kJ/mole for a bending angle of 180° and about 25 kJ/mole for a bend of 360°. Thus antiparallel associations, in which the bend angle is 180°, are favoured over parallel associations where the bend is about 360°.

The rigidity of polypeptide chains also favours right-handed over left-handed connections in β–X–β units. The right-handed twist normally found in β-sheets means that the bending angle for β–X–β units with right-handed connections is less than that for left-handed connections (Fig. 5b). This effect is cooperative, because if connections are not to cross each other in a protein structure (rule 3), all, or nearly all, connections must have the same hand.

The absence of knots in proteins arises from the low probability of the required threading of the polypeptide chain. Calculations show that for chains of 100–300 residues, the probability of knotting is 1–10% if the chains are extremely thin; it is much less if the chains have the actual width of polypeptides or secondary structures (Vologodskii et al 1974, Crippen 1974, Frank-Kamenetskii et al 1975).

These arguments suggest that the origin of the asymmetrical features of the topology of polypeptide chains in protein structures is the relative energetic cost of different folding intermediates. Such intermediates usually have only marginal stabilities, but we might expect rare occasions where the larger energy losses are compensated for by additional interactions within intermediates. Thus, the rules for the topology of polypeptide chains are statistical in nature. Rare exceptions, such as left-handed β–X–β units and chain threading (which resembles unfinished knotting), have been observed.

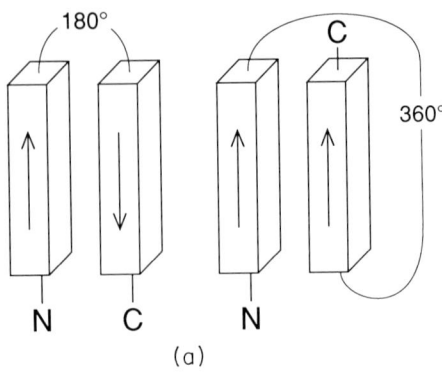

(a)

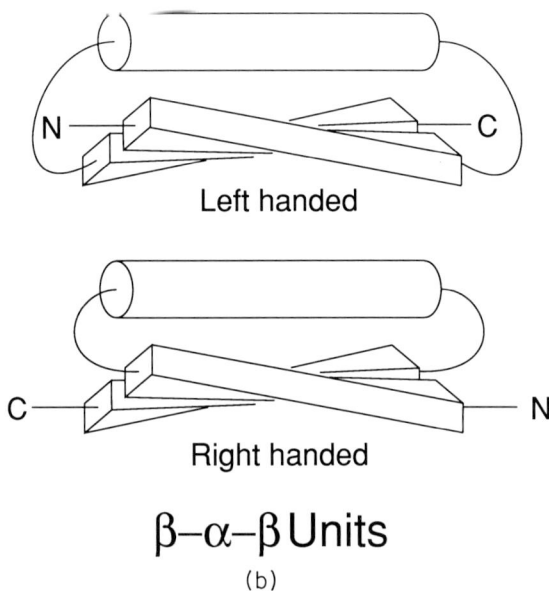

Left handed

Right handed

β–α–β Units

(b)

FIG. 5.   Features of chain topology in protein structures. (a) Pieces of secondary structure adjacent in the sequence tend to be adjacent in the structure and to pack in an antiparallel manner. Antiparallel packing involves the connecting loop bending through about 180°; parallel packing involves bending through about 360°. (b) The connections in β–X–β units (where the βs are parallel strands in the same sheet, though not necessarily adjacent, and X is an α-helix, a strand in a different sheet, or an extended piece of polypeptide) are right handed. Because the strands in the β-sheet are twisted, right-handed connections involve smaller bends by the connecting loop than do left-handed connections.

Stability and the surface area of folded proteins

In the many known protein structures, the packing geometries of α-helices and β-sheets and the topologies of the polypeptide chains are described by rules such as those discussed above (Chothia & Finkelstein 1990). Different combinations of these rules produce an enormous variety of different structures. There is, however, a rule that governs which combinations are permitted. This rule arises from the requirement for the folded structure to be stable.

The free energies of folded proteins are small, usually in the range 20–65 kJ (Privalov 1979). These values represent small differences in the large terms favouring the unfolded or folded state. The main term favouring the unfolded state is the conformational entropy of the polypeptide chain; the value for this depends upon the length of the chain. The main terms favouring the folded state are hydrophobic free energy and hydrogen bonding. The hydrophobic contribution depends on the amount of protein surface that is buried within the folded protein and removed from contact with the solvent (Kauzmann 1959).

The surface area of an unfolded protein is proportional to its molecular weight. The imbalance between the favourable and unfavourable energy terms implies that the buried and accessible surface areas of a folded protein should be correlated with molecular weight. The correlation for proteins built from one polypeptide chain was established some time ago (Chothia 1975). Recently, it has also been established for proteins built of more than one chain (oligomeric proteins) (see Fig. 6; Janin et al 1988). Although the amount of total buried surface is essentially constant for oligomeric proteins of the same molecular weight, the individual contributions of the surfaces buried within and between the subunits vary greatly.

The very different, complex and asymmetrical structures that are found for proteins of given molecular weight are different solutions to the same problem: how to bury the amount of surface required to give the protein sufficient stability to function, within the context of the packing and topology rules.

The symmetry of assemblies of protein molecules

The asymmetrical features of the internal structure of proteins discussed here arise from proteins being formed by L-amino acids. These have conformational properties that result in their polymers forming right-handed α-helices or β-sheets with a right-handed twist. The asymmetrical shape and surface features of these secondary structures are responsible for their asymmetrical arrangement in protein molecules. L-amino acids are also responsible for certain asymmetries in the topology of paths followed by the polypeptide chain in protein structures.

The level of biological structure above that of the individual protein molecule involves assemblies of protein molecules. The structures currently known for

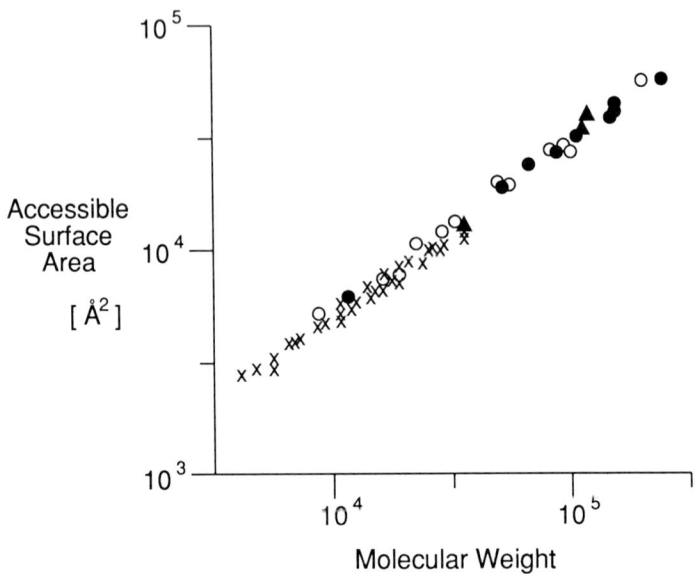

FIG. 6.   The correlation between the surface area that is accessible to solvent and the molecular weight of proteins. X, monomers; ○, dimers; ●, tetramers; ▲, hexamers and octamers. From Janin et al (1988) with permission.

protein assemblies suggest that the asymmetrical nature of amino acids has little or no influence in their general design.

   The reason for this is the different nature of the interacting surfaces. Within proteins, the major interactions involve the close packing of the surfaces of individual α-helices and β-sheets; as I have discussed, the shapes of these surfaces are mainly the consequence of the asymmetrical nature of amino acids. In a few cases, protein assemblies also involve the close packing of the surfaces of individual α-helices and β-sheets; but in most cases, the contact area is formed by adjacent bits of α-helices, β-sheets, and/or connecting loops. Together, these give a quite irregular surface, whose general shape has no direct relation to the asymmetry of amino acid residues.

   Although the surfaces involved in association are irregular, the smaller protein assemblies themselves are usually symmetrical. Specific bonding between identical units necessarily leads to a symmetrical structure since there will be only a limited number of ways in which each unit can be connected to its neighbours to form the maximum number of most stable bonds (Casper 1966). In small assemblies, the components are related by two-, three-, four- or sixfold axes. For example, the enzyme phosphofructokinase has four subunits related by three perpendicular twofold axes (Evans & Hudson 1979). Fig. 7 shows one pair of subunits related by one of the twofold axes. If this protein were made from D-amino acids, the structure of the subunits and their interfaces

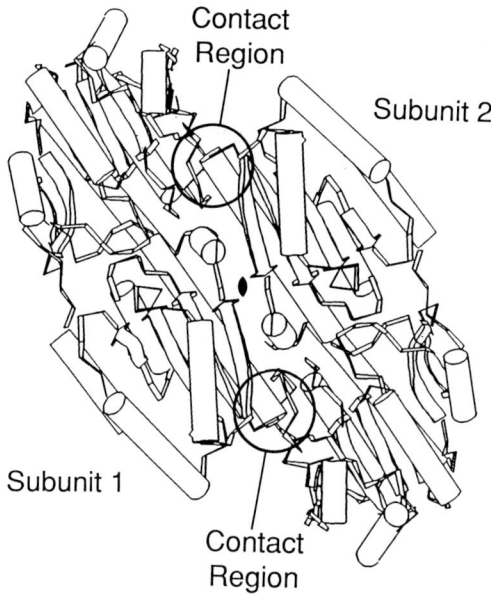

FIG. 7.   Two subunits of the enzyme phosphofructokinase. α-helices are represented by cylinders and β-sheet strands by ribbons. They pack with the α-helix axes approximately parallel to the strands of the β-sheets (see Fig. 4). The contacts between the subunits involve small irregular surfaces and result in the two subunits being related by a twofold axis.

would be inverted but they would still be related by a twofold symmetry axis.

Large assemblies of protein molecules do form asymmetrical structures but, in the cases about which we know, this asymmetry is not related to that of amino acids. Indeed, some protein molecules can, under the appropriate conditions, form structures with different asymmetries. An example of this is the bacterial flagellum discussed by Berg (this volume). This can form a right-handed helical structure with a twist of 7.8° and a left-handed helical structure with a twist of 3.4° (Kamiya et al 1980).

Conclusion

In this paper I have argued that, though the asymmetrical nature of amino acid residues is responsible for the asymmetrical features of α-helices, of β-sheets and of the three-dimensional structure of proteins, it has little or no influence at the next level of biological structure. This concept of the determinants of asymmetry in biological structure operating over a limited range and, consequently, there being different determinants for different ranges of structure may be generally true.

References

Berg HC 1991 Bacterial motility: handedness and symmetry. In: Biological asymmetry and handedness. Wiley, Chichester (Ciba Found Symp 162) p 58–72

Caspar DLD 1966 Design principles in organized biological structures. In: Principles of biomolecular organisation. Churchill, London (Ciba Found Symp) p 7–39

Chothia C 1973 Conformation of twisted β-pleated sheets in proteins. J Mol Biol 75:295–302

Chothia C 1975 Structural invariants in protein folding. Nature (Lond) 254:304–308

Chothia C 1984 Principles that determine the structure of proteins. Annu Rev Biochem 53:537–572

Chothia C, Finkelstein AV 1990 The classification and origins of protein folding patterns. Annu Rev Biochem 53:1007–1039

Chothia C, Janin J 1981 Relative orientation of close-packed β-pleated sheets in proteins. Proc Natl Acad Sci USA 78:4146–4150

Chothia C, Levitt M, Richardson D 1977 Structure of proteins: packing of α-helices and pleated sheets. Proc Natl Acad Sci USA 74:4130–4134

Chothia C, Levitt M, Richardson D 1981 Helix to helix packing in proteins. J Mol Biol 145:215–250

Cohen FE, Sternberg MJE, Taylor WR 1981 Analysis of the tertiary structure of protein β-sheet sandwiches. J Mol Biol 148:253–272

Cohen FE, Sternberg MJE, Taylor WR 1982 Analysis and prediction of the packing of α-helices against a β-sheet in the tertiary structure of globular proteins. J Mol Biol 156:821–862

Crick FHC 1953 The packing of α-helices: simple coiled coils. Acta Crystallogr 6: 689–697

Crippen GM 1974 Topology of globular proteins. J Theor Biol 45:327–338

Evans PR, Hudson PJ 1979 Structure and control of phosphofructokinase from *Bacillus stearothermophilus*. Nature (Lond) 279:500–504

Finkelstein AV, Ptitsyn O 1987 Why do globular proteins fit the limited set of folding patterns? Prog Biophys Mol Biol 50:171–190

Flory PJ 1968 Statistical mechanics of chain molecules. Interscience, New York

Frank-Kamenetskii MD, Lukashin AV, Vologodskii AV 1975 Statistical mechanics and topology of polymer chains. Nature (Lond) 258:398–402

Galloway JW 1991 Macromolecular asymmetry. In: Biological asymmetry and handedness. Wiley, Chichester (Ciba Found Symp 162) p 16–35

Hodgkin DC 1950 X-ray analysis and protein structure. Cold Spring Harbor Symp Quant Biol XIV:65–78

Janin J, Chothia C 1980 Packing of α-helices onto β-pleated sheets and the anatomy of α/β proteins. J Mol Biol 143:95–128

Janin J, Miller S, Chothia C 1988 Surface, subunit interfaces and interior of oligomeric proteins. J Mol Biol 204:155–164

Kamiya R, Asakura S, Yamaguchi S 1980 Formation of helical filaments by copolymerization of two types of 'straight' flagellins. Nature (Lond) 286:628–630

Kauzmann W 1959 Some factors in the interpretation of protein denaturation. Adv Protein Chem 14:1–63

Kendrew J 1961 The three dimensional structure of a protein molecule. Sci Am 205: 96–110

Landau LD, Lifshitz EM 1959 Statistical Physics, Part 1. Pergamon Press, London

Levitt M, Chothia C 1976 Structural patterns in globular proteins. Nature (Lond) 261:552–557

Mason SF 1991 Origins of the handedness of biological molecules. In: Biological asymmetry and handedness. Wiley, Chichester (Ciba Found Symp 162) p 3–15

Murzin AG, Finkelstein AV 1988 General architecture of the α-helical globule. J Mol Biol 204:749–770

Privalov PL 1979 Stability of proteins. Adv Protein Chem 33:167–241

Richardson JS 1981 The anatomy and taxonomy of protein structures. Adv Protein Chem 34:167–339

Vologodskii AV, Lukashin AV, Frank-Kamenetskii MD, Anshelevich VV 1974 Knot problem in statistical mechanics of polymer chains. Zh Eksp Theor Fiz 66:2153–2163

DISCUSSION

Morgan: Does it make any difference if one thinks about this in an evolutionary way? Presumably, the step from the structure of amino acids to a very complex protein like phosphofructokinase is a very large one and must have taken place gradually in many stages.

Chothia: I have simplified things in a sense. I have talked about proteins of approximately 150–200 residues. Very large proteins are usually composed of domains of about this size. Phosphofructokinase consists of four subunits, each of two domains. In the subunit shown on the left of Fig. 7, the bottom domain contains a sheet of four strands with helices packed below and above the sheet. In the top domain there is also a sheet with helices packed on either side. The two domains in the subunit are held together by many different small interactions contributed by several parts of the protein. There are no rules which can relate the symmetrical properties of the secondary structure within the domains to the interaction between the separate protein domains.

Wolpert: I don't understand why you have drawn a barrier between an individual domain and multiprotein structures. It seems to me that if you made a mirror-image protein using D-amino acids, you would have a mirror-image higher order structure. For example, if you had mirror-image tubulin, you would make microtubules with a different helical structure and everything would be mirror image.

Chothia: For helical structures, if you reversed the asymmetry of the amino acids, you would get inverse structures, I agree. But there are no rules which relate the asymmetries of assemblies to the asymmetries of amino acids. I have described the rules that relate the asymmetries of amino acids to the asymmetries of the internal structures of proteins. But the association between globular units depends each time on the particular structure, which is very complicated and an individual characteristic of each protein.

Brown: Are you saying that there can be continued influence beyond your barrier, but handedness is not reliably perpetuated?

Chothia: Yes. There are a few cases where proteins do associate using major elements of secondary structure. For example, the domains in antibodies associate through the packing of β-sheets and the rules I described apply, but that's very unusual.

Kondepudi: Is that also true for structural proteins? You gave phosphofructokinase as an example; enzymes are highly selective in their action as far as chirality is concerned. Only a small part of a functional protein, such as the active site of an enzyme, is involved in its chirally specific activity, for example catalysis. There the chirality comes from the L-amino acids that are the building blocks of the protein.

Chothia: For most structural proteins, the answer is yes. But, even in the rare cases where you do find packing of α-helices, the relations are not simple. Haemagglutinin is a trimeric protein that occurs on the surface of influenza virus. Each subunit contains helices stacked together. Although interactions between α-helices form the contacts between subunits, which is unusual, the particular side chains that occur at the interfaces between the helices override the simple rules that you would expect from the descriptions given in my paper. The subunits in haemagglutinin are related by a threefold axis.

I realize a few examples are not sufficient to prove this, but these examples are typical of what we find. Citrate synthase is a dimer with 440 residues in each subunit; these are about 80% helical. The dimer is formed by four helices in one subunit packing together with four helices of the other subunit. The packing results in the two subunits being related by a twofold axis.

Thwaites: The bacterial wall has a sugar polymer with peptide residues, so the argument for irregular packing in such a structure would be even stronger. The sugar polymer has peptide side chains with both L- and D-amino acids. Nonetheless, many microbiologists feel there is some very regular structure in this, which I can't believe. Dr Chothia's argument supports my idea.

Chothia: There is an interesting story concerning antibodies. You take a protein (P) that has a particular function and raise an antibody (A) which recognizes it. Then you raise an anti-antibody (AA) to the antibody (A). In some cases, when the original protein (P) is involved in recognition, the anti-antibody (AA) will also produce recognition. The established idea in immunology was that the anti-antibody had a site with the same structure as the recognition site of protein P. Recently, the structure has been determined of an antibody and anti-antibody complex (A-AA) (Bentley et al 1990). It is found that there is almost no relationship between the structure of AA and that of P. The protein was lysozyme, and the antibody to it was D13. There are 13 residues on D13 that contact lysozyme. There are also 13 residues in D13 that contact the anti-antibody, but only seven belong to the set that contact lysozyme. Of those seven residues, three have different conformations. So there are only four residues that are involved in similar interactions between the protein or the anti-antibody with the antibody. The binding sites are almost totally different.

Why does this happen? The total binding site of an antibody is about 2300 Å^2. The area involved in recognition is about 700–800 Å^2. In my example, the recognition site and the binding site partially overlap; the side chains on the surface are flexible and three residues change their conformation.

When you first create antibodies to an antigen, you create very many. Then to get an anti-antibody that mimics the original antigen you select one of these— the rare case where there is complete overlap between the recognition sites for the antigen and the anti-antibody. The important point is that recognition doesn't involve the same rules as those that govern protein structure, because you are selecting a small part of a surface which can be made with all sorts of elements of secondary structure. The small amount of surface that is involved in recognition sites means there is an enormous number of possibilities and you have broken free of the constraints of the amino acids' asymmetry.

Galloway: The strict stereoselectivity and specificity that occur as you go from the properties of amino acids to the structure of proteins are very strongly determined. One of the most interesting things that has emerged about proteins is that they have structures which are partly rigid and determined and partly flexible. Is part of what you are saying related to the fact that when you assemble protein structures into bigger structures there is a degree of flexibility?

Chothia: I tried to keep the stereochemistry to a minimum. To answer your question, I need to say something about conformational torsion angles in proteins. The polypeptide chain can be described in terms of the torsion angles about the amino acids. An amino acid within a protein can create a helix or a sheet by rotating around either the ϕ or ψ bond angle. Ramachandran showed 30 years ago that there are only certain permitted values of ϕ and ψ. Helices exist because the conformation allowed in terms of the ϕ and ψ values of the main chain is also the conformation that produces hydrogen bonds between the main chain peptides. Similarly, sheets are formed because the bonds permit an extended conformation in which the residues can hydrogen bond to each other.

There are two types of flexibility. There is that which produces only small modifications in the structure, where, essentially, the residues rock around particular torsion angles. If rocking isn't sufficient, the conformation jumps to a new structure.

For example, in the interface between close-packed helices the side chains intercalate. This doesn't rigidly fix the position of the helices; experimental crystallographic studies show that one helix can move relative to another by 1–2 Å. This is done by rocking on the side chains. Larger movements need a different packing of the side chains.

Frankel: Is the difficulty or impossibility of obtaining inflexible rules for domain–domain or chain–chain interactions related to the practical difficulty of predicting three-dimensional structure from one-dimensional structure?

Chothia: Yes.

Galloway: In pharmacology, you can have a receptor that will bind both enantiomers of a drug. In one case the binding is very much weaker than the other case, but the receptor will bind both enantiomers and produce some activity. The chemical specificity overrides the stereospecificity.

Crow: There are some striking exceptions, such as butaclomol (Bruderlein et al 1975).

Chothia: If we have a surface that is produced by three L-amino acids, A, B and C, one could make exactly the same surface using D-amino acids, but the geometry of the amino acids that created the surface would be different.

Wolpert: So if we built the world from D-amino acids, some multiprotein structures which have asymmetry would maintain the same asymmetry? I don't believe it!

Chothia: No. What I am saying is that one can produce the same surface on a different foundation. One can make buildings with the same outline but quite different insides. The same with proteins: there can be a particular geometry which interacts, but the scaffolding would be very different depending on whether it was made of L- or D-amino acids. Enzymes with quite different folds can recognize the same substrate.

McManus: Let us return to John Galloway's example of architecture. A French chateau typically has towers on the corners in which there are staircases. One recognizes the external conformation of a chateau; it could be built with either right- or left-handed spiral staircases inside without making any difference to the outside.

Wolpert: That's a false analogy, because you are describing a symmetrical structure.

McManus: It needn't be. Think of a cathedral with a tower on one side and a spire on the other.

Weber: We have to consider how proteins are made. In normal protein biosynthesis on ribosomes, only L-amino acids are used. I agree that you can make similar surfaces with L- or D-amino acids, but Nature made the decision a long time ago that normal protein biosynthesis uses only L-amino acids.

Chothia: They are asking intellectually whether I can make similar proteins with D-amino acids.

Weber: It is misleading to ask this question. D-amino acids are not put in by the mechanism of normal protein biosynthesis.

Wolpert: I was doing a thought experiment.

Weber: But you don't have to think about it!

Kondepudi: If you consider a living organism as a lot of proteins brought together, then it's clear that if you built the structure from the complementary, mirror-image amino acids, you might make a cell wall. (So it wouldn't matter whether you used all L-amino acids or all D-amino acids.) But from an evolutionary point of view, a system has to replicate and perpetuate itself. Then it becomes probably impossible that a system could evolve that was able to make two kinds of macromolecules, using all L- or all D-amino acids. Your view would indicate that to be built of proteins consisting of all L-amino acids, as far as big structures are concerned, is incidental not essential.

For evolution of life, I presume you would advocate that forms of life based on one of both L- and D-amino acids arose, then they competed with each other and the L form became dominant.

Chothia: My argument is, given that L-amino acids exist as the natural form, I can explain the consequences for protein secondary structures and their packings.

Weber: We learned earlier that in this universe life is based on L-amino acids. We will get nowhere by thinking that the few cases where D-amino acids occur have anything to do with normal protein biosynthesis and cell structure. D-amino acids are incorporated into proteins by entirely different mechanisms. Since the universe has decided to work with L-amino acids, we don't have to bother about D-amino acids.

Wolpert: If we take microtubules, which have a nice helical structure, the asymmetry or the helicity is a property of the protein subunits. That is the sort of interaction Cyrus Chothia is talking about.

Weber: But if you assemble microtubules on centrioles or freely in solution you can get different things.

Wolpert: But they are dependent on the actual folding of the globular protein.

Weber: I agree.

Wolpert: How many bases in the amino acid sequence would you have to change to change the helicity of a multiprotein structure? What do you know about those rules higher up?

Chothia: If you have an oligomeric protein and you disrupt the interfaces by mutation, you destroy the structure. The question is: can you get an alternative structure?

There may be cases where there are two possible ways of packing proteins with similar but not identical energies. One pattern commonly occurs; it is destabilized by mutation and the second packing with different hand and different geometry then becomes more favourable. I would guess that such cases are rare. Recognition is usually very specific. It is unlikely that there are many proteins which contain alternative hands or alternative surfaces that can come together to create stable structures.

Wolpert: Is it then easier to change the handedness of a multiprotein structure than to change the helicity of a single protein?

Chothia: I think both are extremely difficult!

In his paper, John Galloway mentioned the helical structure of cylindrical viruses. The spherical viruses form symmetrical structures by obeying very precise rules in their geometry that have no connection with the sorts of thing we have been discussing. These viruses are composed of multiples of 60 identical subunits packed to create a closed surface. The rules governing the formation of those structures are driven by the need to form a closed surface and they are totally different from the rules we have discussed so far. For other organelles there are other sets of rules which also have no relationship to what we have been discussing.

Galloway: If you look at some of these big protein assemblies, although the proteins are apparently identical they can obviously interact in different ways with other identical proteins. Otherwise you couldn't form these quasi equivalent shells. There is a flexibility at some stage in the way that the proteins can interact with one another.

Chothia: There are proteins that are hinged and therefore can create different angles. Sometimes identical recognition sites can be hinged to create closed surfaces.

Mason: A spherically symmetrical object may be composed of homochiral units, as is shown by the division of a sphere into two or more chiral components with the same handedness. A postprandial entertainment with an apple, known as *La Coupe du Roi*, illustrates the principle. Two vertical half-cuts are made through the apple—one from the top to the equator, the other, perpendicular to the first, from the bottom up to the equator. Two non-adjacent horizontal quarter cuts are then made along the equator to connect the vertical cuts and the apple separates into two homochiral halves. There are two sets of opposite quadrants through which the horizontal quarter-cuts may be made: one set results in two L-halves, the other in two D-halves. A molecular realization of the principle was reported by Anet et al (1983).

Morgan: Would it help to distinguish, as chaos theory tells us to, between systems that are determinate and systems that are rule governed? Dr Chothia is saying that beyond the cut-off point, the system is not rule governed—at least, we can't discern what the rules are, they are extremely complex. That doesn't mean it's not determinate. You are conceding it is determinate by admitting that if you mirror reverse the amino acids to start with, you mirror reverse the outcome. I think that's the resolution to what seems like a paradox. The transition across the cut-off point is fully determinate but it is not rule governed.

Chothia: There are rules, but they are not related to the handedness of the amino acids. There is a set of rules that applies from L-amino acids to domains in proteins: above that level they cease to be of general importance. There is another set of rules which applies at this level. For example, for the association of folded proteins, you close pack the recognition site and hydrogen bond all residues involved in the interface. The total buried surface area required for the stability of the association of two folded proteins is about 800 Å^2 from each protein.

Galaburda: In a three-dimensional structure, the side chains become more important. It seems to me arbitrary to say that the L-amino acids are not determining the structures of the side chains. How do you draw that arbitrary line?

Chothia: I am saying that if you have a bit of helix, a bit of sheet, a bit of turn, creating the structure, it is clear that you only get that particular structure.

The geometry of the interaction has no simple relationship to the fact that the helix and sheet were composed of L-amino acids.

Wolpert: 'Simple' is the key word.

Frankel: Suppose I travelled to another planet in another solar system and found creatures that looked exactly like us and even spoke English. Then one of these creatures asked me to guess whether it was made of all L-amino acids, all D or mixed L and D. I could answer that it was definitely not mixed L and D, but it could be either all L or all D. Is that right?

Chothia: That is correct.

Kondepudi: It is conceivable that there are big globs of polypeptides, some made of all L-amino acids, some of all D-amino acids, and the L- and D-polymers associate to form large structures. That would be the same as what we see now in bacterial cell walls. By this analogy this creature on the other planet could be composed of both L- and D-amino acids.

Chothia: If a single chain contains both D and L, the structural rules we know no longer apply.

Crow: Is your argument that one can never proceed from the properties of L-amino acids through to brain properties (such as symmetry), or that there could be such a link and that is the unusual circumstance we should be looking for?

Chothia: I think it's probably not useful to think of a direct link from amino acids to the brain. It is more likely that there are sets of rules which come from a very simple origin, in this case L-amino acids, and go through to structures which consist of thousands of amino acids but then stop. Another set of rules takes you further along the route to whole organisms and brains. The relationship between these sets of rules is marginal.

Lawrence: Could there be specially designed proteins (which might be exceptional) that bridge the gap between amino acids and multidomain proteins? If so, could the asymmetry originating with the amino acids be carried through to higher structures?

Chothia: There could be special proteins which do jump the gap. The question is the extent to which they determine the next level in the hierarchy. I think that is very small. The asymmetry can be created by the rules which start at the level of multidomain proteins.

Lawrence: You need something to create the asymmetry to start with.

Chothia: That asymmetry can be created anew at the level of multidomain proteins. It depends on the distribution of the recognition sites.

Burn: Are all the lines representing the extent of a set of rules the same length? Or does the potential for flexibility increase with each step up this ladder? If that were the case, the chance of getting a long chain would diminish as you go higher up. At the beginning, where there is a very simple mechanical relationship, a certain asymmetry would have an effect for quite a long way along the chain. Further up the chain, the complexity and flexibility of the steps would be such that you would not maintain a particular swing for anything like as many steps.

In the part of the chain that you understand, do the steps become more complex and more unpredictable as you go from amino acid to helix to domain?

Chothia: I don't know and I think it's very difficult to answer. I have been talking about the intrinsic properties of structure. L-amino acids combine in a manner dictated by the thermodynamics and kinetics of atoms and molecules to create larger structures. When you go from microscopic structures to macroscopic ones, the properties change. Very few engineers think about the atomic structure: they know the properties of their materials derive from the atomic structure, but they think in far more gross terms. Trying to go all the way along this chain requires an enormous amount of work and information that we don't have. The size of the steps will be discovered only when we know much more. What I have described is the result of 30 years work; longer if we go back to L-amino acids. And there are many problems we still don't understand. The only way we will find out is by looking; *ab initio* calculations have not been very fruitful so far.

Weber: Is part of the reason that this gap exists that you don't yet know the rules to bridge the gap? As you learn more about protein assemblies, will you be able to bridge the gap?

Chothia: Do you expect the new rules to be derived from known rules or that they will be separate?

Weber: I think they will be separate.

McManus: There seem to be two distinct positions. One is that one set of rules applies up to the level of the protein subunit and after that the rules don't apply. Another position is that the rules do apply but the ability to predict gets less and less. Therefore the probability of being able to predict anything, or of there being any effect on the global structure of an organism, becomes diminishingly small.

What impressed me earlier (p 10) was that a difference of probability of 10^{-17} in a physical system can get magnified until we have organisms that are made of only one type of amino acids. So although it may be very difficult to predict and there may be vanishingly small returns, could not that very small difference be magnified and have an effect later on? Does the chain John Burn is talking about get more and more fragile or does the chain somehow manage to strengthen itself?

Chothia: There is no way we can go from amino acids to the three-dimensional structure of proteins without looking. We can understand. Understanding is saying, these are the sorts of things which happen and we know what's involved. But to make a prediction, we need 99.9% knowledge, because structural features are determined by small differences in large terms.

References

Anet FAL, Miura SS, Siegel J, Mislow K 1983 La Coupe du Roi and its relevance to stereochemistry: combination of two homochiral molecules to give an achiral product. J Am Chem Soc 105:1419–1426

Bentley GA, Boulot G, Phottot MM, Poljak RJ 1990 Three-dimensional strutcture of an idiotope–anti-idiotope complex. Nature (Lond) 348:254–257

Bacterial motility: handedness and symmetry

Howard C. Berg

Department of Cellular & Developmental Biology, Harvard University, Biological Laboratories, 16 Divinity Avenue, Cambridge MA 02138, and Rowland Institute for Science, 100 Cambridge Parkway, Cambridge MA 02142, USA

Abstract. Many bacteria swim by rotating thin helical filaments that extend into the external medium, as with common bacteria, or run beneath the outer membrane, as with spirochetes. Each filament is driven at its base by a motor that turns alternately clockwise and counterclockwise. The motor–filament complex is called a flagellum. Other kinds of bacteria glide, but their organelles of locomotion are not known. Since bacteria are microscopic and live in an aqueous environment, they swim at low Reynolds' number; cyclic motion works (e.g. rotation of a helix) but reciprocal motion does not (e.g. stroking of a singly hinged oar). By measuring concentrations of certain chemicals as they move through their environment, making temporal comparisons and modulating the direction of flagellar rotation, bacteria accumulate in regions that they find more favourable. Studies of bacterial chemotaxis are highly advanced, particularly for the peritrichously flagellated species *Escherichia coli*. A great deal is known about chemoreception, receptor–flagellar coupling and adaptation. Recently it has been found that *E. coli* can aggregate in response to signals generated by the cells themselves. Complex patterns form with remarkable symmetries.

1991 Biological asymmetry and handedness. Wiley, Chichester (Ciba Foundation Symposium 162) p 58–72

Most bacteria swim by changing their shape or their orientation. Handedness is fundamental to their motility: it is a constraint imposed by the physics of slow viscous flow. Other bacteria glide or swim without any apparent change in shape: their modes of propulsion remain mysterious. The migrations of cell populations, on the other hand, tend to be symmetrical. In this paper, I will discuss the handedness at length and touch upon the symmetry.

Handedness

Some history

Bütschli (1884) appears to be the first to have realized that one can propel a microorganism by propagating a helical wave along a thin appendage. This

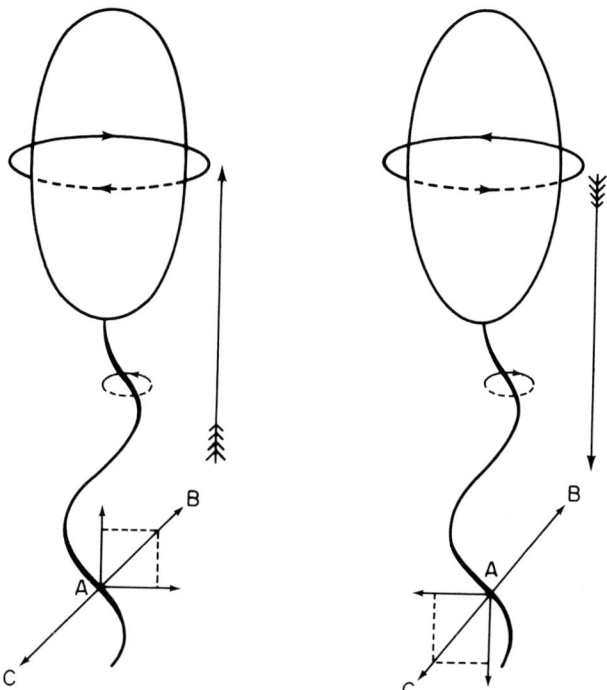

FIG. 1.   A sketch of the forward and backward motion of *Chromatium okenii*, according to Buder (1915). Forward motion (left): the flagellar bundle turns rapidly counterclockwise (as seen by an observer behind the cell) and the cell body rolls more slowly clockwise. Backward motion (right): all directions have reversed. *C. okenii* is a large purple sulphur bacterium, about 6 μm in diameter. It played an important role in early work on bacterial behaviour (reviewed by Berg 1975).

motion generates both torque and thrust. It can be driven, he argued, by the sequential contraction and relaxation of a set of longitudinal lines that run along the sides of the filament, the shortest ones at the inside of the helix, the longest ones at the outside. He developed this theory for protozoa, but it was soon applied to bacteria, for example, by Reichert (1909) and Buder (1915) (Fig. 1). Metzner (1920a) modelled this motion by spinning helical wires in water, but unfortunately at high speed (i.e. at high Reynolds' number). He did not know about the physics of slow viscous flow where Reynolds' number is low. A masterful exposition of this topic was given by Ludwig (1930), who applied it to both ciliary and flagellar propulsion. The hydrodynamicists, in turn, were unaware of his efforts (Taylor 1952). To cut a long story short, a cell as small as *Escherichia coli*, immersed in water, knows a lot about viscosity but nothing about inertia. It generates thrust through viscous shear, not by accelerating fluid.

The flagellum works because the viscous drag on successive segments of the filament, which behave like thin rods, is roughly twice as great when these segments move sideways as when they move lengthwise. One can demonstrate this by dropping a piece of straight wire in a glass of syrup. If the wire is oriented vertically, it falls twice as fast as when it is oriented horizontally. However, if it is oriented obliquely, say downwards to the right, it falls obliquely to the right (Taylor 1967). In a rotating helix, the axial components of the oblique forces add up, generating thrust (cf Fig. 6.3 of Berg 1983). The circumferential components also add up, generating torque. When a microorganism swims at constant speed, all the forces that act upon it must balance, otherwise it would accelerate. Thus, the thrust generated by the flagellum is balanced by the viscous drag due to the translation of the cell body, while the torque generated by the flagellum is balanced by the viscous drag due to its roll.

Life at low Reynolds' number

A microorganism lives in a bizarre world. In this world it does not matter how fast you move, if you retrace your steps, you always return to the same initial position: reciprocal motion is not a viable propulsive strategy. Ludwig illustrated this with an organism propelled by two rigid oars (Fig. 2). Purcell (1977) called this the scallop theorem. He imagined a scallop with one hinge at the edge of its shell and no other degrees of freedom. If macroscopic, it could open its shell slowly and close it rapidly, moving a short distance mouth-first and then a long distance hinge-first. A microscopic scallop, on the other hand, would simply retrace its trajectory. Real scallops turn out to be more sophisticated: they swim mouth-first, not hinge-first (private communication from C. W. McCutchen to E. M. Purcell). Somehow, fluid must be ejected backwards, from the open side of the shell toward the hinge. Also, it turns out that Ludwig's creature (Fig. 2) could make some headway at low Reynolds' number were it to stroke its oars in a non-reciprocal manner, first one and then the other; see Fig. 7 of Purcell (1977). As Ludwig realized, the oars on real ciliates are not rigid: the shape of the power stroke is not the same as that of the recovery stroke. During recovery, the cilium tends to hug the cell body.

Bacterial flagella

The dimensions, handedness and distribution of bacterial flagella vary from species to species. *Chromatium okenii* has about 40 left-handed filaments projecting from one pole. When a cell swims backwards, the filaments change their sense of rotation but not their handedness (Fig. 1). *E. coli* and its close relative *Salmonella typhimurium*, on the other hand, have about six left-handed filaments distributed higgledy-piggledy over the cell surface. One such filament is shown in Fig. 3. The helix diameter and wavelength (pitch)

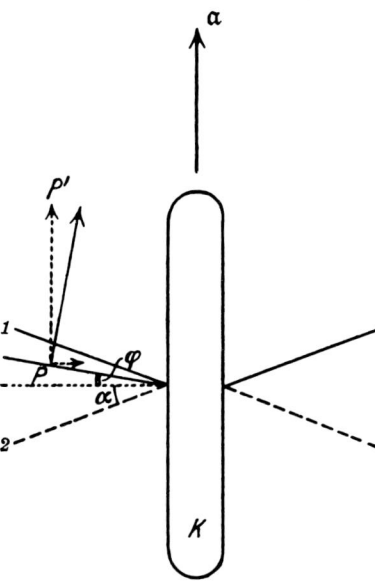

FIG. 2. An organism propelled by two rigid oars, according to Ludwig (1930). A large organism of this kind could make headway upwards by pulling both oars rapidly downwards, then returning them slowly upwards, and repeating this motion. A microorganism of this kind, on the other hand, would move rapidly upwards, then slowly downwards, returning to its initial position. It would just jiggle up and down.

of this filament are 0.63 and 2.65 μm, respectively. When *E. coli* swims smoothly (runs), the filaments form a coherent bundle that pushes the cell steadily forward. For cells grown on a rich medium (tryptone) and observed at 32 °C, the rotation rate averages about 270 Hz and the speed averages about 36 μm/s (Lowe et al 1987). When the direction of rotation switches from counterclockwise to clockwise, the bundle comes apart and the filaments move independently, moving the cell erratically this way and that: the cell is said to tumble. This motion is erratic, partly because the torsion at the base of a filament is large enough to trigger a change in shape: the handedness of the filament switches from left to right and its pitch is halved. Filaments in transition have segments of either handedness that intersect at an angle of 64° (Macnab & Ornston 1977). When the direction of rotation switches again, the filaments revert to their left-handed conformation, the bundle reforms and the cell swims off in a new direction. By alternately running and tumbling, the cell executes a random walk (Berg & Brown 1972). On average, runs last about 1 s, tumbles about 0.1 s. The reason for this difference is not known (see the appendix of Ishihara et al 1983).

The filament in *E. coli* is a self-assembling polymer of a single globular protein, called flagellin, of relative molecular mass about 51 000 (Kuwajima

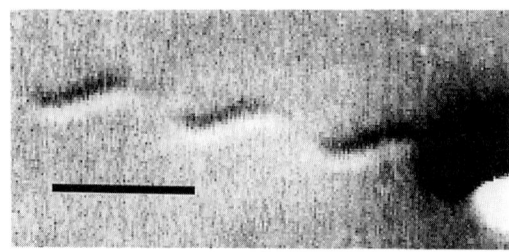

FIG. 3.   Video-enhanced, differential-interference-contrast image of a flagellar filament of *E. coli*. Part of the cell body appears at the right. Bar 2.5 μm. The microscope was focused just above the helical axis, so that segments closer to the observer appear brighter; this demonstrates that the helix is left handed (cf Shimada et al 1975). This is a living cell of wild-type strain AW405 de-energized by treatment with 2,4-dinitrophenol. A different view of the same filament is shown in Fig. 2a of Block et al (1991), which describes the imaging technique.

et al 1986). Some 4000 subunits are required to complete one helical turn (Fig. 3). Only the N- and C-termini are required for assembly (Kuwajima 1988), and only the N-terminus is needed for transport (Kuwajima et al 1989). *In vivo*, the filament is assembled at the distal end (Iino 1969, Emerson et al 1970). The subunits are arranged in 11 nearly longitudinal rows surrounding a central channel about 60 Å in diameter (Namba et al 1989). Each of these protofilaments is a cooperative unit in a two-state model developed to explain polymorphism (cf Calladine 1978, Hotani 1982 and references cited there). It is an interesting problem in crystallography: if each subunit were bonded identically to its neighbours, the filament would be straight; some subunits must be closer to their neighbours than others (cf Klug 1967). What it took time to realize is that one need not contract and relax these protofilaments sequentially, as suggested by Bütschli. All one needs to do is to rotate the filament as a whole; however, that requires a rotary motor and a rotary joint (Berg & Anderson 1973).

Spiral bacteria

There are a number of bacteria whose bodies are helical, e.g. *Spirillum volutans*, with flagellar bundles at either pole. The flagella at the front sweep back over the cell body and contribute to thrust primarily by powering the roll of the body, while those at the back extend behind the cell in a helical conformation and generate thrust in the conventional manner (Metzner 1920b). The cells reverse when the flagella at both poles change their directions of rotation and flip into the other conformation. This strategy is carried to the extreme by spirochetes, where the flagella are intracellular. Thick spirochetes are rigid; most of their thrust is generated by the roll of the cell body (Berg 1976). Thin spirochetes, on the other hand, are flexible; they wave their ends as well as roll (Berg et al 1978, Goldstein & Charon 1988). For example, *Leptospira interrogans* is only about 0.3 µm in diameter by 10 to 20 µm long. Its body is shaped as a tight right-handed helix (Carleton et al 1979). When it swims, the front end of the cell is bent into a helix of much larger diameter and pitch, and the back end is bent into a hook; both ends gyrate. In water, most of the thrust is generated at the front by the backward propagation of a helical wave, driven by the counterclockwise rotation of a left-handed flagellar filament. This filament emerges from the side of the protoplasmic cylinder near the front of the cell and runs backwards between the protoplasmic cylinder and the outer membrane. An identical filament near the back of the cell rotates clockwise (counterclockwise when viewed from behind the cell), but in a hook-shaped conformation. Meanwhile, the body rolls to the right (clockwise when viewed from behind the cell). Once again, the cell reverses when the flagella at both poles change their directions of rotation and their conformation. When the cell is in a gel-like medium, on the other hand, most of the thrust is generated by the roll of the body. This difference is illustrated in Fig. 4, which shows a cell with only one flagellum (easy to find in older cultures). At the top (a), the cell is shown swimming in water. When the flagellum is in a helical conformation the cell moves toward the right, as shown by the arrow. However, when the flagellum is in a hook conformation, the cell stops. The hook gyrates one way and the body rolls the other, but there is little thrust. Things are different when the cell is in a gel-like medium, such as methylcellulose, as shown at the bottom of the figure (b). When the flagellum is in a helical conformation the cell swims to the right, as before. Now, when the flagellum switches to the hook conformation, the cell swims to the left, at about the same speed. Therefore, the thrust must be generated by the roll of the body, not by the propagation of the spiral wave. Evidently, the cell threads its way through the interstices of the gel, like a corkscrew through a cork. The physics has changed; other forces are at work besides those due to viscous shear (Berg & Turner 1979).

Chemotactic signals

In cells like *S. voluntans* or *L. interrogans*, the flagellar motors at one end of the cell normally run clockwise, while those at the other end normally run counterclockwise. There is nothing special about either handedness. The command for reversing is 'switch direction'. In *E. coli*, the motors have a counterclockwise bias and the behavioural signal is 'change the probability of turning clockwise'. The game is different. In cells in which all the known cytoplasmic chemotaxis proteins have been deleted (Wolfe et al 1987) and in lysed-cell preparations (Eisenbach & Matsumura 1988), the motors spin exclusively counterclockwise. Clockwise rotation is thought to be induced by interaction with a phosphorylated derivative of the cytoplasmic protein CheY (cf Stock et al 1989, Bourret et al 1991). Flagellar motors are driven by a proton

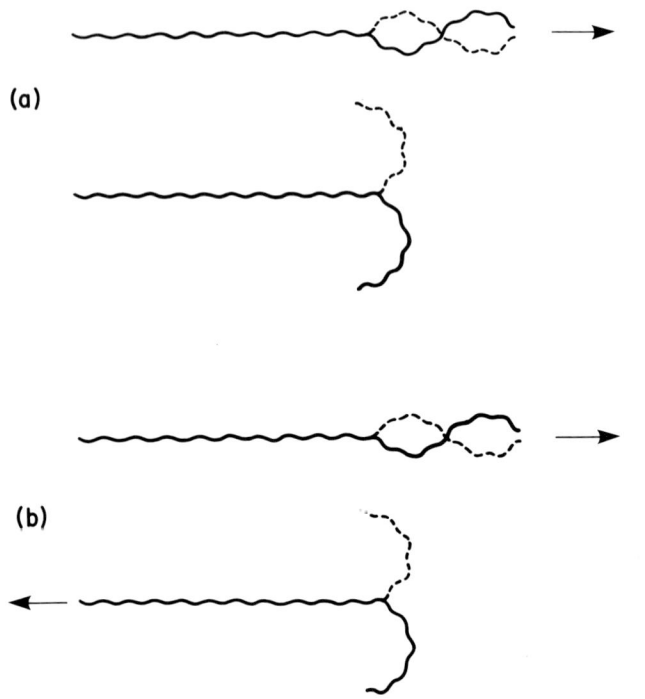

FIG. 4. A schematic drawing of a defective cell of *Leptospira interrogans* serotype *illini*. In water (a) the cell swims to the right (top) or stops (bottom). In 1% methylcellulose (b) it swims in either direction (arrows). This cell has only one intracellular flagellum at its right end, shaped either as a helix (when the motor turns counterclockwise) or as a hook (when it turns clockwise). See the text and Berg et al (1978).

(or, in some species, a sodium) flux, which is directed inwards. We have some ideas about how protons might drive the motor—see, for example Meister et al (1989) and references cited there—but we know less about what controls its direction of rotation. In *E. coli* and *S. typhimurium*, three proteins are implicated in this process: FliG, FliM and FliN. They are thought to reside on the face of the rotor adjacent to the cytoplasm, where they interact with the phosphorylated form of CheY (reviewed by Blair 1990, Macnab 1990).

Gliding bacteria

Other kinds of bacteria glide. They exhibit Brownian movement when in suspension, but move steadily forward or backward along a solid surface, sometimes pivoting clockwise or counterclockwise. No organelles of locomotion have been identified. Work with one species suggests that components of the outer membrane that adsorb to solid surfaces are driven along tracks laid down on the rigid framework of the cell wall (Lapidus & Berg 1982); however, this view is controversial. For possible structural correlates, see Lünsdorf & Reichenbach (1989). It is thought likely that the force-generating elements are similar to those of the rotary motor, since here too the energy is derived from movement of protons or sodium ions down an electrochemical gradient. The best-studied gliding bacterium is *Myxococcus xanthus*. These cells hunt in packs, like wolves, and when starved develop fruiting bodies (cf Shimkets 1990). The most remarkable bacterium in the gliding class is a cyanobacterium of the genus *Synechococcus* that actually swims at speeds of up to 25 μm/s (Waterbury et al 1985). This cell is about the same size and shape as *E. coli*, but it has no external appendages! Somehow, it is able to drive fluid backwards over its surface.

Symmetry

The biased random walk

The motion of cell populations is symmetrical. In *E. coli*, after a tumble there is a slight bias in the direction of the previous run, but otherwise a cell chooses its new direction at random. It tries first one direction, then another, as in a random walk (Berg & Brown 1972). In the presence of a spatial gradient of an attractant, however, runs that happen to carry the cell up the gradient are extended; otherwise the motion remains symmetrical and the cell executes a biased random walk. It sets the bias by measuring concentrations as a function of time and making temporal comparisons. In particular, it counts molecules of a given kind (e.g. aspartate) for a period of about a second, looks up what it found for the previous three seconds, and responds to the difference (reviewed by Berg 1988).

This symmetry can be broken by surfaces. At the bottom of a preparation, near the slide, cells tend to circle clockwise; at the top, near the coverslip, they circle counterclockwise (see Fig. 7 of Berg & Turner 1990). During runs, as noted earlier, the filaments in the flagellar bundle turn counterclockwise and the cell body rolls clockwise (as in Fig. 1, left). Thus, the flagellar bundle tends to roll to the left on the glass, while the cell body rolls to the right. Since the body is in front, the cell steers clockwise (when viewed from above). An extreme and amusing example of wall effects occurs when cells are confined to capillary tubes. If the tube is small (e.g. 10 μm in diameter), the cells do not have room to swim sideways, so the random walk is reduced from three dimensions to one. Cells move through such tubes more efficiently than one would otherwise expect (Berg & Turner 1990).

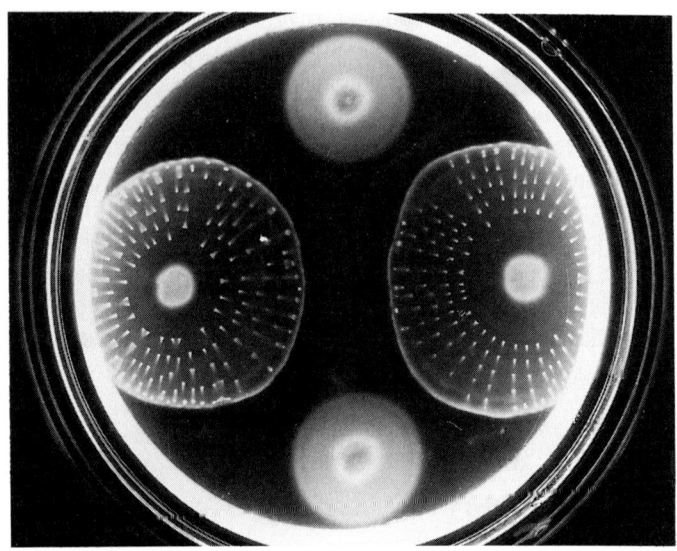

FIG. 5. Examples of patterns formed by *E. coli* on soft agar, visualized by scattered light. The dispersal of cells of an aspartate-blind strain (top and bottom) or an aspartate-sensitive strain (left and right) are shown. The broad white ring near the edge of the plate is an artifact of the illumination; its inside diameter is 7.5 cm. The aspartate-blind cells spread diffusely, as expected for cells that run and tumble but do not follow gradients. The aspartate-sensitive cells formed radial arrays of compact aggregates. The plate contained minimal salts, small amounts of amino acids required for growth, 3 mM succinate (as carbon source) and 50 µg/ml tetrazolium violet (to enhance image contrast). 20 µl of a saturated suspension of cells were inoculated at the centre of each pattern. The plates were incubated at 25 °C for about 40 h. The aspartate-blind strain was RP5854, $\Delta(tar-tap)$5201, from J. S. Parkinson. The aspartate-sensitive strain was HCB317, $\Delta(tsr)$7071, which is serine blind (Wolfe et al 1987).

Pattern formation

When chemotactic cells are inoculated onto soft agar containing nutrients that also serve as chemical attractants, they form a series of concentric rings that migrate outwards at rates of several millimeters per hour (Adler 1966, Nossal 1972). Metabolism of each nutrient generates a spatial gradient, to which the cells respond. Each ring corresponds to the consumption of a different nutrient. In the absence of a chemotactic response cells also migrate outwards, provided that they can tumble as well as run, but they do so more slowly (Wolfe & Berg 1989).

Recently, Elena Budrene has found conditions under which cells form patterns of much greater complexity, as shown in Fig. 5 (Budrene & Berg 1991). Here, cells respond to a chemotactic signal that they generate themselves. Spots of high cell density form in the wake of a band that moves steadily outwards (about

0.7 mm/h). Initially, the cells in these aggregates are highly motile, but after a few hours they stop swimming: thus, the patterns become stable. These patterns do not form in strains missing the aspartate receptor. The patterns are perturbed and finally vanish in the presence of increasing amounts of the non-metabolizable analogue α-methyl-D,L-aspartate. Thus, the attractant is aspartate or an aspartate analogue. Patterns of lower symmetry form in a few minutes when cells are grown in liquid medium, poured in a thin layer in an empty petri plate, and exposed to a fresh aliquot of the carbon source. These patterns are stable for about half an hour, then the cells disperse. It is possible that the attractant is excreted by these cells in response to oxidative stress, which is relieved as the aggregation reduces the local oxygen concentration. If so, bacterial chemotaxis might have evolved not only to enable cells to find food, but also as a means for collective defence.

Acknowledgements

This work was supported by grants from the US National Institutes of Health, the US National Science Foundation and the Rowland Institute for Science.

References

Adler J 1966 Chemotaxis in bacteria. Science (Wash DC) 153:708–716
Berg HC 1975 Chemotaxis in bacteria. Annu Rev Biophys Bioeng 4:119–136
Berg HC 1976 How spirochetes may swim. J Theor Biol 56:269–273
Berg HC 1983 Random walks in biology. Princeton University Press, Princeton, NJ
Berg HC 1988 A physicist looks at bacterial chemotaxis. Cold Spring Harbor Symp Quant Biol 53:1–9
Berg HC, Anderson RA 1973 Bacteria swim by rotating their flagellar filaments. Nature (Lond) 245:380–382
Berg HC, Brown DA 1972 Chemotaxis in *Escherichia coli* analysed by three-dimensional tracking. Nature (Lond) 239:500–504
Berg HC, Turner L 1979 Movement of microorganisms in viscous environments. Nature (Lond) 278:349–351
Berg HC, Turner L 1990 Chemotaxis of bacteria in glass capillary arrays. Biophys J 58:919–930
Berg HC, Bromley DB, Charon NW 1978 Leptospiral motility. Symp Soc Gen Microbiol 28:285–294
Blair DF 1990 The bacterial flagellar motor. Semin Cell Biol 1:75–85
Block SM, Fahrner KA, Berg HC 1991 Visualization of bacterial flagella by video-enhanced light microscopy. J Bacteriol 173:933–936
Bourret RB, Borkovich KA, Simon MI 1991 Signal transduction pathways involving protein phosphorylation in prokaryotes. Annu Rev Biochem 60:401–442
Buder J 1915 Zur Kenntnis des *Thiospirillum jenense* und seiner Reaktionen auf Lichtreize. Jahrb Wiss Bot 56:529–584
Budrene EO, Berg HC 1991 Complex patterns formed by motile cells of *Escherichia coli*. Nature (Lond) 349:630–633
Bütschli O 1884 Protozoa. In: Klassen und Ordnungen des Thier-reichs. Winter, Leipzig vol 1(2):856–857

Calladine CR 1978 Change in waveform in bacterial flagella: the role of mechanics at the molecular level. J Mol Biol 118:457–479

Carleton O, Charon NW, Allender P, O'Brien S 1979 Helix handedness of *Leptospira interrogans* as determined by scanning electron microscopy. J Bacteriol 137: 1413–1416

Eisenbach M, Matsumura P 1988 *In vitro* approach to bacterial chemotaxis. Botanica Acta 101:105–110

Emerson SU, Tokuyasu K, Simon MI 1970 Bacterial flagella: polarity of elongation. Science (Wash DC) 169:190–192

Goldstein SF, Charon NW 1988 Motility of the spirochete *Leptospira*. Cell Motil Cytoskeleton 9:101–110

Hotani H 1982 Micro-video study of moving bacterial flagellar filaments. III. Cyclic transformation induced by mechanical force. J Mol Biol 156:791–806

Iino T 1969 Polarity of flagellar growth in *Salmonella*. J Gen Microbiol 56:227–239

Ishihara A, Segall JE, Block SM, Berg HC 1983 Coordination of flagella on filamentous cells of *Escherichia coli*. J Bacteriol 155:228–237

Klug A 1967 The design of self-assembling systems of equal units. Symp Int Soc Cell Biol 6:1–18

Kuwajima G 1988 Construction of a minimum-size functional flagellin of *Escherichia coli*. J Bacteriol 170:3305–3309

Kuwajima G, Asaka J-I, Fujiwara T, Fujiwara T, Node K, Kondo E 1986 Nucleotide sequence of the *hag* gene encoding flagellin of *Escherichia coli*. J Bacteriol 168:1479–1483

Kuwajima G, Kawagishi I, Homma M, Asaka J-I, Kondo E, Macnab RM 1989 Export of an N-terminal fragment of *Escherichia coli* flagellin by a flagellum-specific pathway. Proc Natl Acad Sci USA 86:4953–4957

Lapidus IR, Berg HC 1982 Gliding motility of *Cytophaga* sp strain U67. J Bacteriol 151:384–398

Lowe G, Meister M, Berg HC 1987 Rapid rotation of flagellar bundles in swimming bacteria. Nature (Lond) 325:637–640

Ludwig W 1930 Zur Theorie der Flimmerbewegung (Dynamik, Nutzeffekt, Energiebilanz). Z Vgl Physiol 13:397–504

Lünsdorf H, Reichenbach H 1989 Ultrastructural details of the apparatus of gliding motility of *Myxococcus fulvus* (Myxobacterales). J Gen Microbiol 135:1633–1641

Macnab RM 1990 Genetics, structure, and assembly of the bacterial flagellum. Symp Soc Gen Microbiol 46:77–106

Macnab RM, Ornston MK 1977 Normal-to-curly flagellar transitions and their role in bacterial tumbling: stabilization of an alternative quaternary structure by mechanical force. J Mol Biol 112:1–30

Meister M, Caplan SR, Berg HC 1989 Dynamics of a tightly coupled mechanism for flagellar rotation. Biophys J 55:905–914

Metzner P 1920a Zur Mechanik der Geisselbewegung. Biol Zentralbl 40:49–87

Metzner P 1920b Die Bewegung und Reizbeantwortung der bipolar begeisselten Spirillen. Jarhb Wiss Bot 59:325–412

Namba K, Yamashita I, Vonderviszt F 1989 Structure of the core and central channel of bacterial flagella. Nature (Lond) 342:648–654

Nossal R 1972 Growth and movement of rings of chemotactic bacteria. Exp Cell Res 75:138–142

Purcell EM 1977 Life at low Reynolds number. Am J Physics 45:3–11

Reichert K 1909 Über die Sichtbarmachung der Geisseln und die Geisselbewegung der Bakterien. Zentralbl Bakteriol Parasitenk Infektionskr Hyg Abt 1 Orig 51:14–94

Shimada K, Kamiya R, Asakura S 1975 Left-handed to right-handed helix conversion in *Salmonella* flagella. Nature (Lond) 254:332–334

Shimkets LJ 1990 Social and developmental biology of the myxobacteria. Microbiol Rev 54:473–501

Stock JB, Ninfa AJ, Stock AM 1989 Protein phosphorylation and regulation of adaptive responses in bacteria. Microbiol Rev 53:450–490

Taylor G 1952 The action of waving cylindrical tails in propelling microscopic organisms. Proc R Soc London A (Phys Sci) 211:225–239

Taylor G 1967 Low-Reynolds-number flows. Encyclopaedia Britannica Educational Corp, Chicago, IL. (This is a 16-mm colour sound film that comes with a set of notes; see there Figs. 13, 14.)

Waterbury JB, Willey JM, Franks DG, Valois FW, Watson SW 1985 A cyanobacterium capable of swimming motility. Science (Wash DC) 230:74–76

Wolfe AJ, Berg HC 1989 Migration of bacteria in semisolid agar. Proc Natl Acad Sci USA 86:6973–6977

Wolfe AJ, Conley MP, Kramer TJ, Berg HC 1987 Reconstitution of signaling in bacterial chemotaxis. J Bacteriol 169:1878–1885

DISCUSSION

Jefferies: It is interesting that you mention Wilhelm Ludwig, because he wrote a very fine book on handedness called 'Das Rechts–Links Problem im Tierreich und beim Menschen'(1932).

Lawrence: What's known about the structure of the flagellin subunit and how it imposes a helix on the filament?

Berg: An X-ray structure of the filament has been published by Namba et al (1989). Image reconstruction has also been done using electron microscopy, by Trachtenberg & De Rosier (1987). As I understand it, the filament is built with the α-helical regions of each subunit lined up along the central pore. The bonding that's crucial for forming the filament is a helix–helix interaction at both the N and C termini. You can delete the centre part of the flagellin molecule and make a perfectly good filament. You can even graft in cholera toxin and still make a filament.

Frankel: There's an old observation by Asakura et al (1966) that if one combines short pieces of filaments from a wild-type cell and flagellin monomers from a curly mutant, or the converse, the filament adopts the form of the seed.

Berg: It depends on the size of the seed (the length of the filament used to nucleate the crystallization). If the seed is small, you get the conformation of the monomer. If the seed is long, you often get the conformation of the seed. I don't know why.

Frankel: Can a seed of one handedness impose its handedness on a filament made of a mutant flagellin which ordinarily forms a filament of the opposite handedness?

Berg: Yes. The normal filament is left handed and the curly filament is right handed. A curly seed about 1 µm long will nucleate the growth of a curly filament from normal monomers. However, a normal seed of this length in a solution of curly monomers also gives a curly filament.

Wolpert: Can you tell us more about the flagellar motor. What's the basis for the reversal of the motor?

Berg: The motor is driven by a proton flux. There is a protein of M_r about 30 000 called MotA that appears to be a proton channel. It has four membrane-spanning domains but most of the protein is in the cytoplasm. Another protein of about the same size, MotB, has one membrane-spanning domain and is mostly in the periplasm. In dominant non-functional mutants of either *motA* or *motB*, the mutant protein binds to the motor and displaces the wild-type protein. In MotA, essentially all the known mutations are in the membrane-spanning region: in MotB, most are in the periplasmic region. We think MotB is a spring that links MotA to the cell wall.

If I look down on the rotor, the force-generating units (MotA, linked to the wall via MotB) appear in a circle around the outside. A proton goes in via MotA and binds to a site on the periphery of the rotor. MotA then moves cir-cumferentially a few Ångströms, the proton leaves the site and enters the cytoplasm. The channel through MotA is in two parts that are offset. One part gets the proton from the periplasm to the rotor; the other gets the proton from the rotor into the cytoplasm. If the motion of this channel complex relative to the rotor is constrained on the basis of which sites are occupied and which are free, one can couple motion of protons through the membrane to angular displacement of the rotor in the plane of the membrane, i.e. one can generate torque (see Meister et al 1989).

Basically, I think the motor is driven by ratchets, like the heads of muscle myosin heavy chains, except that whether a ratchet catches or moves depends on the occupancy of proton-accepting sites. I think the direction of rotation is changed by reversing the sign of the constraints that govern the motion. This could be done by switching to a different row of sites; for example, sites that are neutral when protonated rather than charged. I predict there is some colossal conformational change that moves the periphery of the rotor in a direction normal to the plane of membrane, bringing the channels into contact with a different row of binding sites. We know the motor switches from rotating at top speed in one direction to rotating at top speed in the other direction within a few milliseconds. All the independently moving force generators switch at the same time; I don't know how that can happen unless something occurs on the inside of the motor, not on the outside.

Wolpert: Are there mutants which have a different handedness?

Berg: We have every kind of mutant you can imagine: those that go only clockwise, those that go only counterclockwise, those that don't go, some with straight filaments, any kind of filament you like!

Wolpert: Cyrus, you told us that mutants wouldn't change anything!

Chothia: You don't know the structural basis of this! Native flagellin forms helices with different asymmetries (see Chothia, this volume).

Berg: No one has been able to see, even by X-ray crystallography, differences in the displacement of adjacent flagellin subunits. In the wild-type, it takes about 4000 subunits to complete one turn of the helical filament. So the effects of such displacements are amplified, and the displacements required to convert filaments from one handedness to the other are relatively small.

Layton: Does the movement of these bacteria show a bias towards either counterclockwise or clockwise rotation?

Berg: We have studied the rotation rate of a single cell, tethered to glass by its flagellum, and plotted that as a function of time. The cell was a paralysed *motB* mutant carrying a copy of the wild-type *motB* gene whose expression could be turned on by addition of a chemical inducer. As the wild-type protein was produced, the cell began spinning in a stepwise fashion, first at about 1 Hz, then 2 Hz, 3 Hz and so on. These experiments show that each motor has eight independent force-generating units, each composed of both *motA*- and *motB*-encoded proteins (Blair & Berg 1988).

Each recording was a separate measurement of the rotation direction and speed made using a stop-watch while looking at a video screen. At the lowest level, the motor tended to spend most of its time going counterclockwise; after that it rotated approximately equal times in each direction. We do not know why a motor with a single force generator has a different rotational bias than one with several force generators.

Morgan: How periodic is the pattern of bacteria shown in Fig. 5? If you do a 2D Fourier transformation, do you get big power peaks?

Berg: I haven't done it, but you would. You can see by eye that it is periodic.

Yost: How do you distinguish between motility and cells replicating in some areas more than in other areas?

Berg: By doing the experiment differently. We grow the cells in a flask, then spread them on a petri dish and add more growth medium. A few minutes later they have formed distinct aggregates. These patterns are not periodic on a grand scale like those in Fig. 5; the spacing between the spots is only roughly constant and the spots are a variety of sizes. There are some conditions under which these patterns start forming in seconds. The replication time of the bacteria under these conditions is about three hours.

Wood: Are the patterns medium dependent?

Berg: Yes, subtle differences in the medium produce different patterns. If we use 3 mM succinate instead of 2 mM, we get a different pattern. These patterns are robust in the sense that if we repeat the experiment under the same conditions, we get the same kind of pattern. There are no genetic changes involved, because we can pick bacteria from anywhere on the plate, inoculate them onto a fresh plate and get the same pattern.

Morgan: If each cell had around it an attractive region and then further out an inhibitory region and this profile looked like a Mexican hat, it would be band limited in spatial-frequency terms. Wouldn't you then get periodic structure?

Berg: Yes. The region further out (away from the centre of an aggregate) is inhibitory, because the cells have migrated away from that region and, therefore, are no longer excreting an attractant there. A mechanochemical model of this kind has been developed by Oster & Murray (1989).

References

Asakura S, Eguchi G, Iino T 1966 *Salmonella* flagella: *in vitro* reconstitution and overall shapes of flagellar filaments. J Mol Biol 16:302–316

Blair DF, Berg HC 1988 Restoration of torque in defective flagellar motors. Science (Wash DC) 242:1678–1681

Chothia C 1991 Asymmetry in protein structure. In: Biological asymmetry and handedness. Wiley, Chichester (Ciba Found Symp 162) p 36–57

Ludwig W 1932 Das Rechts–Links Problem im Tierreich und beim Menschen. Springer-Verlag, Berlin & Heidelberg (Facsimile edition 1970)

Meister M, Caplan SR, Berg HC 1989 Dynamics of a tightly coupled mechanism for flagellar rotation. Biophys J 55:905–914

Namba K, Yamashita I, Vonderviszt F 1989 Structure of the core and central channel of bacterial flagella. Nature (Lond) 342:648–654

Oster GF, Murray JD 1989 Pattern formation models and developmental constraints. J Exp Zool 251:186–202

Trachtenberg S, DeRosier DJ 1987 Three-dimensional structure of the frozen-hydrated flagellar filament: the left-handed filament of *Salmonella typhimurium*. J Mol Biol 195:581–601

Intracellular handedness in ciliates

Joseph Frankel

Department of Biology, University of Iowa, Iowa City, IA 52242, USA

Abstract. Ciliated protozoa have intrinsically asymmetrical ciliary structures that are asymmetrically arranged over the cell surface. These structures can be arranged in two enantiomorphic configurations, 'right-handed' (RH) and 'left-handed' (LH). Whereas one of these configurations (arbitrarily, RH) is apparently universal in Nature and predominant in the laboratory, mirror-image (RH–LH) doublets and reverse (LH) singlets have been generated and studied in eight different ciliate genera. In all these, the *internal* asymmetry of individual ciliary structures remains normal even when the asymmetry of *arrangement* of these structures is reversed. The individual structures may sometimes become inverted (rotationally permuted). LH forms reproduce themselves if they are able to feed, or reorganize periodically before starving to death if they are not. Changes of cellular handedness depend upon unusual geometric configurations and in most cases are unrelated to genic changes. In hypotrich ciliates changes of handedness can be provoked by artificially generated juxtapositions of anterior and posterior cell regions or of right and left cell margins. Reversal of handedness in ciliates can be visualized as a consequence of (re-)establishment of a normal sequence of normally spaced positional values following geometric disturbances created by the experimenter or by the regulating cell.

1991 Biological asymmetry and handedness. Wiley, Chichester (Ciba Foundation Symposium 162) p 73–93

Form and pattern are commonly uncoupled in those protistan cells whose principal patterned elements are sets of cilia or flagella. This is nowhere clearer than in the ciliates, in which cell form is frequently radial and occasionally bilateral, yet the abundant ciliary structures are always asymmetrical. Their asymmetry pervades every level of structural organization, from the ultrastructure of the individual cilium to the global arrangement of ciliary complexes over the cell surface.

An asymmetrical structure can exhibit two enantiomorphic forms. Such dual forms exist in Nature in small structures, such as amino acids, and in large structures, such as hands. Can they also be observed at levels in between, such as in intracellular structural organization? The answer depends upon which intermediate level one examines.

Ciliates typically possess numerous structural complexes, technically known as kinetids, that are organized around individual ciliary basal bodies or small

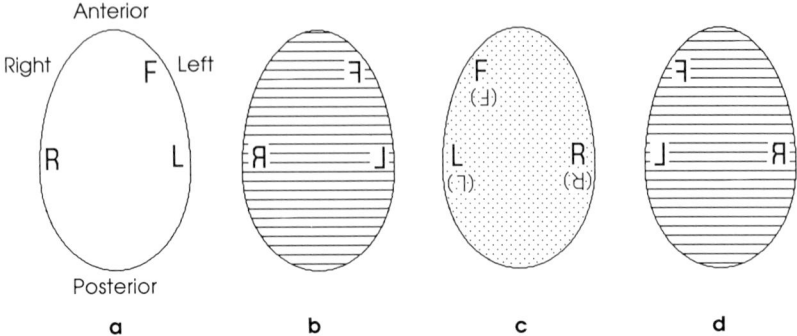

FIG. 1. Abstract representations of arrangements of ciliary complexes in a ciliate. Configurations corresponding to those actually found are (a) normal and (c) global reversal; configurations not observed are (b) local reversal and (d) local plus global reversal. In (c), the upside-down letters in parentheses represent inverted (rotationally permuted) structures, as have commonly been observed. For further explanation, see the text.

groups of basal bodies. These complexes have a definite arrangement over the cell surface, illustrated schematically in Fig. 1. Reversal of asymmetry is possible at two levels. The structures (represented abstractly by letters) could be individually (locally) reversed or their global arrangement could be reversed. Four combinations of local and global reversal are then logically possible, as shown in Fig. 1 (this assumes that all structures are either normally oriented or coordinately reversed). Of these four combinations, only the normal (Fig. 1a) and the globally reversed (Fig. 1c) ones have actually been observed. These two arrangements will be called *right-handed* (RH) and *left-handed* (LH), respectively. As normal ('right-handed') ciliates have ciliary patterns that are systematically related in different genera, the RH and LH forms can be distinguished easily and non-arbitrarily.

Propagation of handedness

How can the handedness of arrangement of structural complexes persist through cell division? The answer relies on the topology of ciliate propagation. Ciliates grow longitudinally and divide transversely. New ciliary structures are formed before cytokinesis begins, producing two tandem sets of structures in identical arrangements (Fig. 2b), which are then partitioned to the two daughter cells (Fig. 2c).

The precise way in which this happens is different in different ciliates (see chapter 3 of Frankel 1989). I will concentrate on one particular group of ciliates, the hypotrichs, and notably on the genus *Stylonychia*, which will serve as my

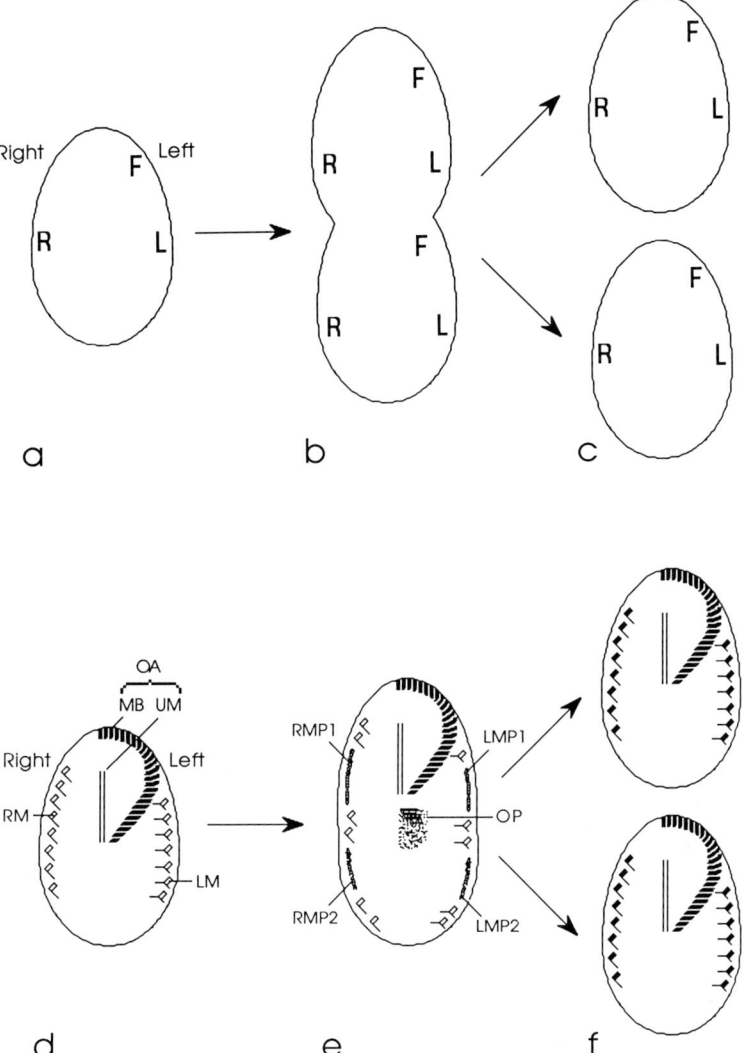

FIG. 2. Propagation of ciliary structures through division, (a–c) represented abstractly and (d–f) drawn more concretely for the ventral surface of *Stylonychia*. Here and elsewhere, the cell's left corresponds to the observer's right. In (d), left marginal (LM) and right marginal (RM) cirri are labelled, as are the membranelle band (MB) and undulating membrane (UM) of the oral apparatus (OA). In (e), the right and left marginal cirral primordia for the anterior division product (LMP1, RMP1) and for the posterior division product (LMP2, RMP2) are labelled, as is the oral primordium (OP). Old cirri are shown in outline, new cirri are filled in. For further explanation, see the text.

primary example throughout this essay. Fig. 2d is an abridged diagram of the ventral ciliary structures of this ciliate (for the true picture, see Fig. 4a). The three structural complexes shown, the left marginal cirral row, the right marginal cirral row and the oral apparatus, can be taken as more concrete versions of the schematic L, R and F of Fig. 2a. The cirral rows are composed of compound ciliary units (cirri) that have prominent accessory fibres. The arrangement of accessory fibres is different in the cirri of the right and left marginal rows (see Fig. 4a); this difference distinguishes these cirral rows even when they are in abnormal locations. The oral apparatus is more complex, with a band of compound ciliary structures known as membranelles on the left and two parallel undulating membranes on the right. These two sets of distinctive structures together form a lop-sided 'V', with the cell mouth, or buccal opening (Fig. 4a), near their point of posterior convergence.

In *Stylonychia*, only the old oral apparatus is passed intact to one of the daughter cells (the anterior daughter). The other ciliary structures are all replaced before cell division. The oral apparatus of the posterior daughter cell is derived from an oral primordium (Fig. 2e) that differentiates posterior to the old oral apparatus. The new marginal cirri are formed from four marginal cirral primordia (Fig. 2e), with LMP1 and RMP1 destined for the anterior division product, and LMP2 and RMP2 destined for the posterior division product. The other ciliary primordia, notably those for the prominent cirri located on the midventral surface, are not shown here; they appear at the same cell latitudes as the marginal ciliary primordia. The old cirri that did not participate in the formation of the cirral primordia are then resorbed. The newly formed ciliary primordia spread over the surfaces of the daughter cells and replace the old ciliary structures. The positioning of the ciliary primordia is precise and regular, so the preexisting arrangement of ciliary structures is perpetuated in the daughter cells.

Mirror-image doublets

With few exceptions, left-handed ciliates are derived from mirror-image doublets (Table 1). Therefore, to understand how reversals of handedness take place, one must first learn about ciliate doublets.

In doublets, two cells are fused together, with no internal partitions. Ciliate doublets may be either homopolar, with the anteroposterior axes of the two component cells similarly aligned (head-to-head and tail-to-tail), or heteropolar, with the anteroposterior axes oppositely aligned (head-to-tail). Homopolar doublets, but not heteropolar doublets, can divide and propagate their doublet condition (Fauré-Fremiet 1945).

The types of homopolar doublets observed in *Stylonychia* and other hypotrichs are illustrated in schematic cross-sectional views in Fig. 3c, following conventions introduced in Fig. 3a and 3b. The doublets are classified according to the

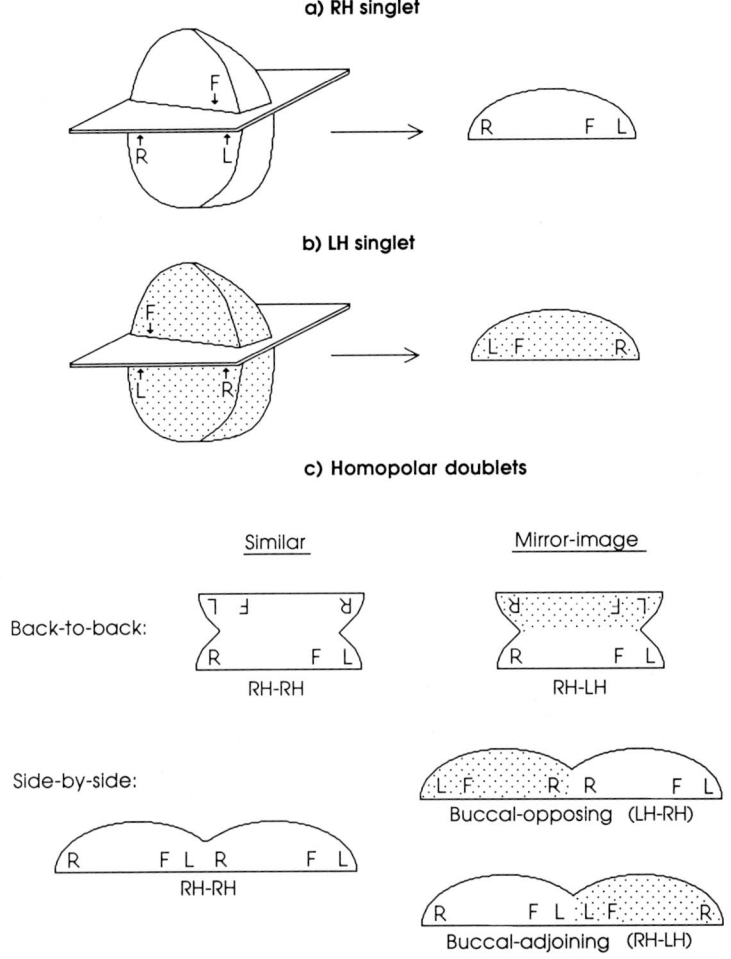

FIG. 3.   The types of homopolar doublets observed in hypotrich ciliates. Right-handed (RH) cells or cell components are open, left-handed (LH) cells or cell components are stippled. (a,b) Abstract cross-sections of (a) RH and (b) LH cells. (c) The types of doublets actually observed, classified according to geometry (back-to-back versus side-by-side) and topology (similar versus mirror-image). For further explanation, see the text.

relative positions of the components (back-to-back versus side-by-side) and their symmetry relations [similar (RH–RH) versus mirror-image (RH–LH)].

Back-to-back RH–RH homopolar doublets are relatively common in hypotrichs and are capable of indefinite vegetative self-reproduction (Fauré-Fremiet 1945, Grimes 1973). In contrast, back-to-back mirror-image (RH–LH) doublets have been described only once in a hypotrich, by Fauré-Fremiet (1945),

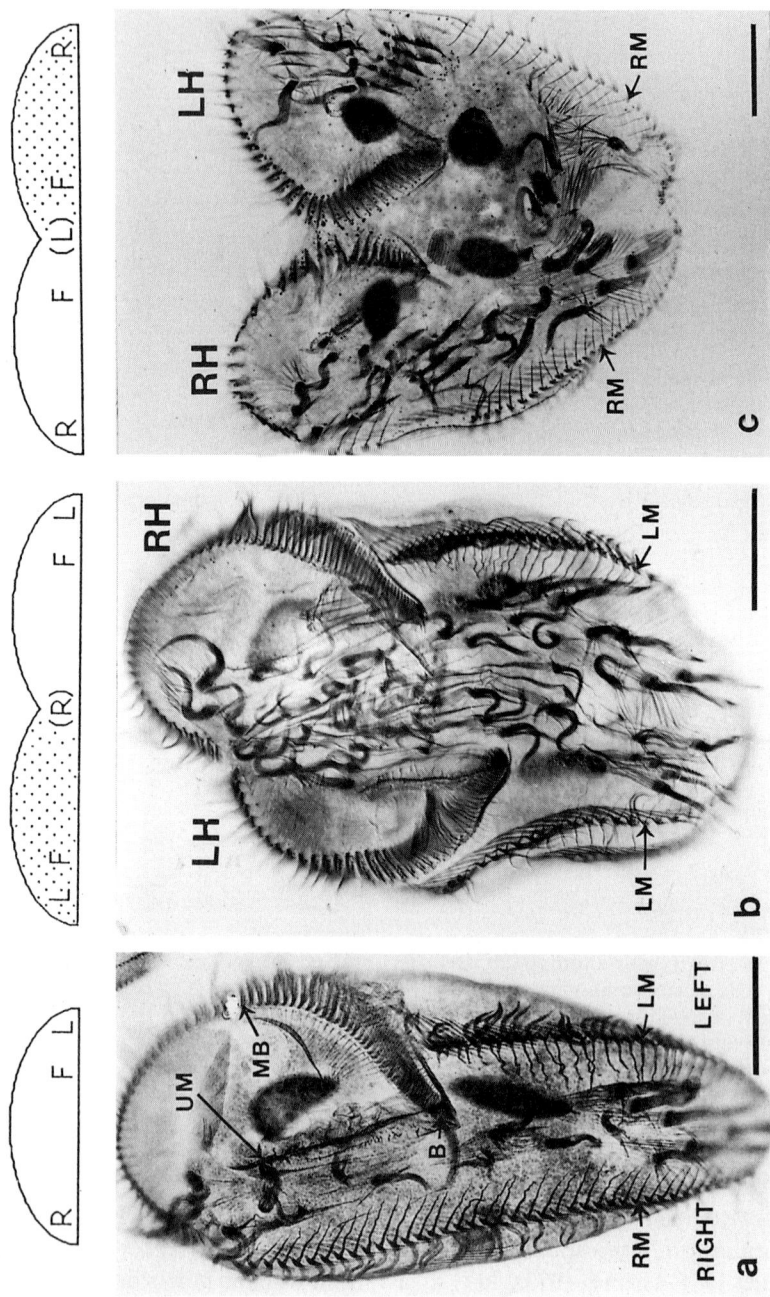

who observed transient propagation of this configuration in a clone of cells that split apart to form viable RH and inviable LH singlets after a few weeks of observation.

Side-by-side homopolar RH–RH doublets (Fig. 3c, lower left), unlike their back-to-back RH–RH counterparts, are unstable. In contrast, side-by-side LH–RH doublets of the buccal-opposing variety (Figs. 3c, 4b) are stable. Such mirror-image doublets were first experimentally produced in *Stylonychia* and shown to be self-reproducing by Tchang et al (1964); they have since been generated and asexually propagated in four other laboratories (Table 1). These doublets can maintain their original organization after conjugating with singlets (Shi & Qiu 1989).

Side-by-side buccal-adjoining RH–LH doublets (Figs. 3c, 4c) are unstable, largely because of a tendency to lose oral structures at the plane of symmetry (Grimes & L'Hernault 1979, Shi et al 1991).

Mirror-image doublets have also been observed in other groups of ciliates: they were described in the heterotrich ciliate *Blepharisma* by Suzuki (1957), and later observed in the hymenostomes *Glaucoma* and *Tetrahymena* by Suhama (1982) and by Frankel & Nelsen (1986), respectively (Table 1). In *Tetrahymena*, there are also forms in which the dorsal surface is variably transformed to a reversed (LH) ventral surface; these forms are dependent on the action of *janus* mutant genes and are not considered further here (see Frankel 1989 for review and references).

The origin of left-handed organization

To convert an RH to an LH organization, the cell must somehow break across a major topological barrier. It is thus not surprising that naturally occurring

---

FIG. 4. Schematic cross-sections (above) and protein-silver preparations of the ventral surfaces (below) of (a) a normal right-handed cell of the hypotrich *Stylonychia mytilus*, (b) an established buccal-opposing (LH–RH) mirror-image doublet, and (c) a newly formed buccal adjoining (RH–LH) mirror-image doublet. The structures labelled in (a) are the same as those in Fig. 2d, plus the buccal opening (B). In the doublets, the right-handed (RH) and left-handed (LH) components are indicated. In the buccal-opposing mirror-image doublet (b), both cirral rows at the outer margins are of the left marginal (LM) type, with prominent accessory fibres extending transversely to the right, whereas in the buccal-adjoining mirror-image doublet (c), both outer marginal cirral rows are of the right marginal (RM) type, with accessory fibres extending diagonally to the left on the RH side and to the right (inverted) on the LH side. Marginal cirral rows are absent from the boundary between RH and LH components in the central region of the doublet. The doublet in (c) is undergoing morphogenetic reorganization. Scale bars = 50 µm. These photographs were kindly supplied by Dr Xinbai Shi; Fig. 5c, labelled differently, was published as Fig. 9H of Shi et al (1991).

TABLE 1 Occurrence of mirror-image doublets and left-handed forms in ciliates

Ciliate group	Genus	Mirror-image (MI) doublets Stability	References	Left-handed singlets Origin from MI doublets	Capacity to divide	References
Hetero-trichida	*Stentor*	—	—	No[e]	No	14
	Blepharisma	Transient	13	—	—	—
Hypotrichida	*Stylonychia*	Stable[a]	2,5,8, 15,17,18	Yes	No[g]	2,9,16
	Stylonychia	Transient[b]	3,10	—	—	—
	Pleurotricha	Stable[a]	4	Yes	No	2
	Paraurostyla	Stable[a]	5	—	—	—
	Urostyla	Transient[c]	1	Yes	No	1
Hymenosto-matida	*Glaucoma*	Stable	11	Yes	Yes	12
	Tetrahymena	Transient[d]	6	Likely[f]	Yes	7

[a]Buccal-opposing side-by-side mirror-image (LH–RH) doublets. [b]Buccal-adjoining side-by-side mirror-image (RH–LH) doublets. [c]Back-to-back mirror-image (RH–LH) doublets. [d]Does not include double-ventral forms arising from the action of *janus* genes. [e]LH singlets resulted from regulation after sequential transections and rotations of halves of the cell around both the vertical and horizontal axes. [f]LH singlets were selected from cultures containing regulating homopolar doublets. [g]In one case, three successive divisions occurred before death (Grimes 1990).
References: (1) Fauré-Fremiet 1945; (2) Grimes 1990; (3) Grimes & L'Hernault 1979; (4) Grimes et al 1980; (5) Jerka-Dziadosz 1983; (6) Frankel & Nelsen 1986; (7) Nelsen & Frankel 1989; (8) Shi & Frankel 1990; (9) Shi et al 1990; (10) Shi et al 1991; (11) Suhama 1982; (12) Suhama, 1985; (13) Suzuki 1957; (14) Tartar 1966; (15) Tchang et al 1964; (16) Tchang & Pang 1977; (17) Tuffrau & Totwen-Nowakowska 1988; (18) Yano & Suhama 1991.

mirror-image doublets and LH singlets have not been reported. The topological breakthrough has been accomplished in three circumstances in *Stylonychia*. One mode involved extensive folding of a central disc removed from a dividing cell, and was the original means by which self-reproducing mirror-image doublets were obtained (Tchang et al 1964); this mode is not thoroughly understood (Frankel 1989, p 199–200). Here I will summarize the two better-investigated modes of origin of mirror-image doublets, which involve respecification along the anteroposterior and lateral axes, respectively.

Respecification of the anteroposterior axis

Shi and his collaborators have devised an operation that provokes a reversal of the anteroposterior axis in *Stylonychia mytilus* (Shi et al 1991) and other hypotrich ciliates (X.-b. Shi, L. Lu, Z. Qiu, W. He, J. Frankel, in preparation). Fig. 5 provides a simplified description of this operation and its consequences in *Stylonychia*. The operation brings about an exchange of most of the anterior and posterior halves of the cell, with preservation of the original polarity and asymmetry of the two halves (Fig. 5a,b). The practical necessities of performing the operation on these small cells (about 0.2 mm in length) required removal of both ends and of one margin (removal of the cell's right margin is shown here). One consequence of the operation is that a zone originally near the posterior end of the cell ('F' in Fig. 5a,b) becomes juxtaposed on another zone originally near the anterior end of the cell ('A').

Stylonychia has great capacities for morphallactic regeneration. If the two transposed pieces had been totally separated, each would have produced complete sets of ciliary primordia to regenerate the total ciliary pattern with unchanged polarity (Dembowska 1925). In the transposed situation, often one of the regions did exactly that, whereas the other underwent a reversal of anteroposterior polarity.

The anteriorly transposed posterior region (P→A) of *Stylonychia* regenerated its ciliary organization as it would have if it had been isolated, with an oral primordium (cf Fig. 5c, OP1) differentiating at the normal primordium site (E in Fig. 5a,b). The new oral structures produced by this oral primordium then

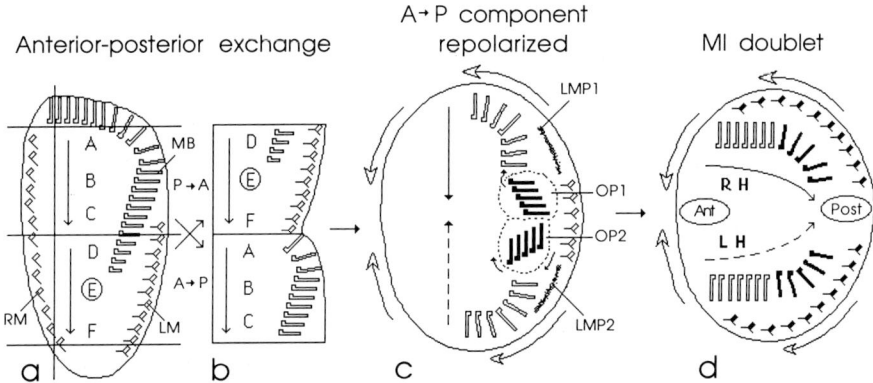

FIG. 5. The anterior-posterior exchange operation in *Stylonychia mytilus*. All diagrams include the membranelle band (MB) and marginal cirri: left marginal (LM) and right marginal (RM) cirri are structurally different in *Stylonychia*. Old structures are shown in outline, new structures are filled in. (a) The cell before the operation. The lines cutting across the cell indicate the planes of transection, the vertical arrows inside the cell indicate cell polarity, and the large letters indicate latitudinal positions; the approximate latitude at which the oral primordium differentiates is circled. (b) The cell immediately after the operation, with the posterior-to-anterior (P→A) and anterior-to-posterior (A→P) transpositions indicated. (c) The cell approximately 2 h after the operation, showing the sites of oral primordia (OP1, OP2) and left marginal cirral primordia (LMP1, LMP2). The dashed vertical arrow inside each cell indicates the region of reversal of polarity; the open-headed arrows outside the cells show the direction of folding. (d) The eventual configuration attained about 4 h after the operation, showing the right-handed (RH) and left-handed (LH) components and the anterior (Ant) and posterior (Post) ends of the mirror-image doublet. For further explanation, see the text.

moved anteriorly to merge with the remaining old oral structures of the P→A region (only membranelles are shown) (Fig. 5d), retaining their original polarity. Other ciliary primordia, including the left marginal cirral primordium (LMP1 in Fig. 5c), appeared at latitudes similar to that of the oral primordium, regenerating sets of cell structures with the normal RH pattern.

The posteriorly transposed anterior fragment (A→P) also regenerated a complete set of ciliary structures, but with reversed polarity. A second oral primordium (OP2) was formed, with a set of membranelles that were rotationally permuted relative to those in OP1 (Fig. 5c). This set of new membranelles then underwent a clockwise rotation to seat itself on the remaining old oral membranelles of the A→P region, but now with a reversal of global polarity: the former posterior portion of the membranelle band became anterior, whereas the presumptive anterior end became posterior. Other ciliary primordia appeared (only LMP2 is shown) in an even more obviously reversed arrangement (see photographs in Shi et al 1991). At the same time, the complex folded toward the wounded right edge, converting the tail-to tail linear configuration shown in Fig. 5c into the parallel configuration of Fig. 5d. The P→A region thus became the RH component of a buccal-opposing mirror-image doublet, whereas the reversed A→P region became the LH component of the doublet. Marginal cirri of the *left* marginal type enveloped both sides of the doublet. This entire reorganization took about four hours (Shi et al 1991).

When the left rather than the right edge was excised during the original operation, a similar reversal of polarity took place in the A→P region, but folding occurred toward the left and the outcome was a buccal-adjoining rather than a buccal-opposing mirror-image doublet (not shown here; see Shi et al 1991).

A consequence of the reversal of global polarity operating with normal ciliary units is that many of these units became inverted (rotationally permuted) in the LH component of the mirror-image doublet. Some structures, such as the individual membranelles, remained individually inverted in the self-reproducing mirror-image doublet (Grimes et al 1980, Jerka-Dziadosz 1983). Other structures, such as the inverted left marginal cirri, commonly were replaced by normally oriented cirri, which were still of the left marginal structural type (cf Fig. 4b) (Jerka-Dziadosz 1985, Shi et al 1991).

Respecification of the lateral axis

There is another route to the formation of mirror-image doublets. This sequence, described in three different investigations on three related species of hypotrichs, always begins with side-by-side (RH–RH) doublets generated by translocation of daughter cells following division arrest, and ends with stable buccal-opposing (LH–RH) mirror-image doublets (Jerka-Dziadosz 1983, Tuffrau & Nowakowska 1988, Yano & Suhama 1991). Some of the events in between have not been fully elucidated: Yano & Suhama's summary figure, based on continuous observation

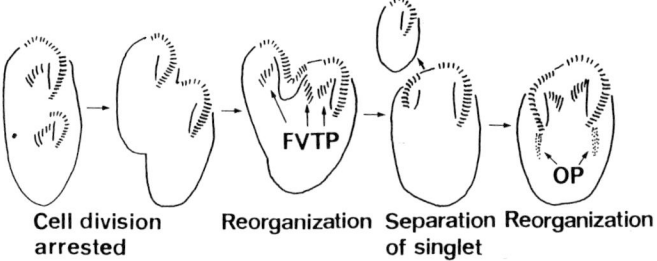

| Cell division | Reorganization | Separation | Reorganization |
| arrested | | of singlet | |

FIG. 6. Schematic illustrations of the transformation into a mirror-image doublet of a *Stylonychia pustulata* cell that had been arrested in division by heat shock, showing the oral structures (unlabelled), the frontal-ventral-transverse primordia (FVTP) and the oral primordia (OP). For further explanation, see the text. From Yano & Suhama (1991), with permission. This figure was kindly provided by Dr Mikio Suhama prior to its publication.

of living cells of *Stylonychia pustulata* going through the transition from side-by-side RH–RH doublets to mirror-image doublets (Fig. 6), represents a fair summary of what is known about this process. The critical event is the 'reorganization' step, which seems to involve a U-shaped bend of the oral structures of the right component, and the formation of *three* sets of FVTP (frontal-ventral-transverse primordia). These are primordia for the prominent central cirri (which I did not introduce earlier). The two outer sets of FVT cirral primordia develop normally within the two original components of the side-by-side doublet, whereas the middle set develops in between and appears to be a mirror image of both of the outer primordia. After this reorganization step, a small normal (RH) singlet cell buds off on the right side, leaving an imperfect mirror-image doublet, which becomes perfected in subsequent reorganization(s). The mirror-image doublet produced by this means is in most ways identical to that generated by the anteroposterior interaction described earlier.

My interpretation of these events is shown in a stylized form in Fig. 7, with the 'reorganization' and 'separation of singlet' step combined to save space. In my view, the side-by-side RH–RH doublet that results from the division arrest is unstable solely because it contains a major positional discontinuity between the right and left marginal structures along the line of juncture, thereby violating the 'principle of normal neighbours' postulated by Mittenthal (1981) for control of patterning in multicellular organisms. This discontinuity provokes the system to restore positional continuity by formation of an LH domain, resulting in a RH–LH–RH 'triplet' similar to those previously generated in *Blepharisma* (Kumazawa 1979) and in *Tetrahymena* (Fauré-Fremiet 1948, Nelsen & Frankel 1986). The *Stylonychia* triplet is resolved into an LH–RH mirror-image doublet by budding off the right RH component. Since the details of this pathway have not been worked out fully, this scheme should be regarded as provisional.

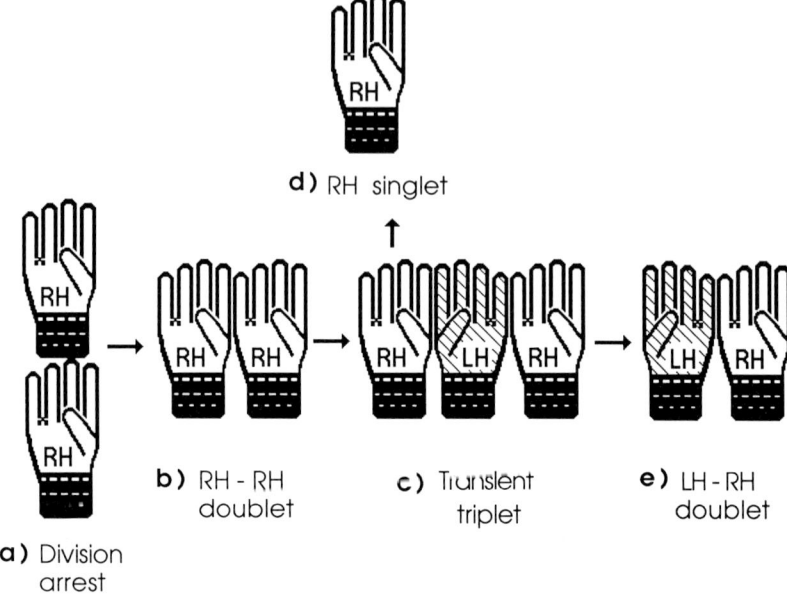

d) RH singlet

a) Division arrest

b) RH - RH doublet

c) Transient triplet

e) LH - RH doublet

FIG. 7.   A stylized interpretation of the events illustrated in Fig. 6, representing cells as if they were hands. Right-handed (RH) components are open, left-handed (LH) components shaded. (a) A normal RH cell arrested in division becomes converted into (b) a side-by-side RH–RH doublet through the sliding anteriorly of the posterior daughter cell. The right edge of one cell ('thumb') thereby comes to lie adjacent to the left edge of the other ('little finger'). To fill in this discontinuity, (c) an LH component becomes interpolated between the two RH components. This 'triplet' form is transient, being resolved by (d) the budding off of the RH component on the right side of the triplet, leaving behind (e) a side-by-side LH–RH mirror-image doublet.

Despite great differences in detail between the two well-investigated pathways of formation of mirror-image doublets, they share two similarities. One is that in both situations the starting point involves the juxtaposition of territories that are normally distant from each other: anterior and posterior extremities in one case, right and left margins in the other. The new neighbours are highly abnormal in both cases. The other similarity is the formation of a reversed set of ciliary structures near the site of positional discontinuity, thereby resolving that discontinuity. In both cases, intercalation might be taking place to restore normal sequences of positional values, which would imply the existence of an intra-cellular version of the process originally postulated for multicellular organisms by French et al (1976). This generalization, however, is subject to one caveat for each of the two experimental situations: in the case of respecification along the anteroposterior axis, an alternative explanation based on the idea of a 'polarizing zone' (Tickle et al 1975) may be tenable (Shi et al 1991), whereas in the case of respecification of the lateral axis, the LH organization interposed between the two original RH components appears to be partial rather than complete.

Left-handed singlets

The reader will undoubtedly have guessed how one can obtain a left-handed (LH) singlet from a mirror-image doublet—by longitudinal bisection along the plane of symmetry of a mirror-image doublet to separate the RH and LH components. If both components of the original mirror-image doublet are nucleated (as generally is the case), then both derived singlet forms ought to be capable of growth and division. This has been shown to be true in LH singlets of the hymenostome ciliates *Glaucoma scintillans* (Suhama 1985) and *Tetrahymena thermophila* (Nelsen & Frankel 1989) (Table 1).

In the hypotrich *Stylonychia*, however, an LH singlet cannot grow, almost certainly because it cannot feed (Fauré-Fremiet 1945, Tchang & Pang 1977, Grimes 1990, Shi et al 1990). This inability to feed is a consequence of the 180° inversion of the individual membranelles of the oral apparatus in the LH component: a buccal opening (cell mouth) forms at the posterior end of the oral apparatus where the membranelles and undulating membranes converge (Fig. 4b), but the inverted membranelles in the LH component sweep food particles anteriorly, away from that mouth. This does not affect the viability of a mirror-image doublet, in which the LH component is supported by the normal RH component with which it is in cytoplasmic continuity, but is fatal for an isolated LH cell. This isolated LH cell undergoes futile morphogenetic reorganizations, always ending up with the same lethal topology despite the presence of a normal macronucleus that doubtless produces normal gene products (Grimes 1990, Shi et al 1990, cf Tartar 1966). The only known way that an LH cell can be 'rescued' is by converting it into a mirror-image doublet through a transposition operation analogous to the one described above (Shi et al 1987).

Conclusions

The phenomena outlined in the previous sections and parallel observations in *Tetrahymena* and *Glaucoma* (Frankel 1990) challenge our general understanding of pattern formation in development. The capacity of simple operations to elicit the rapid conversion of a heritable normal ciliate organization to an equally heritable mirror-image organization indicates that the difference between patterns of opposite handedness cannot in this case be specified by genic differences [a conclusion that was also demonstrated for LH cells of *Tetrahymena* by different means (Nelsen et al 1989)]. In my view, the results described here indicate that intracellular positional information must exist in the ciliate surface region; different surface regions are non-equivalent (cf Lewis

& Wolpert 1976) in some way that is developmentally important. The dynamics of pattern reversal based on intracellular non-equivalence has been formulated precisely in a theoretical model for *Tetrahymena* (Brandts & Trainor 1990a,b). Although this model is entirely abstract, it none the less suggests a system in which at least two components interact to specify positional values (Frankel 1990). The molecular basis of this system remains elusive.

Despite our inability (as yet) to provide a molecular basis for handedness in ciliates, the similarity in the dynamics underlying changes in handedness of ciliates to those encountered during pattern regulation in eggs and multicellular organisms indicates that the genic interactions and cell-to-cell signalling that are usually invoked to explain these dynamics probably do not provide a *sufficient* explanation. The cell cortex—perhaps in eggs as well as ciliates—may still hold mysteries that lie beyond current molecular biological understanding.

Acknowledgements

I would like to thank especially Dr Xinbai Shi and Dr Mikio Suhama for communicating the information that made this essay possible. I would also like to thank Dr Shi for providing the photographs, Dr Suhama for providing a copy of Fig. 6 prior to publication, and Dr E. Marlo Nelsen for painstakingly designing the other line illustrations on his MacIntosh computer. An earlier version of this manuscript was read by Drs Anne W. K. Frankel, Gary W. Grimes, Maria Jerka-Dziadosz, E. Marlo Nelsen, Xinbai Shi and Norman E. Williams. The author's research is supported by NIH grant HD-08485.

References

Brandts WAM, Trainor LEH 1990a A non-linear field model of pattern formation: intercalation and morphallactic regulation. J Theor Biol 146:37–56

Brandts WAM, Trainor LEH 1990b A non-linear field model of pattern formation: application to intracellular pattern reversal in *Tetrahymena*. J Theor Biol 146:57–86

Dembowska WS 1925 Studien über die Regeneration von *Stylonychia mytilus*. Wilhelm Roux's Arch Entwicklungsmech Org 104:185–209

Fauré-Fremiet E 1945 Symétrie et polarité chez les Ciliés bi-ou multicomposites. Bull Biol Fr Belg 79:106–150

Fauré-Fremiet E 1948 Doublets homopolaires et régulation morphogénétique chez le cilié *Leucophrys patula*. Arch Anat Microscop Morphol Exp 37:183–203

Frankel J 1989 Pattern formation: ciliate studies and models. Oxford University Press, New York

Frankel J 1990 Positional order and cellular handedness. J Cell Sci 97:205–211

French V, Bryant PJ, Bryant SV 1976 Pattern regulation in epimorphic fields. Science (Wash DC) 193:969–981

Frankel J, Nelsen EM 1986 Intracellular pattern reversal in *Tetrahymena thermophila* II. Transient expression of a *janus* phenocopy in balanced doublets. Dev Biol 114:72–96

Grimes GW 1973 An analysis of the determinative difference between singlets and doublets of *Oxytricha fallax*. Genet Res 21:57–66

Grimes GW 1990 Inheritance of cortical patterns in ciliated protozoa. In: Malacinski GM (ed) Cytoplasmic information systems. (Primers in developmental biology, vol IV) Macmillan, New York, p 23–43

Grimes GW, L'Hernault SW 1979 Cytogeometrical determination of ciliary pattern formation in the hypotrich ciliate *Stylonychia mytilus*. Dev Biol 70:372–395

Grimes GW, McKenna ME, Goldsmith-Spoegler CM, Knaupp EA 1980 Patterning and assembly of ciliature are independent processes in hypotrich ciliates. Science (Wash DC) 209:281–283

Jerka-Dziadosz M 1983 The origin of mirror-image symmetry doublet cells in the hypotrich ciliate *Paraurostyla weissei*. Roux's Arch Dev Biol 192:179–188

Jerka-Dziadosz M 1985 Mirror-image configuration of the cortical pattern causes modifications in propagation of microtubular structures in the hypotrich ciliate *Paraurostyla weissei*. Roux's Arch Dev Biol 194:311–324

Kumazawa H 1979 Homopolar grafting in *Blepharisma japonicum*. J Exp Zool 207: 1–16

Lewis JH, Wolpert L 1976 The principle of non-equivalence in development. J Theor Biol 62:479–490

Mittenthal JE 1981 The rule of normal neighbors: a hypothesis for morphogenetic pattern regulation. Dev Biol 88:15–26

Nelsen EM, Frankel J 1986 Intracellular pattern reversal in *Tetrahymena thermophila* I. Evidence for reverse intercalation in unbalanced doublets. Dev Biol 114:53–71

Nelsen EM, Frankel J 1989 Maintenance and regulation of cellular handedness in *Tetrahymena*. Development 105:457–471

Nelsen EM, Frankel J, Jenkins LM 1989 Non-genic inheritance of cellular handedness. Development 105:447–456

Shi X-b, Frankel J 1990 Morphology and development of mirror-image doublets of *Stylonychia mytilus*. J Protozool 37:1–13

Shi X-b, Qiu Z 1989 Induction of amicronucleate mirror-imaged doublet in the ciliate *Stylonychia mytilus* and observation of its reproductive behavior. Acta Zool Sin 35:364–369 (In Chinese, with English summary)

Shi X-b, Qiu Z, Lu L 1987 The correction of the reversed half in mirror-imaged doublet of *Stylonychia mytilus* and its genetic implications. Symp Chin Protozool Soc 4:38–39 (abstract, in Chinese)

Shi X-b, Lu L, Qiu Z, Frankel J 1990 Morphology and development of left-handed singlets derived from mirror-image doublets of *Stylonychia mytilus*. J Protozool 37:14–19

Shi X-b, Lu L, Qiu Z, He W, Frankel J 1991 Microsurgically generated discontinuities provoke heritable changes in cellular handedness in a ciliate, *Stylonychia mytilus*. Development 111:337–356

Suhama M 1982 Homopolar doublets of the ciliate *Glaucoma scintillans* with a reversed oral apparatus. I. Development of the oral primordium. J Sci Hiroshima Univ Ser B Div 1 (Zool) 30:51–65

Suhama M 1985 Reproducing singlets with an inverted oral apparatus in *Glaucoma scintillans* (Ciliophora, Hymenostomatida). J Protozool 32:454–459

Suzuki S 1957 Morphogenesis in the regeneration of *Blepharisma undulans japonicus* Suzuki. Bull Yamagata Univ (Nat Sci) 4:85–192

Tartar V 1966 Stentors in dilemmas. Z Allg Mikrobiol 6:125–134

Tchang T-r, Pang Y-b 1977 The cytoplasmic differentiation of jumelle *Stylonychia*. Sci Sin 20:234–243

Tchang T-r, Shi X-b, Pang Y-b 1964 An induced monster ciliate transmitted through three hundred and more generations. Sci Sin 13:850–853

Tickle C, Summerbell D, Wolpert L 1975 Positional signalling and specification of digits in chick limb morphogenesis. Nature (Lond) 254:199–202

Tuffrau M, Totwen-Nowakowska I 1988 Différents types de doublets chez *Stylonychia mytilus* (Cilié Hypotriche): genese, morphologie, reorganisation, polarité. Ann Sci Nat Zool Paris 13e ser 9:21–36

Yano JJ, Suhama M 1991 Pattern formation of mirror-image doublets of the ciliate *Stylonychia pustulata*. J Protozool 38:111–121

DISCUSSION

Fujinaga: Is global reversal the equivalent of what we call complete situs inversus?

Frankel: Global reversal and situs inversus are really two different things. Take my right hand and my left hand: suppose they were cut off and put on the table, we could then speak of the handedness of each individual hand like we can speak of the handedness of an amino acid. By global reversal, I mean the reversal from the asymmetry of one hand to the asymmetry of the other hand in terms of internal arrangement of its parts. Situs inversus refers to an organism that already has a basic bilateral symmetry. We are talking not about what makes the right hand different from the left hand, but about what makes one prefer a right hand or left hand, or what perhaps creates a slight asymmetry superimposed on a basic bilateral symmetry.

Brown: You said that you never saw local inversions.

Frankel: That is correct, if by 'inversions' you really mean 'reversals of asymmetry'.

Brown: Do you ever get global plus local reversals?

Frankel: Never.

Fujinaga: Is this terminology of local and global reversal well accepted?

Frankel: It's well accepted among the small band of people working on ciliates. They speak about local and global or about small scale and large scale patterning.

Morgan: Psychologists use similar terminology. In perceptual experiments, for example, one can use large letters composed of smaller elements that are themselves letters. For example, a large 'F' could be composed of small 'g's. There is a lot of perceptual work looking at the facilitation and inhibition between local and global processing in these composite figures (Navon 1977).

Lawrence: Does the process of intercalation in ciliates occur in a particular order? If you go from F through to A (Fig. 5), do you start with E and B then bridge the remaining gap (D and C) step by step?

Frankel: In *Stylonychia* and other hypotrichs, all new ciliary structures arise in a stereotyped sequence, with an oral primordium appearing first, FVT cirral primordia next, then marginal cirral primordia (Shi et al 1990). Rather than developing step by step from the boundaries inward to fill the gap, the entire territory reorganizes in some as yet undetectable manner: then the new ciliary structures appear not far from the centre of the reversed territory and continue to develop in their usual sequence but with reversed polarity.

There is indirect evidence for the initial process of 'invisible' repolarization. Normally, when there are two sets of ciliary primordia in the same ciliate cell, even when they are of opposite handedness, their development is synchronous (Grimes et al 1981, Shi et al 1990). However, where global polarity is in the process of being reversed, the development of the set of ciliary primordia undergoing reversal is delayed relative to that of the set developing with the original polarity (Grimes & L'Hernault 1979, Shi et al 1991). In subsequent rounds of reorganization and division, after the reversal has become established, the two sets of primordia develop synchronously. The delay associated with reversal of polarity, and hence with handedness, occurs regardless of whether the reversal involves a change from the normal RH organization to the abnormal LH organization (as described here), or the less common reversal from LH to RH (X.-b. Shi, Z. Qiu, L. Lu, unpublished). Therefore, the delay is inherent in the process of reorganization. Unfortunately, we do not know how this initial reorganization proceeds spatially.

Lewis: When the ciliate is cut during these microsurgical manipulations, why don't the contents of the cell spill out?

Frankel: In some ciliates, for example *Euplotes*, they do! In *Stentor*, a big ciliate that has been operated on extensively, not only do the intracellular contents not spill, but the halves of the cells are very hard to graft together. The cytoplasm congeals immediately the ciliate is cut. I would bet this is Ca^{2+} dependent. When you bisect a *Stentor*, the halves become like two balls that repel each other. Vance Tartar grafted them by cutting and annealing at virtually the same time. I think that is why Shi Xin-bai had to 'graft' the smaller *Stylonychia* cell by cutting off the edges and then transposing parts while maintaining cytoplasmic continuity (see Figs 1, 2 in Shi et al 1991).

Lewis: Is this healing of membrane or healing of a mass of microtubules or what?

Frankel: No one has studied this ultrastructurally. It would be interesting to see whether membrane forms instantaneously. I doubt it would.

Lawrence: In multicellular systems, people often emphasize the importance of cell division in the generation of normal neighbours. The situation in ciliates seems similar with respect to intercalation, even though there is no division of the buds into cells. Perhaps the mechanism of patterning does not depend on cell division but involves something else that we know little about.

Stern: Except that in unicellular organisms the generation of new membrane may be analogous to the generation of new cells in multicellular organisms.

Frankel: We don't know what is reorganizing in ciliates. In the cases I described, there is no evident growth; the cells are starving, not feeding. It looks like classical morphallaxis—development through reorganization of what is there rather than through intervening growth. This phenomenon can be described only at a rather gross level; we don't know the molecular mechanism. We can never be sure whether ciliate intercalation is truly homologous to this process in multicellular organisms or whether it has evolved independently. I would like to believe that in some sense it's homologous, but I don't have any convincing arguments for that.

Wolpert: Am I right in thinking that every asymmetrical cell organelle is related to a basal body?

Frankel: In the ciliate cell surface region, yes.

Wolpert: So is the only asymmetrical organelle, as distinct from molecules like myosin and actin, in the cell the centriole or basal body?

Lewis: Isolated smooth muscle cells are supposed to adopt a corkscrew shape, which probably reflects asymmetry in the bundles of actin and myosin.

Wolpert: Those are cells as distinct from an organelle.

Layton: In some sporozoans there is a helical structure. What is that and what does it have to do with this?

Frankel: Spores of the Myxozoa and Microspora, two distantly related phyla within what used to be called the Sporozoa, possess helically wound polar filaments located within polar capsules (Myxozoa; Lom 1990) or polar tubes free within the spore (Microspora; Canning 1990). The polar capsules of myxozoans are remarkably similar to the nematocytes of coelenterates (Lom 1990). I do not know of any information on the direction of winding of either structure. Morphogenesis of these helical structures seems to be different in the two phyla. In the Myxozoa it occurs by outgrowth from the base (Lom 1990); in Microspora by coalescence of Golgi vesicles (Vinckier 1975). In both cases, the structures are utterly unlike basal bodies or centrioles.

Layton: In the midpiece of spermatozoa of many species, there is a helical arrangement of the mitochondria (Zamboni 1991). I have been unable to find out whether this helix has a consistent handedness.

Wolpert: That may be a counter example. Certainly, the centriole is the obvious cellular organelle with structural asymmetry at a multiprotein level. Klaus, do we know anything about assembly of the centriole?

Weber: I am not aware of anything that is firmly established.

Morgan: What is thought now about centrioles being autonomous in the sense of self-replicating? Are the only organelles that show asymmetry also autonomous?

Frankel: Normally, new centrioles or basal bodies form near old ones, but there are some well documented exceptions. One is in the amoeboflagellate *Naegleria* (Fulton & Dingle 1971); another is in *Oxytricha*, a relative of *Stylonichia*. It forms cysts that lose all of their basal bodies (Grimes 1973). Gary Grimes serially sectioned the cysts but couldn't find any basal bodies. He claimed there were no microtubules either; he might have missed a few microtubules, I doubt he missed a basal body. So you don't need an old basal body to make a new one. There could be something left that's asymmetrical that you simply cannot distinguish using the electron microscope.

There were claims around 1970 that there was DNA in basal bodies. These were pretty well refuted (Fulton 1971, Hartman et al 1974). It was recently claimed that the *uni* linkage group, which contains a number of flagellar markers, was located in the basal body, as shown by molecular biological techniques (Hall et al 1989). This has also been disputed (Johnson & Rosenbaum 1990).

There is fairly reasonable evidence that there is RNA in basal bodies. In ciliates, Ruth Dippell (1976) showed that the dense material in the lumen of the basal bodies is dissipated by RNAse but not by DNAse or proteases. There were other demonstrations in this and other systems (e.g. Hartman et al 1974). But there are no ribosomes in the basal body and nobody knows what that RNA does. Even when new basal bodies form near old ones, they don't divide: a new one forms at a right angle to the old one, several hundred Ångströms away (Dippell 1968). So I do not think that one can call basal bodies self-replicating.

Weber: I don't think anyone has cloned the RNA in basal bodies. It really is not clear if it is a contaminant or if it is a truly structural feature, in which case you would expect it to have a unique sequence.

Jefferies: Ribosomes are asymmetrical organelles.

Morgan: Could someone explain what sort of asymmetry ribosomes have?

Chothia: In eukaryotes, it is a large complex of RNA and 83 different polypeptides. It has a complicated shape which is asymmetrical.

Kondepudi: All the speakers have talked about an asymmetrical creature of one kind or another. From an evolutionary point of view, at the level of molecules we can say we need autocatalysis and competition for one molecular species to dominate (see p 11–12). No one has commented on what kinds of mechanisms at a higher level create these asymmetries and dominance of, say, a left-handed ciliate over a right-handed one.

Frankel: In ciliates, the mode of creation of higher-level asymmetry is unknown. The dominance of the right-handed form is because the left-handed form either cannot feed and divide at all, for example in *Stylonychia*, or in certain other cases such as *Tetrahymena* the left-handed cells can feed and divide but they divide more slowly than right-handed forms. A left-handed ciliate never survives in Nature because the individual organelles are not reversed and there is an inescapable conflict between the internal structure and the arrangement of organelles.

Berg: In bacteria, if you look at the wavelength and amplitude of a flagellum compared to the size of the cell body and consider the hydrodynamics, you find these are optimal for generating thrust. The dominance of a particular structure depends on its utility. From basic physical principles, if you ask how well an object that size can measure changes in concentration, you find that *E. coli* does as well as anything could. These are highly evolved creatures; they have been optimized over very many generations.

Thwaites: There is a bacterium, *B. subtilis*, mutant forms of which exist that grow into long multicellular filaments. Short filaments can be produced in cultures of wild-type cells but they are transitory. The mutants lack a lytic enzyme; this prevents the cells from separating into daughter cells after cell duplication. Filaments up to several millimetres long then form, containing thousands of individual cells. The cells usually divide when they are 3–4 µm long. In a filament, they are distinct; but the filament doesn't look like a string of sausages, it has no invaginations of any kind. It is like a cylindrical fibre.

By considering the properties of the filament, we infer that the cell doesn't grow by simple elongation, it grows with a twist within a spectrum which, in terms of surface helix angle, extends about $\pm 10°$. Whether this has any function in the normal parting of daughter cells, we are not sure.

If a filament is allowed to get long enough, it forms a prototype multicellular organism, the shape of which is governed solely by the elementary rules of structural mechanics. A single filament of perhaps a thousand cells during its lengthening not only twists, but because of Brownian motion or because of stochastic differences in growth rates at different points in its cell wall, it waves about. It does this in a characteristic manner that Mendelson called 'writhing'. A filament twists and bends in a fairly random manner, until one part of it touches another part. It touches in precisely the right way to form a helix of the same hand as that in which it is growing. There is a reasonable argument as to why this should be so. If a filament growing like this is restrained in torsion, it develops a contrary torque. If the tension in a filament under torque is reduced, it folds into a 2-ply structure in the opposite hand to that of the torque. One can show this with a piece of string. For a growing filament in which twisting is prevented, this is the same hand as the original twist. The 2-ply structure behaves in the same way as a single filament, so that 4-ply and eventually *n*-ply structures are obtained, where *n* can be as much as 1000. This is an example of a single 'rule': blocked twist causes plying, leading to the formation of a large multicellular structure (Mendelson & Thwaites 1989).

Mendelson first observed these structures in 1976 and called them 'macrofibres'. They look like little bits of textile yarn in which the helix angle increases with the number of filaments. This is because, for a given environment, the macrofibres from a given mutant have the same twist, no matter how many filaments they contain (Mendelson et al 1984). They also have a characteristic average length of folding, with some statistical variation. The folding process, unlike those of structural instability, is a dynamic one. It isn't a classical buckling problem, like that of a column with end thrust, although that can be induced in a very viscous growth medium. The plying then proceeds somewhat differently but the same twist is observed. Both the twisting rate and the folding length can be measured using time-lapse microcinematography.

The origin of twisting-with-growth could be the way in which the cell wall is built—that's one model, but it requires changes in the pattern with changes in environment. I have been working on the mechanics of what happens if there is a slight helical anisotropy in the mechanical properties of the cell wall. The combined effect of the turgor pressure within the bacterium and the upwelling of the wall material from the inside (the bacterial cell wall is continuously added to at the membrane) causes twisting. In addition, the net charge on the wall has the effect of a non-uniform pressure. Depending on the relative magnitude of this and the turgor pressure, twist could be of either hand without changes in wall arrangement or material properties. It is a reasonable hypothesis that this alone is sufficient to cause twisting-with-growth (Thwaites & Mendelson 1991).

References

Canning EU 1990 Phylum Microspora. In: Margulis L, Corliss JO, Melkonian M, Chapman DJ (eds) Handbook of Protoctista. Jones & Bartlett, Boston p 53–72

Dippell RV 1968 The development of basal bodies in *Paramecium*. Proc Natl Acad Sci USA 61:461–468

Dippell RV 1976 Effects of nuclease and protease digestion on the ultrastructure of *Paramecium* basal bodies. J Cell Biol 69:622–637

Fulton C 1971 Centrioles. In: Reinhardt J, Ursprung H (eds) Origin and continuity of cell organelles. vol 2: Results and problems in cell differentiation. Springer-Verlag, Berlin, p 170–221

Fulton C, Dingle AD 1971 Basal bodies, but not centrioles, in *Naegleria*. J Cell Biol 51:826–836

Grimes GW 1973 Morphological discontinuity of kinetosomes during the life cycle of *Oxytricha fallax*. J Cell Biol 57:229–232

Grimes GW, Knaupp-Waldvogel EA, Goldsmith-Spoegler CM 1981 Cytogeometrical determination of ciliary pattern formation in the hypotrich ciliate *Stylonychia mytilus*. II. Stability and field regulation. Dev Biol 84:477–480

Hall JL, Ramanis Z, Luck DJL 1989 Basal body/centriolar DNA: molecular genetic studies in *Chlamydomonas*. Cell 59:121–132

Hartman H, Puma JP, Gurney TL Jr 1974 Evidence for the association of RNA with the ciliary basal bodies of *Paramecium*. J Cell Sci 16:241–259

Johnson KA, Rosenbaum JL 1990 The basal bodies of *Chlamydomonas reinhardtii* do not contain immunologically detectable DNA. Cell 62:615–619

Lom J 1990 Phylum Myxozoa. In: Margulis L, Corliss JO, Melkonian M, Chapman DJ (eds) Handbook of Protoctista. Jones & Bartlett, Boston, p 36–52

Mendelson NH, Thwaites JJ 1989 Do forces and the physical nature of cellular materials govern biological processes? Comments Theor Biol 1:217–236

Mendelson NH, Favre D, Thwaites JJ 1984 Twisted states of *Bacillus subtilis* macrofibers reflect structural states of the cell wall. Proc Natl Acad Sci USA 81:3562–3566

Navon D 1977 Forest before trees: the precedence of global features in visual perception. Cognit Psychol 9:353–383

Thwaites JJ, Mendelson NH 1991 Mechanical behaviour of bacterial cell walls. Adv Microb Physiol 32:173–222

Vinckier D 1975 Nosemoides gen n., *N. vivieri* (Vinckier, Devauchelle, & Prensier, 1970) comb. nov. (Microsporidie): étude de la différenciation sporoblastique et genèse des différentes structures de la spore. J Protozool 22:170–184

Zamboni L 1991 Physiology and pathophysiology of the human spermatozoan: the role of electron microscopy. J Electron Microsc Tech 17:412–436

Two types of bilateral symmetry in the Metazoa: chordate and bilaterian

R. P. S. Jefferies

The Natural History Museum, Department of Palaeontology, Cromwell Road, London SW7 5BD, UK

Abstract. The chordate sagittal plane is perpendicular to the sagittal plane primitive for the bilaterally symmetrical metazoans (Bilateria). The earliest metazoans, when symmetrical at all, were probably radial in symmetry. The axis of symmetry was vertical and the mouth, when present, opened either upward or downward. The Bilateria evolved from the primitive metazoan condition by acquiring bilateral symmetry, mesoderm, a brain at the anterior end and protonephridia. Perhaps in the stem lineage of the Bilateria a hydroid-like or medusoid-like ancestor fell over on one side onto a substrate (pleurothetism). If so, the anteroposterior axis of Bilateria would be homologous with the vertical axis of radial symmetry in coelenterates. The bilaterian plane of symmetry arose to include the anteroposterior axis. The Deuterostomia (the Hemichordata, Echinodermata and Chordata) evolved within the Bilateria by producing the mouth as a secondary perforation. Within the deuterostomes the echinoderms and chordates constitute a monophyletic group named Dexiothetica. Hemichordates retain the primitive bilaterian sagittal plane. The Dexiothetica derive from an ancestor like the present-day hemichordate *Cephalodiscus* which had lain down on the primitive right side (dexiothetism) and acquired a calcite skeleton. The echinoderms evolved from this ancestor by losing the ancestral locomotory tail and gill slit, becoming static, moving the mouth to the centre of the new upper surface and developing radial pentameral symmetry. The chordates evolved from the same ancestor by developing a notochord in the tail, losing the water vascular system, evolving a filter-feeding pharynx and developing a new vertical plane of bilateral symmetry perpendicular to the old bilaterian plane. Evidence derived from certain bizarre Palaeozoic marine fossils (calcichordates) gives a detailed history of the early evolution of echinoderms and chordates and shows how the new bilateral symmetry was gradually acquired in chordates. This symmetry began in the tail (which contained the notochord and was also the leading end in locomotion) and advanced forward into the head.

1991 Biological asymmetry and handedness. Wiley, Chichester (Ciba Foundation Symposium 162) p 94–127

This paper deals with two problems: (1) the origin of bilateral symmetry in the Bilateria as a great subgroup of the Metazoa; and (2) the origin of bilateral symmetry in the Chordata, as a rather small subgroup of the Bilateria. These problems are not identical, because the sagittal plane of a human is perpendicular to that of a beetle.

Cladistic methodology

Cladistics attempts to classify organisms phylogenetically, using recency of common ancestry as the fundamental criterion for the recognition of groups (Fig. 1). For general treatments see Hennig (1966, 1969), Wiley (1981) and Ax (1985, 1987).

The central concept of cladistics is the *monophyletic group*, which, by definition, comprises all organisms descended from a single stem species (latest common ancestor). Extant monophyletic groups can be joined in pairs called *sister groups*, each pair sharing a latest common ancestor that was not ancestral to anything else still extant. Two sister groups constitute a monophyletic group of higher rank than themselves; by joining monophyletic groups in pairs of sister groups, a complete system of extant organisms can be built. The results can be presented as a dichotomously branching tree (a cladogram) with the extant groups at the ends of the branches.

Monophyletic groups are distinguished by advanced features shared by all members (unless secondarily lost); these features are called the *autapomorphies* of the group. They evolved, one after another, in the *stem lineage* of the group (the exact lineage of ancestors and descendants not shared by the extant members with any other extant monophyletic group). An advanced feature which joins two sister groups is called a *synapomorphy* of the two; it is an autapomorphy of the higher group formed by the two sister groups.

When extinct forms are taken into account, there are two obvious ways of delimiting a monophyletic group. The narrower delimitation, called the *crown group*, comprises the latest common ancestor of the extant members of the group and all descendants of that stem species, whether extant or extinct. The wider delimitation, called the *total group*, comprises all forms that are more closely related to the living members of the group than to anything else still extant. Subtraction of the crown group from the total group gives an assemblage of extinct forms called the *stem group*. The stem lineage passes through the stem group but most members of the stem group will belong to side branches from that lineage.

A stem group is subdivided into *plesions* (Patterson & Rosen 1977, Craske & Jefferies 1989), each comprising all those members of a stem group which, so far as can be discerned, are equally closely related to the relevant crown group. One plesion is said to be less or more *crownward* than another.

Traditional categorial ranks (e.g. family, order) mostly disappear in cladistics, because it is logically impossible to devise rules for assigning them. The only exceptions are the species, which can be objectively defined as a group of

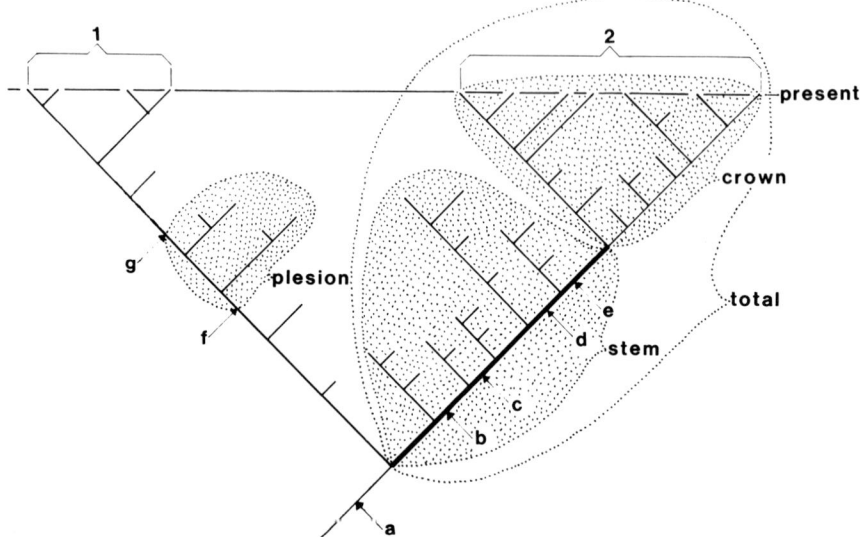

FIG. 1.   Cladistic terminology. Groups 1 and 2 are extant sister groups. The thick line is the stem lineage of group 2. Innovations *b*, *c*, *d* and *e* are autapomorphies of group 2. Innovation *a* is an autapomorphy of the group (1 + 2) and a synapomorphy of groups 1 and 2. Successive innovations *f* and *g*, acquired in the stem lineage of group 1, define a plesion in the stem group of 1.

potentially interbreeding individuals, and the genus, which has to be retained for nomenclatorial reasons.

The above is a brief summary of a complicated and controversial subject.

The origin of bilateral symmetry in the Bilateria

The Metazoa, comprising the sponges, the strange marine animal *Trichoplax adhaerens*, the coelenterates and the Bilateria, are likely to be a monophyletic group because they share the following autapomorphies: multicellularity; diploidy with gametic meiosis; oogenesis with the formation of tiny polar bodies; sperm with a conical head, a middle piece containing a few mitochondria and two centrioles, of which the distal one forms the basal body of the cilium; and collagen. These autapomorphies would have been acquired in segment 1 of Fig. 2.

The Metazoa can be divided into the sponges and all other metazoans. Likely autapomorphies of the sponges (Porifera) include: sessility; the dermal pores and osculum, allowing the flow of a feeding current through the sponge; and perhaps a planktic larva. These autapomorphies would have arisen in segment 2 of Fig. 2.

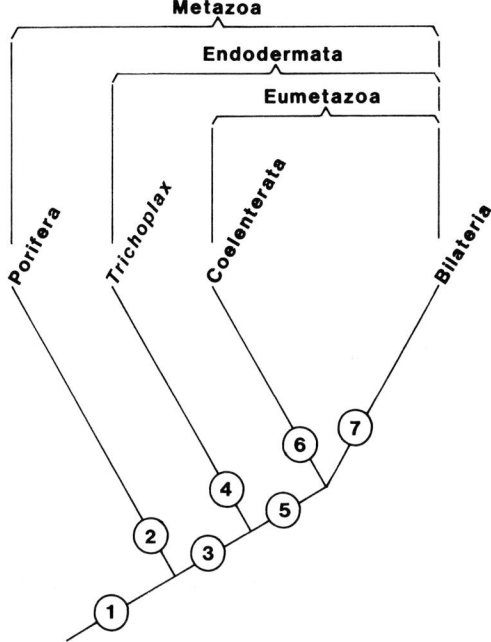

FIG. 2.   Basic subdivisions within the Metazoa. For explanation of innovations 1–7, see text.

The probably monophyletic group of all metazoans except sponges has never been named. I propose to call it the Endodermata, reflecting the presence of a true enzyme-secreting endoderm. It comprises *Trichoplax adhaerens* (the sole representative of the 'phylum' Placozoa) and the Eumetazoa.

Trichoplax is a small, flexible, approximately disc-shaped marine animal found resting on, and moving over, hard substrates. It was first described, and its significance as a primitive metazoan was first recognized, by Schulze (1883) but his work was ignored until Grell's recent studies (1981) of the animal. It consists of an endoderm, an ectoderm and a layer of fibrous contractile cells in between, which are probably homologous with muscle cells. The endodermal cells are connected by ultrastructural features called belt desmosomes that do not occur in sponges. *Trichoplax* has no nerve cells. Most of the ectodermal and endodermal cells are monociliate and the cilia are diplosomal (i.e. they have an extra centrosome at the base) like those of sponges, coelenterates and the bilaterian group Gnathostomulida (Ax 1987, p 258). *Trichoplax* digests food particles by doming up from the substrate to form a temporary gut into which digestive enzymes are secreted by glandular endodermal cells: the opening of the gut (the temporary 'mouth') faces downwards. *Trichoplax* is presumed to reproduce sexually, for ova and cleaved eggs have been seen, though not sperm. It can also reproduce asexually by budding and by binary fission.

The autapomorphies of the group Endodermata, which would have evolved in segment 3 of Fig. 2, include belt desmosomes, muscles, gland cells in the endodermis secreting digestive enzymes, and the power to invaginate to form a gut (though only temporarily in *Trichoplax*).

The coelenterates and the bilaterally symmetrical metazoans (Bilateria) constitute a monophyletic group called the Eumetazoa (Fig. 2). The striking autapomorphies for this group are: permanent gut and blastopore (= mouth); sensory cells; and nerve cells. Autapomorphies of the Coelenterata include tetraradial symmetry and the presence of nematoblasts (stinging cells). There are two main body forms in the Coelenterata—the planktic, mouth-down medusoid and the fixed, mouth-up hydroid. Nobody knows which was the primitive coelenterate condition.

Autapomorphies of the Bilateria are: bilateral symmetry associated with unidirectional locomotion and a horizontal anteroposterior axis; brain and longitudinal nerve tracts; mesoderm, probably arising in ontogeny from the 4d cell of spiral cleavage; and protonephridia, each built of three standard cells. Presumably, the most primitive Bilateria had no anus, thus resembling the coelenterates. The Plathelminthomorpha (flatworms) are a large monophyletic group of Bilateria without an anus, and may be primitive for Bilateria in this respect.

Another possible autapomorphy of the Bilateria involves a change in orientation, though this seems never to have been suggested before (Fig. 3). The bilaterian longitudinal axis, which commonly has the mouth near one end, is primitively horizontal. But the axis of radial symmetry in Coelenterata, which likewise has the mouth at one end, is usually vertical, whether mouth-up or mouth-down. The fact that *Trichoplax* is mouth-down suggests that such a vertical axis is primitive for Endodermata. I suggest, therefore, that the anteroposterior axis of Bilateria is homologous with the vertical axis of Coelenterata but has rotated through 90°. Thus the primitive bilaterian could have resembled a hydroid or medusoid which had taken to lying on one side. This proposed rotation I call pleurothetism (Gr. *pleura*, a side; *thetos*, adapted for resting on).

The main argument against the pleurothetic rotation is the fact that many plathelminths, such as the turbellarian *Planaria*, have the mouth near the middle of the ventral surface. The latter might thus correspond to the unrotated lower surface of a medusoid. However, according to Ax (1987, p 250f) the Plathelminthomorpha can be divided into two sister groups—the Gnathostomulida and the Plathelminthes. The Gnathostomulida have the mouth near the anterior end of the body, as do the Catenulida which in many respects are the most primitive Plathelminthes. This suggests that the mouth was primitively anterior in the Plathelminthomorpha, and therefore perhaps in the

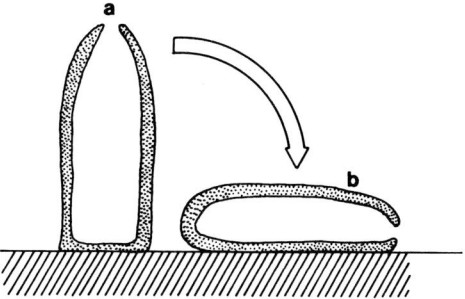

FIG. 3. It is suggested that a non-bilaterian ancestor (a), living in a mouth-up position, may have given rise to the Bilateria (b) by falling over on one side (the process of pleurothetism). Equally, pleurothetism may have involved descent from a mouth-downward ancestor.

Bilateria as a whole. This is consistent with a pleurothetic rotation in the stem lineage of the Bilateria.

The hypothesis of pleurothetic rotation in the origin of the Bilateria is testable. The anteroposterior axis is presumed to be homologous in *Drosophila* and the mouse, because the genes specifying positional information along this axis in both organisms have the same sequence along the chromosome, which corresponds to the spatial sequence of their domains of activity along the anteroposterior axis (Holland 1990). This axis is presumably homologous throughout the Bilateria. If the positional information along the vertical axis of symmetry in a *Hydra*, for example, proves to be specified by homologous genes in the same sequence, then the pleurothetic rotation probably took place.

The vertical plane of bilateral symmetry primitive for the Bilateria presumably has to do with unidirectional motion. Pushing an asymmetrical object is directionally unstable, requiring continual corrections to move it in one particular direction. Pulling such an object is directionally stable, because the object will yaw until forces resisting the movement exert an equal moment on both sides of the pulling point. However, the total force resisting the pull can be decreased by diminishing the projected frontal area and slimming the body on either side of the longitudinal axis. This will tend to produce a plane of symmetry including that axis. Symmetry of external shape and of the locomotor organs will be more important than internal symmetry.

The origin of bilateral symmetry in the Chordata

Within the Bilateria is the probably monophyletic group of the Radialia (Fig. 4). It comprises the so-called tentaculate phyla of the Bryozoa, Phoronida and Brachiopoda; perhaps the Chaetognatha or arrow worms; and the Deuterostomia, including the Chordata.

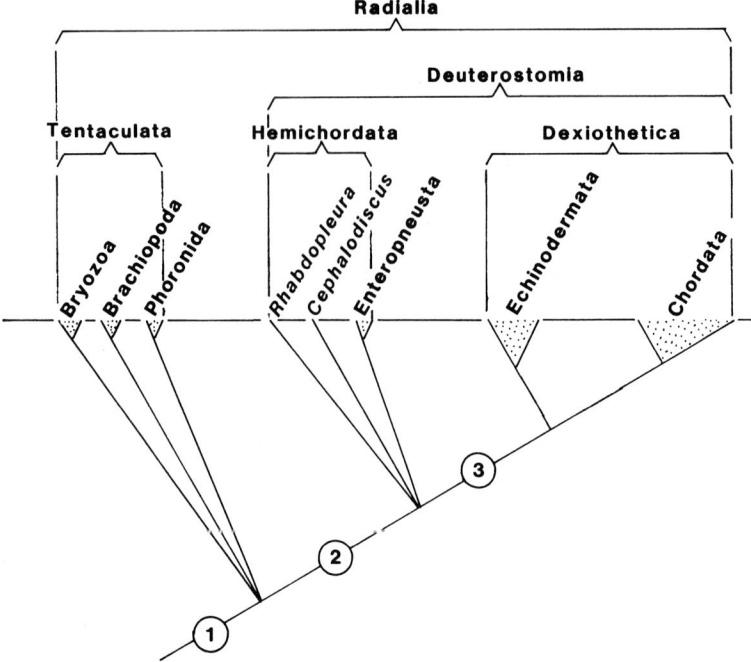

FIG. 4. Phylogenetic relationships among the Radialia. (1) Trimery; (2) deuterostomy; (3) dexiothetism.

Probable autapomorphies of the Radialia are radial cleavage of the eggs and trimery, i.e. the body consists of three regions, the protosome, mesosome and metasome; the mesosome is modified to form ciliated tentacles which produce a feeding current; and the protosome contains an unpaired protocoel, the mesosome a pair of mesocoels, and the metasome a pair of metacoels.

In the Deuterostomia the blastopore becomes the anus in ontogeny, while the mouth arises as a secondary perforation. The deuterostomes consist of the Hemichordata, the Echinodermata and the Chordata, and are sometimes extended to include the Chaetognatha.

Dexiothetism

The Echinodermata and Chordata constitute the monophyletic group Dexio-thetica (Jefferies 1979) with the following autapomorphies: a calcareous skeleton, mesodermal in origin; a rotation of the animal's body through 90° to fall onto the original right side (new ventral side) (dexiothetism), and various changes resulting from this rotation; and the enclosure of the protosome inside the body. The evidence that these features are autapomorphies of a group Dexiothetica comes largely from fossils.

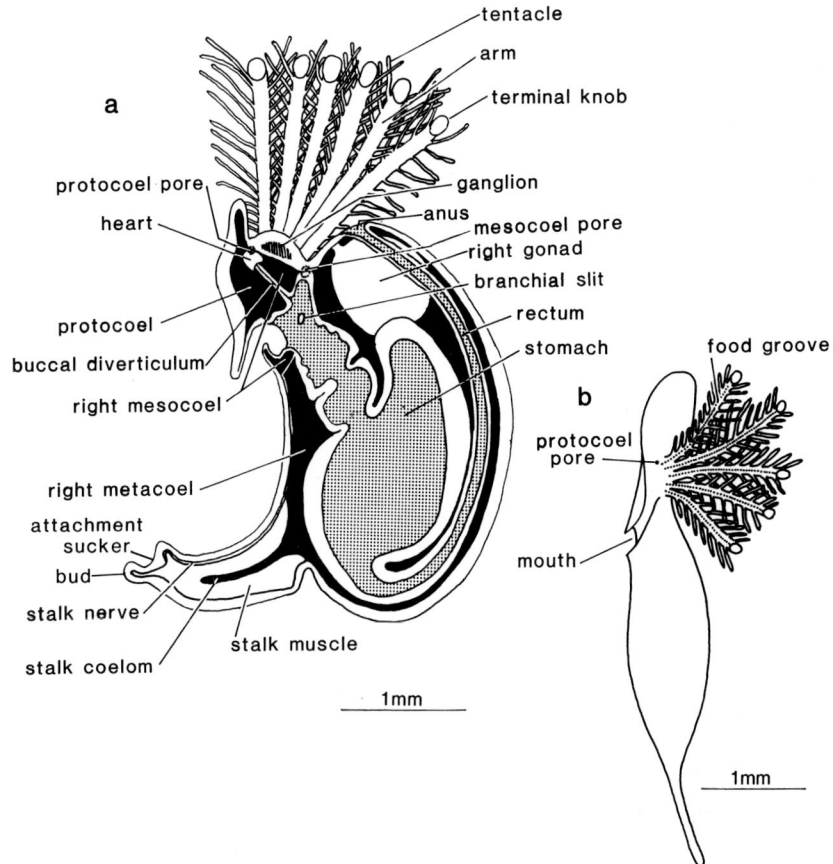

FIG. 5.   The anatomy of *Cephalodiscus*. (a) The right half of a zooid, in left aspect. (b) A complete zooid seen from the left.

The present-day hemichordate *Cephalodiscus* probably resembles the immediate ancestors of the Dexiothetica (Fig. 5). It lives on the sea floor. Each individual (zooid) secretes a horny tube in which it lives and the tubes are joined in bundles so that the animals are semi-colonial. Each zooid is trimerous, with a protosome, a tentaculate mesosome and a metasome. The metasome is divided into the swollen trunk anteriorly and the thin, extensile, locomotory stalk posteriorly. There is a sucking disc on the end of the stalk and new zooids are budded off around the edge of this disc. There is an unpaired protocoel in the protosome, a pair of mesocoels in the mesosome, and a pair of metacoels in the metasome. There are various body openings: a pair of protocoel pores; a pair of mesocoel pores; the mouth just ventral to the mesosome; and in the trunk a pair of gill slits, a pair of gonopores and an anus. Inside the protosome

Cephalodiscus-like ancestor

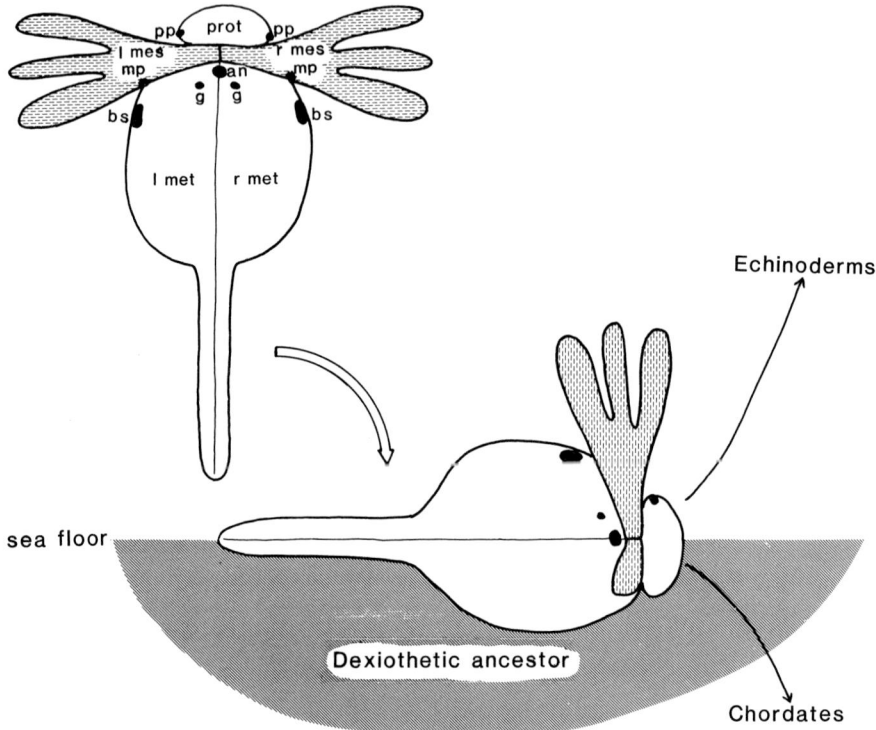

FIG. 6.   Dexiothetism in the stem lineage of the Dexiothetica (chordates + echinoderms). an, anus; bs, branchial slit; g, gonopore; 1 mes and r mes, left and right mesocoel; 1 met, r met, left and right metacoel; pp, protocoel pore (= hydropore); prot, protocoel.

is the heart, an excretory organ called the glomerulus and a projection of the gut called the stomochord. A large ganglion exists in the dorsal midline of the mesosome. The trunk contains the gut and a pair of gonads. The gut consists of the pharynx, with a pair of gill slits opening through its wall, a large stomach and a long narrow intestine running up to the anus. The stalk is highly muscular and abundantly innervated. The nervous system of *Cephalodiscus* seems to be confined entirely to the base of the epidermis.

   As to habits, the cups of *Cephalodiscus* are extended into horny 'flagpoles', which the zooids climb when feeding. In climbing, the zooid clings to the flagpole alternately by the stalk, which it uses as a prehensile extensible tail with a sucking disc on the end, and by the ventral face of the protosome. When it feeds, the ciliated tentacles are held up curved with the convex side outwards and their ends touching, to form a sphere open at the top. Cilia on the fine side-branches of

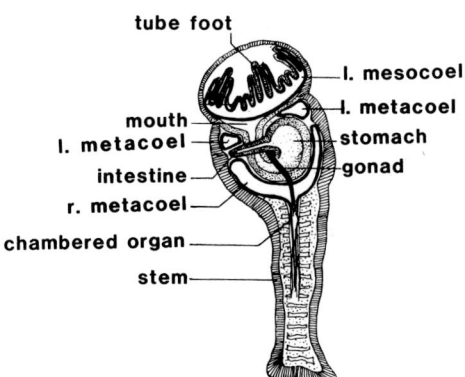

FIG. 7.   An attached larva of the crinoid *Antedon* to show the effects of dexiothetism in the embryology of echinoderms. Compare with Fig. 6. The gonad is the primary gonad whose position corresponds to the axial organ and the axocoel (= protocoel). For ease of comparison, the diagram uses hemichordate anatomical nomenclature: thus in echinoderms the metacoels are usually called somatocoels, etc.

the tentacles drive a flow of water into the sphere which emerges through the top opening (Lester 1985). Mucus, flowing down grooves on the outside of the sphere in the main axis of the tentacles, traps food particles from this water current and conveys them to the mouth.

Dexiothetism (Fig. 6) implies that an animal resembling *Cephalodiscus* took to crawling right-side-downwards on the mud of the sea floor, pulling itself along by the stalk. There were several results: the right metacoel came to be under the left one; the openings of the original right side became injurious and were lost; and the right tentacles were thrust into the mud, became useless and were lost. At about the same time as the dexiothetic rotation, the left mesocoel pore was lost and replaced by a communication between the left mesocoel and the protocoel (the so-called stone canal). The protosome, from being an externally visible part of the body, became encircled by the metasome. A calcite skeleton of echinoderm type was acquired, though the plates would at first have been uniform, small and spicular. The resulting animal would have been the latest common ancestor of the echinoderms and chordates.

This hypothetical animal can be compared with an early larva of the crinoid *Antedon* among the echinoderms (Fig. 7). In both cases the right metacoel is beneath the left metacoel, with the gut largely carried in the horizontal mesentery between them. The tentacles and water vascular system of *Antedon* and the equivalent left mesocoel of the hypothetical animal face upward. Neither has an equivalent of the right mesocoel. The left protocoel pore opens upward and there is no right protocoel pore. Both would have had a calcareous skeleton. The protocoel of *Antedon*, which is associated with the primary gonad of Fig. 7,

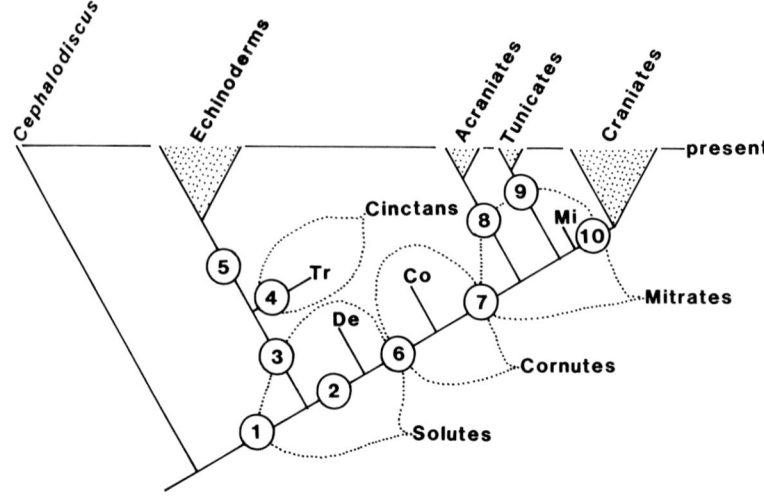

FIG. 8.   Phylogenetic position of primitive fossil chordates and echinoderms (solutes, cornutes, mitrates and cinctans). De, *Dendrocystoides*; Co, *Cothurnocystis*; Mi, *Mitrocystella*; Tr, *Trochocystites*. Evolutionary innovations are: (1) dexiothetism, enclosure of protocoel in body, calcite skeleton; (2) notochord; (3) loss of tail; (4) anchoring stele; (5) mouth in centre of upper surface, pentameral symmetry, loss of remaining gill slit; (6) loss of arm, feeding by endostylar mucous filter in pharynx; (7) right gill slits and pharynx; (8, 9 and 10) loss, three times, of calcite skeleton.

is enclosed in the body, rather than being the coelom of a separate body region. The gonad of echinoderms originates in ontogeny from the wall of the left metacoel (left somatocoel) (Hyman 1955, Gemmill 1914, MacBride 1914) and is therefore homologous with the left gonad of *Cephalodiscus*.

The most important differences are that the larva of *Antedon* is attached to a hard substrate by a stem; shows pentameral symmetry induced by the five-rayed water vascular system; has the mouth near the centre of the upper surface; has no gill slit; and has no stalk-tail, for it is not likely that the stem of larval *Antedon* is homologous with the stalk of *Cephalodiscus*.

The resemblances between the hypothetical dexiothetic ancestor and larval *Antedon* would have been acquired in the stem lineage of the Dexiothetica. Most of the differences probably evolved in the echinoderm stem lineage. The view that echinoderms are descended from a dexiothetic ancestor is thus based on embryology. The evidence that the same was true of chordates comes from fossils.

Dexiothetism and the fossil record

The fossils in question used to be called 'carpoid echinoderms'; the interrelationships of the forms discussed here are summarized in Fig. 8. They have an echinoderm-like calcite skeleton but show no radial symmetry. All carpoids which are chordates can be called 'calcichordates'. The first person to argue in detail that the 'carpoids' were ancestral to the chordates was the Swedish zoologist Torsten Gislén (1930).

Four groups of 'carpoids' will be discussed: the Soluta, the Cincta, the Cornuta and the Mitrata. They are all marine and range in known age from Lower Cambrian to Upper Carboniferous, i.e. from about 550 to 320 million years ago. I shall use the terms right, left, dorsal, ventral, anterior and posterior by homology with fishes.

A solute

An example of the Soluta is *Dendrocystoides scoticus* (Bather) (Fig. 9) which I have recently reconstructed in detail (Jefferies 1990). It is about 440 million years old (Upper Ordovician). It consisted of a head, an arm and a tail and was highly asymmetrical except the posterior part of the tail. The head was covered with numerous, mostly irregular plates and was somewhat flatter ventrally than dorsally. It is fairly certain that the head lay, ventral surface down, on the mud of the sea floor and the animal habitually pulled itself rearward by the tail. There is strong evidence that cornutes and mitrates likewise moved rearward.

The tail was divided into fore, mid and hind tail. The hind tail was long, slender and rigid with a skeleton made from a dorsal and a ventral series of hemicylindrical plates. Seen from the side, most of its length was slightly convex dorsally, but near the end it turned downwards as a rigid hook. The hind tail was exactly bilaterally symmetrical. A canal along its axis probably carried a notochord with a central longitudinal blood vessel.

The fore tail was short and flexible and the skeleton had a large lumen. Details of the skeleton show that it mainly flexed from side to side, wagging the mid and hind tail as a single unit. Such flexure would require an anti-compressional structure in the fore tail for it to bend without telescoping, i.e. a notochord. This would have been a forward continuation of the organ in the axial canal of the hind tail. There must have been powerful muscles in the fore tail, presumably on either side of the notochord. The mid tail had a highly asymmetrical skeleton which formed a hollow cone to accommodate the posterior end of these muscles. I suggest that *Dendrocystoides* crawled rearwards by waving the tail from side to side, thrusting the terminal hook into the mud during the power strokes as a punt pole and lifting it clear of the mud during the return strokes. The skeleton of the foretail was not symmetrical because the plates of right and left sides tend to alternate.

The arm skeleton could open ventrally along its length, to expose the tube feet of the water vascular system. These probably picked food particles out of the sediment. There is skeletal evidence that they were arranged in groups

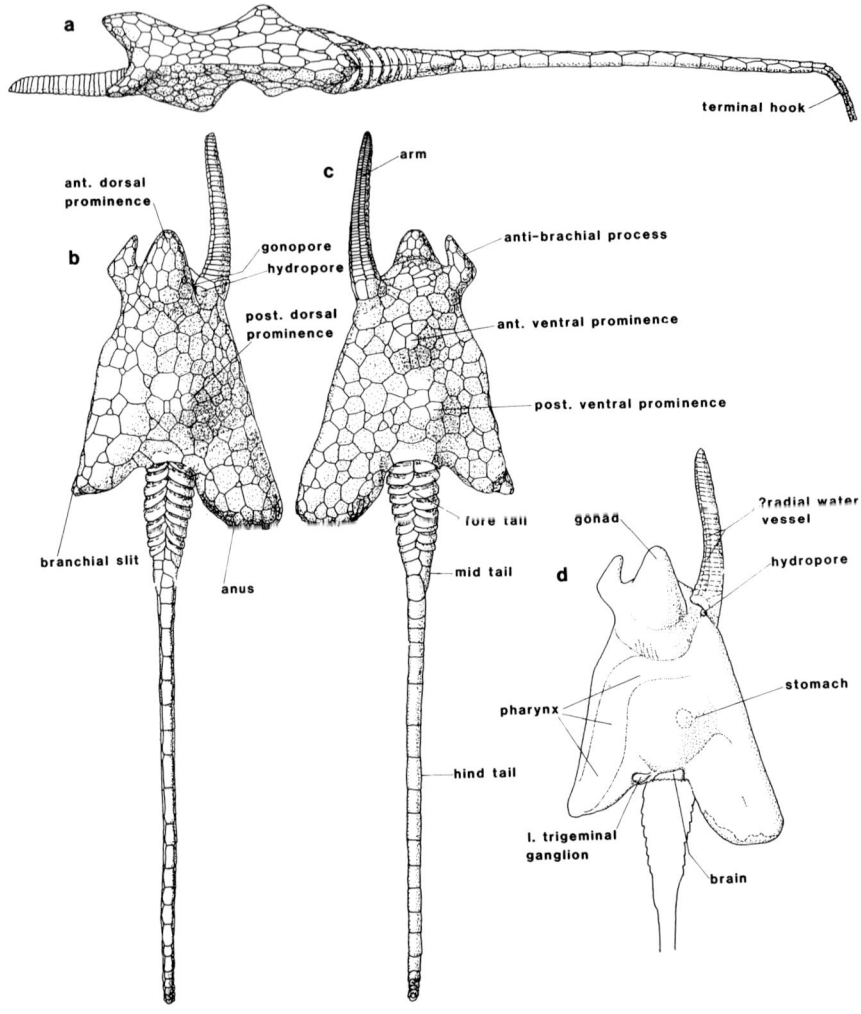

FIG. 9.   The solute *Dendrocystoides scoticus* (Bather). (a) Left lateral aspect; (b) dorsal; (c) ventral; (d) dorsal aspect of soft parts of the head.

of three like the tube feet of crinoids. The arm could wave from side to side so that the tube feet could reach downwards and graze food from the mud.

The mouth of *Dendrocystoides* was at the proximal end of the arm, at anterior right of the head. The hydropore, which would have been the left opening of the protocoel (axial sinus or axocoel) and indirectly of the left mesocoel (= water vascular system = left hydrocoel), was just behind and right of the mouth. The gonopore was left of the mouth and farther from it than the hydropore.

The anus was at posterior right in the head, while at posterior left was a pinhole-sized branchial slit. The latter would probably have released surplus water from the pharynx.

Inside the head (Fig. 9d) the anatomy of various organs is suggested by ridges and grooves on the internal surface of the skeleton and the positions of the head openings. The gonad was probably in the anterior dorsal prominence, near the gonopore. The pharynx approached the gill opening from anterior right. I cannot determine whether there was a buccal cavity. The posterior dorsal prominence probably carried the stomach. The brain was situated where the head joined the tail and was accompanied by a ganglion, probably the left trigeminal ganglion, just left of it. Perhaps a right ganglion existed but without leaving any trace.

The position of the brain just in front of the tail and at the front end of the notochord recurs in cornutes and mitrates and is therefore primitive for the chordates. This position, far from the mouth and corresponding to the proximal end of the stalk in *Cephalodiscus*, shows that the chordate brain is not homologous with the mesocoel ganglion of *Cephalodiscus* or the enteropneusts (acorn worms; Hemichordata). Nor can it be homologous with the brain, in its primitive bilaterian anterior position, of plathelminths, nematodes, molluscs, annelids or arthropods.

Thus, there was perfect bilateral symmetry in the hind tail of *Dendrocystoides*, imperfect bilateral symmetry in the fore tail and almost none in the head, except that the outline approximates an isosceles triangle.

Dendrocystoides can be compared with a *Cephalodiscus* that has undergone dexiothetism. Fig. 10a, b shows *Cephalodiscus* in left view and *Dendrocystoides* in dorsal view. In a clockwise journey around the two sketches we pass in *Cephalodiscus*: mouth, hydropore, left mesocoel pore, left gonopore, anus, stalk (tail), gill slit. In *Dendrocystoides* we pass: mouth, hydropore, anus, tail, gill slit, gonopore. The sequence is the same, except that *Dendrocystoides* has no left mesocoel pore and the gonopore is out of place. The arm and tentacles have similar anterior positions in both animals. The comparison is greatly strengthened by echinoderm embryology which, as mentioned above, implies that the gonad of *Dendrocystoides* is homologous to the left gonad of *Cephalodiscus*; the arm of *Dendrocystoides* is homologous to one of the left tentacles of *Cephalodiscus*; and the hydropore of *Dendrocystoides* is homologous with the left protocoel pore of *Cephalodiscus*. The surviving left arm of the *Cephalodiscus*-like ancestor would have rotated through 180° in *Dendrocystoides*, so that the food-collecting mucous stream would face down rather than up.

The implied homology of stalk and tail makes functional sense because both are locomotory and muscular and are extended parallel to the substrate when in use. Moreover, in the Middle Cambrian solute *Castericystis vali*, juveniles are often seen attached to the adult by the ends of their tails (Ubaghs & Robison

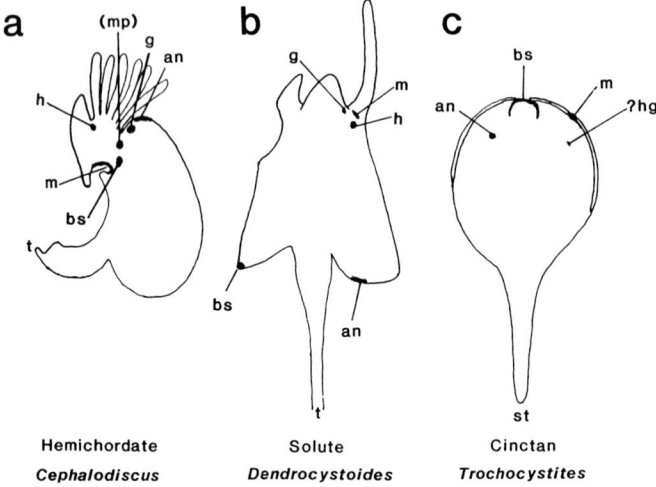

FIG. 10.   Evidence for dexiothetism. The hemichordate *Cephalodiscus* (a) in left aspect, compared with the solute *Dendrocystoides* (b) and the cinctan *Trochocystites* (c), both in dorsal aspect, to show the identical topology of all three. an, anus; bs, branchial slit; g, gonopore; h, hydropore (= left protocoel pore); ?hg, probable hydropore–gonopore in *Trochocystites*; m, mouth; mp, mesocoel pore; st, cinctan stele; t, stalk of *Cephalodiscus* = tail of *Dendrocystoides*.

1985). This suggests that juvenile *Castericystis* had an attachment sucker on the end of the tail, comparable with the sucker at the end of the stalk in the adult *Cephalodiscus* (Fig. 5). The most important difference between the stalk and the tail lies in the presumption that *Dendrocystoides* had a notochord. Other differences include the brain at the proximal end of the tail, probably a proximal concentration of muscles in the tail, and a complicated tail skeleton.

Several likely synapomorphies of *Dendrocystoides* with more crownward chordates, including the presumed notochord, the brain, the left trigeminal ganglion, the posterior position of the gill slit on the left and of the anus on the right, suggest that it was a chordate near the base of the chordate stem group.

A cinctan

Trochocystites bohemicus Barrande (Fig. 11) was a cinctan that lived about 530 million years ago (Middle Cambrian). It was shaped like a tennis racquet, the 'handle' being called the stele and the 'head' the theca (Jefferies 1990). There was a frame of large marginal plates around the theca. A dorsal and a ventral integument, each of many plates, were attached to the edges of the thecal frame. The dorsal integument may have been slightly flexible peripherally, but its central parts and the whole of the ventral integument were stiff. The skeleton of the

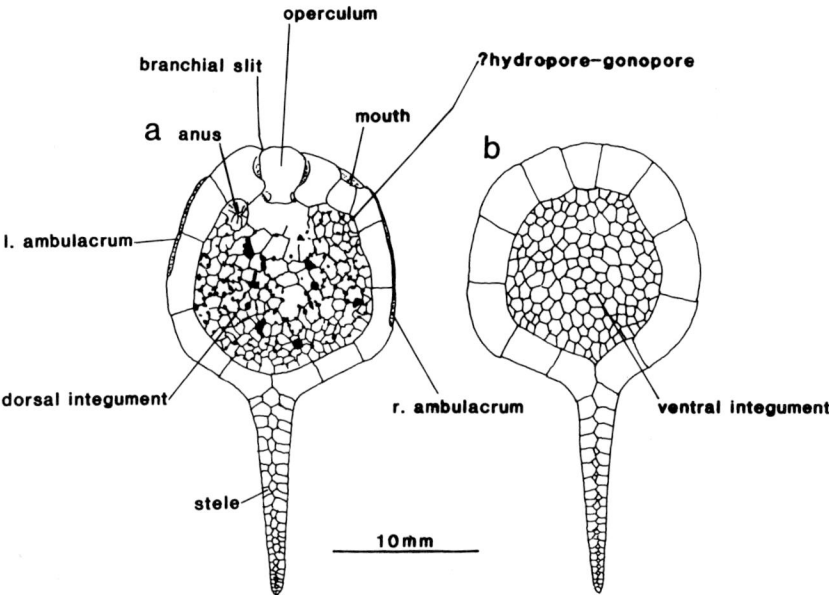

FIG. 11.   The cinctan *Trochocystites bohemicus* Barrande. (a) dorsal aspect; (b) ventral.

stele consisted mainly of two series of plates which passed forwards at right
and left into the marginal plates of the theca. It is fairly certain that the theca
habitually lay, ventral surface downwards, on the mud of the sea floor. Tubercles
on the ventral surface of the frame would lift the theca above the sea floor.
The stele, which would have been slightly flexible up and down but not sideways,
probably sloped downwards into the sea floor and acted as an anchor, so that
the theca, and the thecal openings, would take a downstream position in any
bottom current. The stele was probably not homologous with the tail of solutes,
since it served as an anchor rather than being locomotory and resembles an
outward protrusion of the frame.

The openings are again crucial. At anterior right was the mouth, identifiable
because two grooves converge on it from right and left. By comparison with
echinoderms these would be ambulacral grooves with the mouth situated where
they met. The grooves have flexible plated upper and lower lips comparable
with the cover plates on the ambulacral grooves of many primitive fossil five-
rayed echinoderms. The anus was at anterior left, just inside the frame. There
was a gill slit at the very anterior end, and there was probably a hydropore-
gonopore, just behind and right of the mouth on the inside of the frame
(W.-P. Friedrich, personal communication).

The cinctans were probably stem-group echinoderms, because the ambulacra
were recumbent on the surface of the body, as was primitively true of five-rayed

echinoderms (Paul & Smith 1985); the ambulacra were biradial, not uniradial as in solutes, and there are signs of vestigial biradiality (and of vestigial triradiality) in the imperfectly quinqueradial ambulacral system of primitive five-rayed echinoderms (Jefferies 1986, p 319); the cinctans were probably passive filter feeders, extracting food particles from water on the lee-side of the ambulacra, like modern crinoids (Macurda & Meyer 1974), rather than being sediment grazers like solutes or active filter feeders like *Cephalodiscus*; and there was no tail (if the stele was not a tail).

Trochocystites conforms to the dexiothetic model, though not so persuasively as *Dendrocystoides* does (Fig. 10a, c). It takes a very basal position in the echinoderm stem group, because it retains a gill slit; the mouth is in the primitive, now horizontal, sagittal plane rather than in the centre of the upper surface; and the ambulacral system has two rays rather than five.

Trochocystites is strongly asymmetrical in the position of the mouth, anus and hydropore–gonopore; this asymmetry was probably inherited from the dexiothetic ancestor. There is, however, almost perfect bilateral symmetry of outline and body shape about a new vertical longitudinal mirror plane which bisected the branchial slit, theca and stele. This new plane of symmetry was perpendicular to the hemichordate sagittal plane, which had become horizontal. It adapted the theca to taking up a hydrodynamically stable position, down-stream of the anchoring stele, in a bottom current. So far as can be deter-mined, the new plane coincided with the sagittal plane of chordates, but it must have evolved independently.

A cornute

Cothurnocystis elizae Bather (Fig. 12) is an example of the Cornuta and comes from the same bed as *Dendrocystoides scoticus*, being thus about 440 million years old (Upper Ordovician). It has been studied in great detail (Bather 1913, Jefferies 1968, 1986, Jefferies & Lewis 1978, Ubaghs 1967, Woods & Jefferies 1991).

It consisted of a head and a tail; unlike solutes it had no arm. Seen from above, the head was like a mediaeval long-toed boot in shape (Gr. *cothurnos* = boot). It lay, ventral surface down, on the sea floor and there are ventral spikes which would raise the head above the sea floor and make it move easily in a rearward direction, pulled by the tail. The head was edged by a frame of large marginal plates, to which a dorsal and a ventral integument were attached. The integuments were plated and flexible and the ventral one was crossed by an anteroposterior strut.

A line of about 16 branchial slits penetrated the left part of the dorsal integument, just in front of the left posterior frame, in the same general position as the single branchial slit of *Dendrocystoides*. Each slit had an outlet-valve structure (Jefferies 1986, p 193). The larva of amphioxus (*Branchiostoma*)

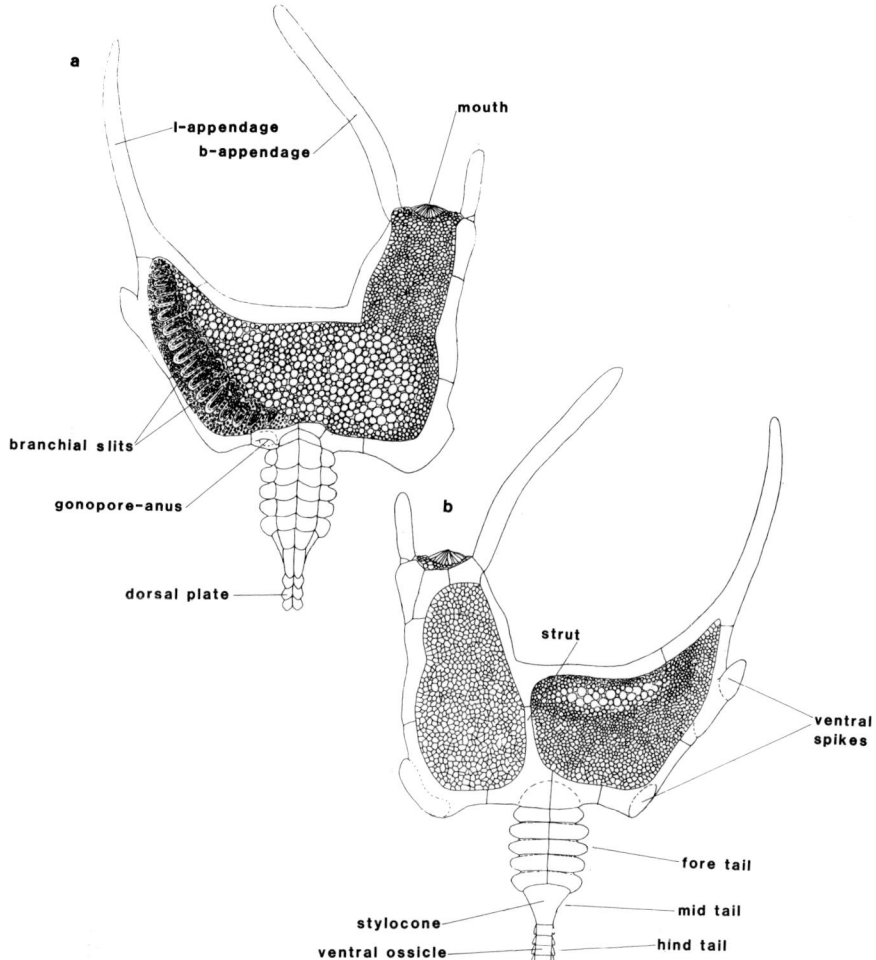

FIG. 12.   The cornute *Cothurnocystis elizae* Bather. (a) dorsal aspect; (b) ventral. The hind tail was much longer than shown here.

likewise has left gill slits only, and in this respect recapitulates the cornute and solute condition.

The gonopore-anus in *Cothurnocystis* lay just left of the tail. Two tubes in the soft parts, the narrow gonoduct and the broad rectum, approached it from the right, inside the head. Some cornutes show evidence of an acusticolateralis receptor near the gonopore-anus (Jefferies et al 1987); this probably existed in *Cothurnocystis* also. The mouth in *Cothurnocystis* lay at anterior right in the head.

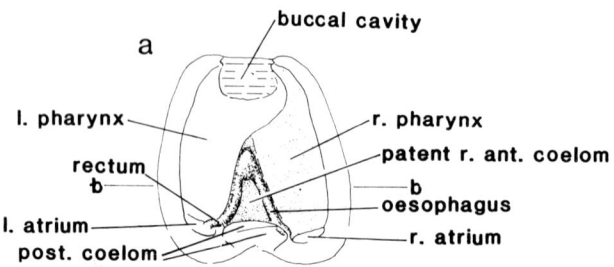

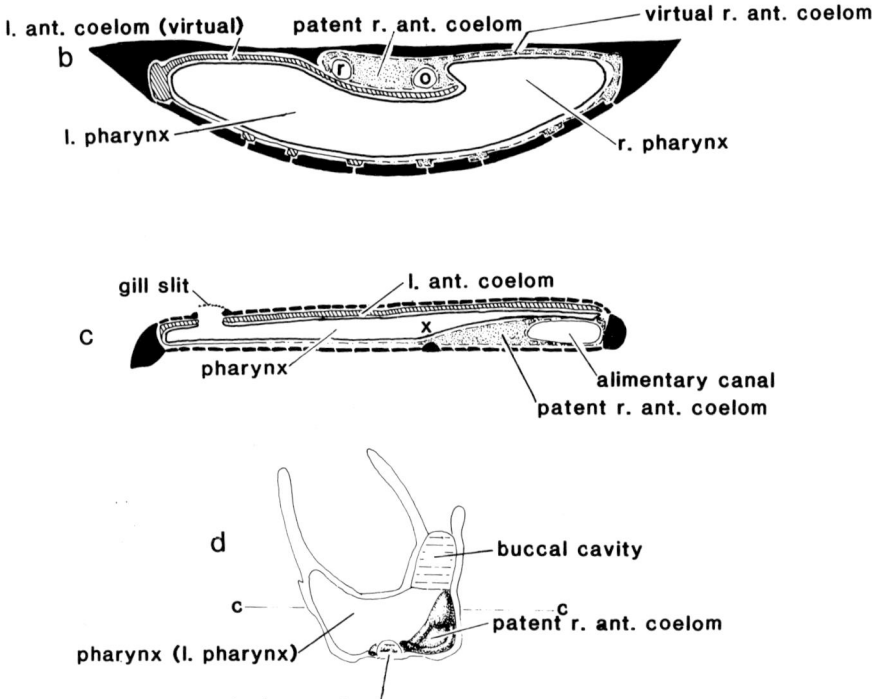

FIG. 13. Head chambers in a mitrate and a cornute, to show the basic similarity.
(a) Dorsal aspect of the mitrate *Mitrocystites*. (b) Transverse section through b-b in (a).
(c) Transverse section through the cornute *Cothurnocystis* at c-c in (d). (d) Dorsal aspect
of *Cothurnocystis*. o, oesophagus; r, rectum; ×, point at which the right pharynx would
sprout out of the left pharynx towards posterior right, to produce the mitrate condition.

The head contained several chambers, recognizable by details of sculpture on
the internal surfaces of the marginal plates and variations in the integument
plates (Fig. 13c, d). The buccal cavity filled the 'ankle' part of the 'boot', just
behind the mouth. The posterior coelom lay just anterior to the tail with the
rectum and gonoduct running across its floor; it was homologous with the left

epicardium of tunicates. The cavity of the right anterior coelom lay on the floor, mainly in the right posterior part of the head (the 'heel' part of the 'boot'). It would have contained the gonad, the non-pharyngeal part of the gut and probably the heart, among other organs. Virtual extensions (i.e. without any cavities) may have passed out from it to the rest of the ventral surface. The right anterior coelom, being beneath the pharynx, would correspond to the right metacoel of *Cephalodiscus*. The pharynx extended from the buccal cavity at anterior right to the gill slits at posterior left. On purely comparative grounds, there was probably a virtual left anterior coelom overlying the pharynx and buccal cavity and corresponding to the left metacoel of *Cephalodiscus*.

The head chambers of *Cothurnocystis* thus resembled those of *Dendrocystoides* in their general lay-out (cf. Fig. 9d). However, *Dendrocystoides* had no posterior coelom; it perhaps had no buccal cavity; the pharynx was much smaller and with only one tiny gill slit; the gonad was anterior; and the heart was probably anterior in the head, near the hydropore. The differences in the pharynx probably relate to a change in the mode of feeding, for *Dendrocystoides* would have picked up food particles using the tube feet of the arm, while *Cothurnocystis*, having no arm, presumably filtered food particles from a feeding current, using a mucous filter inside the pharynx secreted by an endostyle. As mentioned below, there is strong evidence that mitrates fed in this manner, but with right gill slits as well as left ones. Among extant organisms, larval amphioxus similarly has left gill slits only and produces a mucous filter from the endostyle (Olsson 1983). The feeding current of *Cothurnocystis* would have been propelled by cilia lining the pharynx, perhaps helped by muscular contraction of the pharyngeal wall.

The tail of *Cothurnocystis* was divided into fore, mid and hind tail. Unlike *Dendrocystoides*, it was perfectly symmetrical. The fore tail had a large lumen and a skeleton adapted to right–left flexion. It probably contained a notochord and powerful muscles.

The hind tail was less rigid than in *Dendrocystoides*. It had an unpaired series of hemicylindrical ossicles ventrally and paired imbricate plates dorsally. The notochord, extending rearward from the fore tail, occupied a median groove in the dorsal surface of the ventral ossicles. The hind tail could flex up and down, and in particular the end could be bent downward, as a hook, to penetrate and grip the sea floor. The space above the notochord must therefore have contained muscles.

The mid tail had a large conical, anteriorly excavate ossicle which, like the mid tail skeleton of *Dendrocystoides*, would receive the distal muscles of the fore tail, so that the fore tail could wag the mid and hind tail as a unit.

Studies with a working model show how *Cothurnocystis* could crawl rearwards across the sea floor, by waving the tail from side to side and pushing the terminal hook into the mud during the power strokes (Woods & Jefferies 1991). There would have been much yaw, with the head sliding alternately along two long anterior curved plates of the head (l-appendage and b-appendage) like curved sledge runners. This suggested method of crawling is basically like that of *Dendrocystoides*.

The brain of *Cothurnocystis* lay in a cerebral basin in the skeleton just anterior to the tail. There is evidence of two ganglia, probably the right and left trigeminal ganglia, just anterior to the brain.

The anatomy of *Cothurnocystis* is, therefore, basically like that of *Dendrocystoides*. The differences are mostly synapomorphies of *Cothurnocystis* with more crownward cornutes or with crown-group chordates. They show that *Cothurnocystis* was more crownward in the chordate stem group than *Dendrocystoides*.

As to bilateral symmetry, the head of *Cothurnocystis* shows none, except that the brain with its paired trigeminal ganglia is symmetrical. Indeed, the head is even less symmetrical than that of *Dendrocystoides*, probably because of the increased size of the pharynx and perhaps of the buccal cavity. The tail shows perfect bilateral symmetry, however.

A mitrate

The mitrate *Mitrocystella incipiens* (Barrande) (Fig. 14) is about 460 million years old (Middle Ordovician) and belongs to the stem group of the Craniata. All known mitrates are primitive members of the chordate crown group.

Mitrocystella consisted of a head and a tail. The tail was perfectly bilaterally symmetrical and the head almost so, at least externally, except for one plate at anterior left on the dorsal surface. The head was almost flat dorsally and strongly convex ventrally. The dorsal surface was rigid, whereas the ventral surface was flexible, at least anteriorly. The tail had fore, mid and hind parts, but these are probably not homologous with the like-named parts of cornutes and solutes. The head would have lain hull-down in the sea floor, perhaps lightly covered with mud on the dorsal surface.

Transverse ribs on the posteroventral surface of the head consistently have the anterior slope steeper than the posterior one. Similar ribs in recent crabs and bivalves are used to grip sediment during locomotion. The ribs of *Mitrocystella* would have had the same function, allowing the head to slide rearwards through the mud, pulled by the tail. Thus *Mitrocystella* moved much as the cornutes and solutes did, except that the tail acted in ventral flexion, with the muscles of the fore tail pushing the hind tail forwards against the mud and so pulling the head rearwards. There is evidence of a notochord, dorsal nerve cord, spinal ganglia and muscle blocks in the hind tail.

The head had four openings: the mouth anteriorly, a pair of atrial openings and a lateral line. The mouth was asymmetrical; it opened a few degrees leftward and its mid-point was left of the midline of the head. The atrial openings were situated at posterior right and left. At them, a slight gape could occur between

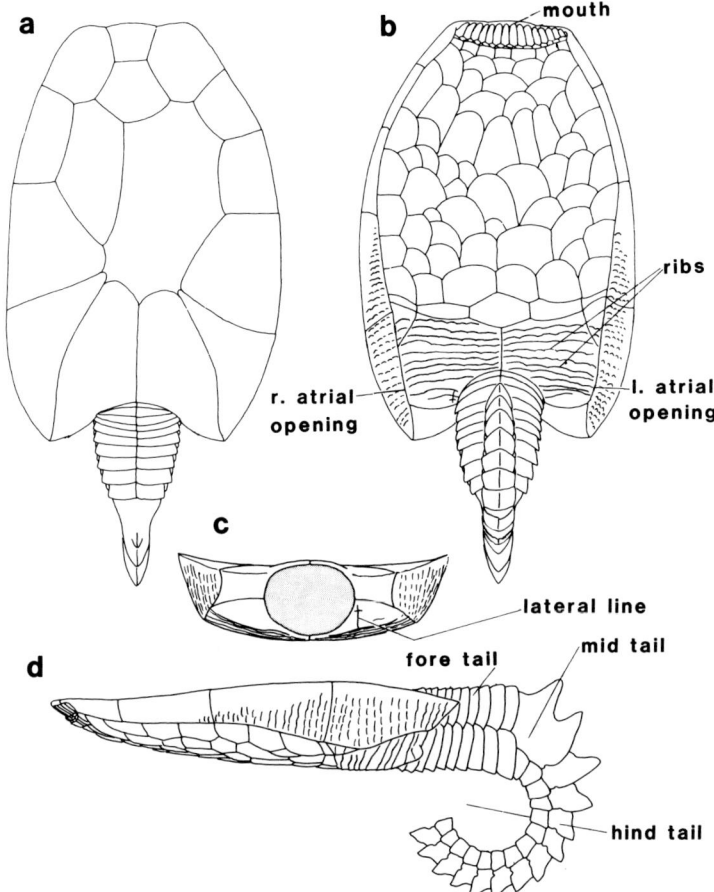

FIG. 14.   The mitrate *Mitrocystella incipiens* (Barrande). (a) dorsal aspect; (b) ventral; (c) posterior; (d) left.

adjacent plates by which water on both sides, and faeces and gametes on the left, were released from atrial chambers inside the head. The lateral line, a branched groove on the right posterior surface of the head, probably carried a series of neuromasts.

The chambers inside the head of *Mitrocystella*, which were similar to those of the mitrate *Mitrocystites* shown in Fig. 13a, b, can be reconstructed from the anatomy of the internal surface of the skeleton. They are fundamentally as in a cornute, but with modifications towards an imperfect bilateral symmetry. Thus the fossils suggest that there was a primary or left pharynx, homologous with that of cornutes, and running from anterior right to posterior left. Out of this, both in ontogeny and phylogeny, a new right pharynx sprouted to

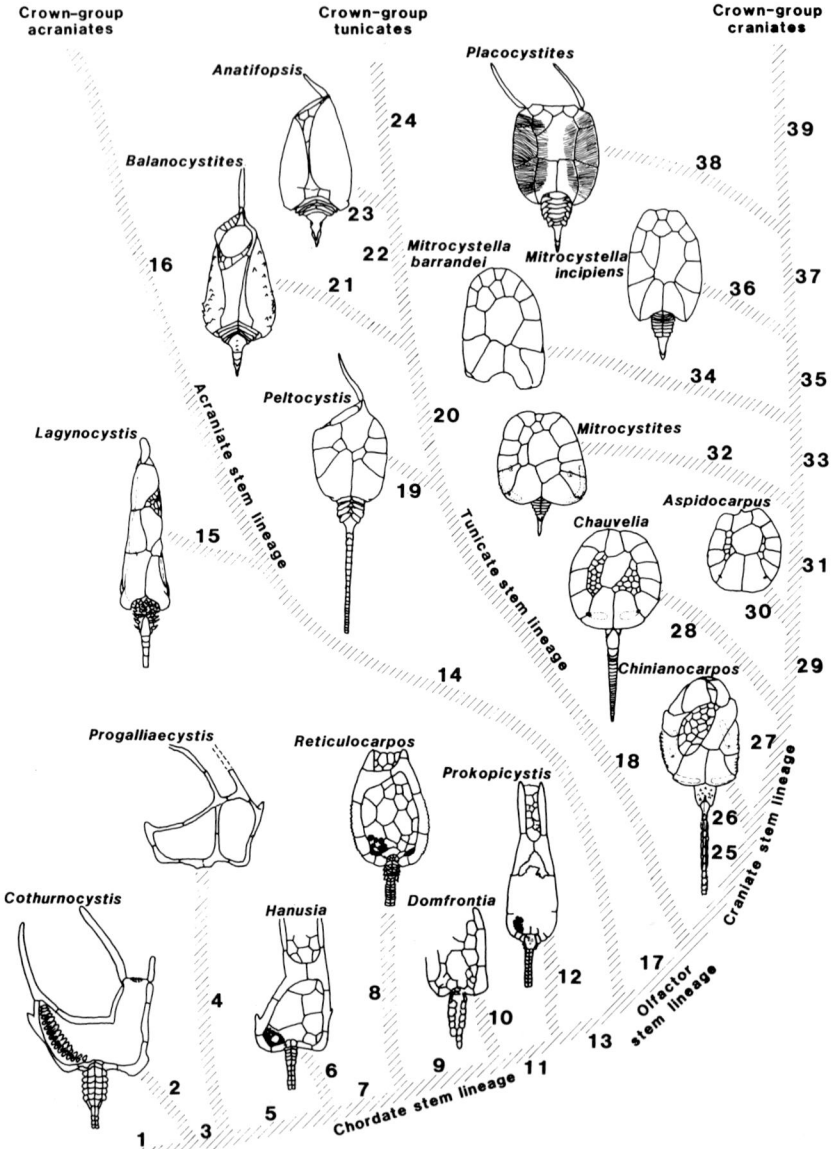

FIG. 15. The phylogeny of mitrates and crownward cornutes. The animals are shown
in dorsal aspect. The tails are sometimes shown fully, sometimes omitted, sometimes
artificially abbreviated and sometimes not completely visible. Outline symmetry of the
head was acquired in segments 3, 5 and 7; right gill slits and the right pharynx were
acquired in segment 13; the autapomorphies of the here-proposed group Olfactores
(= tunicates + craniates) were acquired in segment 17.

the posterior right at the point marked × in Fig. 13c. (In amphioxus the left or primary larval gill slits are followed suddenly, at metamorphosis, by right or secondary gill slits.) Both in ontogeny and phylogeny the right pharynx inflated beneath the cavity and contents of the right anterior coelom, lifted them up and squashed them into a median position, hanging from the ceiling of the head. A virtual left anterior coelom, homologous with the left metacoel of *Cephalodiscus* and with the left mandibular somite of vertebrates, probably covered the dorsal surface of the buccal cavity and left pharynx. Similarly, a virtual extension of the right anterior coelom, equivalent to the right metacoel of *Cephalodiscus* and to the right mandibular somite of vertebrates, probably extended over the dorsal surface of the right pharynx and spread to an unknown extent ventral to the pharynx. Gill slits would have existed where the left and right pharynges met their respective atria. A buccal cavity was present anterior to the left pharynx; there was also a posterior coelom, which, however, was double, equivalent to both the left and the right epicardia of tunicates. There is strong evidence that the rectum opened into the left atrium, as it does in a tunicate tadpole. The gonoduct would also have opened into the left atrium.

Mitrocystella and other mitrates would have fed, like recent tunicates, acraniates (amphioxus) and larval lampreys, using an endostylar mucous filter. As in recent tunicates, there were many asymmetries in this apparatus and several features which recur in the tunicate pharynx can be located on skeletal evidence (Jefferies 1986, p 270–284).

The nervous system in the head of *Mitrocystella* can be reconstructed in detail because of various grooves and canals in the skeleton. There is evidence of a brain just in front of the tail, divided into prosencephalon and a deuterencephalon, and of many different cranial nerves. The nervous system was bilaterally symmetrical, in most respects, except for the acusticolateralis system. Asymmetries of the latter may be connected with fundamental asymmetries in the branchial system because the atria are homologous with the auditory capsules of vertebrates.

In the craniate stem group seven plesions can now be recognized among mitrates (Fig. 15), with *Mitrocystella incipiens* in the sixth. In the crownward direction, however, the record stops with entire loss of the calcite skeleton. Many important craniate autapomorphies would have arisen in the craniate stem lineage after the skeleton disappeared (segment 39 of Fig. 15). The fossils demonstrate clearly that the tunicates, and not the acraniates such as amphioxus, are the sister group of the craniates. All mitrates show signs that the left pharynx preceded the right pharynx in ontogeny, and recent acraniates retain this ontogenetic asymmetry, whereas tunicates and craniates have lost it independently. Mitrates resemble giant, adult, calcite-plated tunicate tadpoles and represent the tunicate tadpole-like ancestors which many authors have postulated in the ancestry of the vertebrates (e.g. Romer 1972).

Situs inversus is known in one species of mitrate, the stem-group tunicate *Peltocystis cornuta* Thoral (Lower Ordovician) (Fig. 15; Ubaghs 1969, p 82). Perhaps this species had an echinoderm-like larva which, like some flatfishes, could lie down either on the right or on the left. So far as known, no cornutes, solutes or cinctans show situs inversus.

Evolutionary conclusions

The fossils show that partial bilateral symmetry of the head was acquired at least twice in the calcichordates: in cornutes related to the genus *Phyllocystis* (Jefferies 1986, p 326) and in the most crownward cornutes and the mitrates (Fig. 15). The latter development is especially important because it represents the first stage in the evolution of the bilateral symmetry of the head in modern chordates. Beginning with *Cothurnocystis*, the first step, as seen in *Progalliaecystis*, would have been a reduction in the size of the l-appendage on the left of the head and development of a spike on the right side of the head (the e-spike), almost equal in size to what was now the l-spike. The animal probably crawled rearwards by pivoting the head alternately about these two spikes, using the tail in ventral flexion—thus the locomotion would have been more bilaterally symmetrical than in *Cothurnocystis*. The next step was a drastic reduction in the size of the left or primary pharynx, as seen in *Hanusia* (Cripps 1989). Then almost perfect symmetry of the head outline was acquired, along with ventral flexion of the tail approximately in the sagittal plane, as in *Reticulocarpos* (this was accompanied by a drastic reduction in body size). A right pharynx and right gill slits arose in the origin of the mitrates. If the gill slits were ciliated to pump the feeding current, this would have doubled the pumping rate of the pharynx. Fig. 15 also shows that the most primitive craniate mitrate *Chinianocarpos* was externally less symmetrical than more crownward craniate mitrates such as *Mitrocystella*.

The fossils show that bilateral symmetry evolved in the chordates from the posterior forwards and from the outside inwards. In solutes the only strictly bilateral part of the animal is the hind tail. In *Cothurnocystis* the whole tail is symmetrical, but the head is entirely asymmetrical. In more crownward cornutes such as *Reticulocarpos* the tail is symmetrical, the outline of the head is symmetrical, but the interior of the head remains asymmetrical. In mitrates the tail is symmetrical, the posterior part of the head is externally symmetrical, but the anterior part of the head is externally often asymmetrical and the inside of the head is likewise asymmetrical, though less so than in cornutes because there were right gill slits as well as left ones.

There is a functional interpretation of these generalizations. In solutes and primitive cornutes, the tail tended to become symmetrical because it was waved from side to side in locomotion, and thus used equally on right and left, while movement of the head probably involved yaw, with different non-symmetrical structures active in rotating clockwise and anti-clockwise. In more crownward cornutes, the yaw first became symmetrical and then was lost because the tail flexed ventrally in the sagittal plane of the head. The surface of the head came in contact with the bottom mud during locomotion and therefore became symmetrical, whereas the inside of the head, in its protected situation, remained asymmetrical. The acquisition of right gill slits in the origin of mitrates would have no locomotory advantage but would double the throughput of the feeding current through the head. The posterior parts of the head in mitrates, being the leading parts in locomotion, came in contact with relatively strong undisturbed mud and therefore the selection pressure towards bilateral symmetry was more intense than for the trailing anterior parts of the head in contact with weaker mud.

A subtler embryological cause may also have acted in producing bilateral symmetry. The tail contained the notochord, and this is an enormously important constraint over embryological development in general and possibly over bilateral symmetry in particular. Perhap bilateral symmetry in part results from an influence spreading out from the notochord?

In summary, the origin of the chordate plane of bilateral symmetry can now be traced using fossil evidence. This plane is not homologous with the primitive bilaterian plane of bilateral symmetry, but perpendicular to it. The sagittal plane of a human is perpendicular to that of a beetle.

Acknowledgement

I wish to thank Fritz Friedrich, of Würzburg University, Germany, for the use of some of his recent results on the Cincta.

References

Ax P 1985 Das Phylogenetische System. Fischer, Stuttgart

Ax P 1987 The phylogenetic system. Wiley, Chichester (Translation of Ax 1985)

Ax P 1989 Basic phylogenetic systematization of the Metazoa. In: Fernholm B, Bremer K, Jörnvall H (eds) The hierarchy of life. Elsevier, Amsterdam, p 229–245

Bather FA 1913 Caradocian Cystidea from Girvan. Trans R Soc Edinb 49:359–529

Craske AJ, Jefferies RPS 1989 A new mitrate from the Upper Ordovician of Norway, and a new approach to subdividing a plesion. Palaeontology (Lond) 32:9–99

Cripps AP 1989 A new genus of stem chordate (Cornuta) from the Lower and Middle Ordovician of Czechoslovakia and the origin of bilateral symmetry in the chordates. Géobios 22:215–245

Gemmill JF 1914 The development and certain points in the adult structure of the starfish Asterias rubens L. Philos Trans R Soc Lond B Biol Sci 205:213–294

Gislén TRE 1930 Affinities between the Echinodermata, Enteropneusta and Chordonia. Zool Bidr Upps 12:99–304

Grell KG 1981 Trichoplax adhaerens and the origin of the Metazoa. In: Ranzi L (ed) Origine dei grandi phyla dei Metazoa. Atti dei Convegni dei Lincei vol 49:107–121

Hennig W 1966 Phylogenetic systematics. University of Illinois, Urbana

Hennig W 1969 Die Stammesgeschichte der Insekten. Kramer, Frankfurt-am-Main

Holland PWH 1990 Homeobox genes and segmentation: co-option, co-evolution, and convergence. Semin Dev Biol 1:135–145

Hyman LH 1955 The Invertebrates, vol 4: Echinodermata, the coelomate Bilateria. McGraw-Hill, New York

Jefferies RPS 1968 The subphylum Calcichordata (Jefferies 1967)—primitive fossil chordates with echinoderm affinities. Bull Br Mus (Nat Hist) Geol 16:243–339

Jefferies RPS 1979 The origin of chordates—a methodological essay. In: House MR (ed) The origin of major invertebrate groups. Syst Assoc Spec vol 12:443–477

Jefferies RPS 1986 The ancestry of the vertebrates. Br Mus (Nat Hist), London

Jefferies RPS 1990 The solute *Dendrocystoides scoticus* from the Upper Ordovician of Scotland and the ancestry of chordates and echinoderms. Palaeontology (Lond) 33:631–679

Jefferies RPS, Lewis DN 1978 The English Silurian fossil *Placocystites forbesianus* and the ancestry of the vertebrates. Philos Trans R Soc Lond B Biol Sci 282: 205–323

Jefferies RPS, Lewis M, Donovan SK 1987 *Protocystites menevensis*—a stem-group chordate (Cornuta) from the Middle Cambrian of South Wales. Palaeontology (Lond) 40:420–484

Lester SM 1985 *Cephalodiscus* sp. (Hemichordata: Pterobranchia): observations of functional morphology, behavior and occurrence in shallow water around Bermuda. Mar Biol (New York) 85:263–268

MacBride EW 1914 Textbook of embryology. vol 1: Invertebrata. Macmillan, London

Macurda DB, Meyer DL 1974 Feeding posture of modern stalked crinoids. Nature (Lond) 247:394–396

Olsson R 1983 Club-shaped gland and endostyle in larval *Branchiostoma lanceolatum* (Cephalochordata). Zoomorphology, 103:1–13

Patterson C, Rosen DE 1977 Review of the ichthyodectiform and other Mesozoic teleost fishes and the theory and practice of classifying fossils. Bull Am Mus Nat Hist 158:85–172

Paul CRC, Smith AB 1985 The early radiation and phylogeny of echinoderms. Biol Rev Camb Philos Soc 59:443–481

Romer AS 1972 The vertebrate as dual animal–somatic and visceral. Evol Biol 6: 121–156

Schulze FE 1883 *Tricoplax adhaerens* nov. gen., nov. spec. Zool Anz 6:92–97

Ubaghs G 1967 Homostelea. In: Moore RC (ed) Treatise on invertebrate paleontology. Part S. Echinodermata 1. University of Kansas and Geological Society of America, Lawrence, Kansas, vol 2:S565–S581

Ubaghs G 1969 Les echinodermes carpoides de l'ordovicien inférieur de la Montagne Noire. Cah Paléontol CNRS, Paris

Ubaghs G, Robison RA 1985 A new homoiostelean and a new eocrinoid from the Middle Cambrian of Utah. Univ Kans Paleontol Contrib 115:1–24

Wiley EO 1981 Phylogenetics—the theory and practice of phylogenetic systematics. Wiley, New York

Woods IS, Jefferies RPS 1991 A new stem-group chordate from the Lower Ordovician of South Wales, and the problem of locomotion in boot-shaped cornutes. Palaeontology (Lond), in press

DISCUSSION

Chothia: Could you give us an idea of the time scale for these evolutionary events?

Jefferies: The earliest radiate echinoderm, *Arkarua adami*, is known from the late preCambrian, about 600 million years ago (Gehling 1987). The split between the echinoderms and the chordates would be older than that, but I suppose not much older (Jefferies 1990). There are enormous gaps in the sequence which geologists always point out to me. These fossils don't come in the right stratigraphical order, by any means.

The earliest known cornute dates from the middle Cambrian era, which was about 530 million years ago. The earliest craniates with a phosphatic skeleton come from the upper Cambrian, about 520 million years ago.

Chothia: So the whole transformation that you described took about 100 million years.

Jefferies: Yes, perhaps. Evolution could have happened well before the time from which we have the earliest known relevant fossils.

Layton: How big were these creatures?

Jefferies: The head of *Cothurnocystis* was about 2 cm across. The earliest mitrates were about 1 cm in head length.

Morgan: You said the original dexiothetic ancestor settled onto its right-hand side. That implies there was a left and a right, so presumably it was a bilateral animal already.

Jefferies: The *Cephalodiscus*-like ancestor presumably had the same bilateral asymmetry as is primitive to bilaterians and persists in most bilaterians today. The sagittal plane of a beetle, I believe, is the primitive bilaterian sagittal plane. Ours is perpendicular to that and is the chordate sagittal plane.

Wolpert: Are you saying that the anteroposterior axis of the insect is not the same as ours?

Jefferies: No, the anteroposterior axis is the same; the plane of bilateral symmetry is different. I think the anteroposterior axis is homologous throughout the Bilateria. It may be homologous with the vertical axis of coelenterates.

Lawrence: So the anteroposterior axis could be homologous but the dorsoventral axis is a new one. The mechanism for forming the dorsoventral axis in the chordates and echinoderms could then be fundamentally different from that which is responsible for the dorsoventral axis in an insect.

Jefferies: I wonder whether the the old embryological system for producing right and left was somehow resuscitated to produce the new right and left.

Lawrence: That is possible. Some genes involved in axis formation seem to be used again and again during development.

Wolpert: The moment you become dorsoventral, embryologically, it presents no difficulty to become bilateral. Take a ball, representing the embryo, and choose any point you like to be the front. Set up a gradient with the tail at the other end. Then you can choose any other point to be the dorsal side, and bilaterality, with left and right, is given automatically.

Morgan: So our plane of symmetry has arisen secondarily after this flop over to the right which led to an atrophy of many structures on the right-hand side of the original plane of bilateral symmetry? What are the implications of this for later vertebrate evolution?

Jefferies: I can only go as far as the mitrates in evolution. After that, the calcite skeleton disappears, so from about *Mitrocystella* it's guess work with no help from the fossil record.

What seems to have happened independently in the tunicate and the craniate stem lineages is that the left and right pharynges, for example, became equated —they appear at the same time in ontogeny without any trace of asymmetry. Amphioxus keeps the old system: first it produces the primary gill slits, which are left gill slits; then the secondary gill slits, which are right gill slits, appear at metamorphosis opposite the primaries; then tertiary gill slits arise at right and left behind those. All those three sets of gill slits are found in the mitrate *Lagynocystis*. There is good evidence that the gill bars and gill slits were present in that organism (Jefferies 1986).

Frankel: The left gill slits of *Cothurnocystis* would correspond to the dorsal side of its ancestor, is that correct?

Jefferies: The *Cephalodiscus*-like ancestor of *Cothurnocystis* had only a pair of gill slits, and the right one was lost as a result of dexiothetism. The original left gill slit, that had become dorsal, was also on the left side of the new organism . . . .

Frankel: New left or old dorsal?

Jefferies: New left and new dorsal, and old left as well. The single gill slit of solutes, as seen in *Dendrocystoides*, corresponds to the left gill slit of *Cephalodiscus*. In going from solutes to cornutes there seems to have been some sort of replication, so that one gill slit gave rise to several. In this transition to cornutes, the feeding arm was lost, as one can see by comparing Fig. 9 (*Dendrocystoides*) with Fig. 12 (*Cothurnocystis*). The organism still had to feed. Presumably, cornutes fed by an endostylar filter and endostylar pharynx of the sort which occurs in tunicates and amphioxus, but with left gill slits only, as in larval amphioxus. So the endostylar filter was invented at the transition between solutes and cornutes. The replication of the gill slit from one to several was presumably to increase the flow of water through the pharynx to improve this feeding method.

Corballis: The *Cephalodiscus*-like ancestor that fell over was bilaterally symmetrical. At some point you linked bilateral symmetry to locomotion but the original animal didn't locomote, did it?

Jefferies: Cephalodiscus does locomote and, presumably, the *Cephalodiscus*-like ancestor of the Dexiothetica did so too. There were two falling-overs: pleurothetism, at the beginning of the Bilateria, and dexiothetism, at the beginning of the Dexiothetica. Bilateral symmetry can arise as a result of locomotion. Also, many active filter-feeders, those that produce a feeding current, have a bilateral symmetry related to that current. *Rhabdopleura*, a hemichordate, has a bilateral symmetry related to its feeding current; it has two tentacles. So feeding may be involved as well as locomotion.

Galaburda: One gets the impression that proceeding toward the mitrates, because of this 'falling over', the symmetry that develops is quite phenotypic. Different basic structures will grow on one side to become symmetrical to older structures that were present before in a kind of convergent evolution of the two sides of a bilaterally symmetrical structure. Is there any evidence that in extant similar organisms some genes are expressed on one side and not on the other side?

Jefferies: I can't say much about genes but I can mention a natural historical parallel. Flatfish start with a bilaterally symmetrical larva. The adult flops over onto one side—in some species it doesn't matter which side—the upper surface becomes coloured, the lower surface becomes white and a new plane of symmetry develops perpendicular to the old one. I think that is analogous to what happened in the origin of Dexiothetica. How it happens, I don't know. I can see the selective advantage in becoming bilaterally symmetrical.

Burn: In flatfish embryos, you can trace the migration of the eye by the relationship of the optic nerves. Do you see that in palaeontology? Did organs that had become buried migrate to the new right or left side? If you adopt a new left–right symmetry on what was the left side, does that induce lateralized structures to move to a new position? Or could these organs have moved simply to get back on the top?

Jefferies: There is one case of a structure that has moved to get back on the top. The right anterior coelom, which was homologous with the right metacoel of *Cephalodiscus* and in cornutes was ventral beneath the pharynx, extended dorsalwards and leftwards in the origin of mitrates to a position dorsal to the new right pharynx at posterior right of the head. Thus, in this case, old right (new ventral) was converted to new right. The right anterior coelom of mitrates is homologous with the right mandibular somite of extant vertebrates and in these it is exactly symmetrical to the left mandibular somite.

In *Cothurnocystis*, the left gill slits are homologous with the old left gill slit of *Cephalodiscus*. When the *Cephalodiscus*-like ancestor fell onto its right side, the old left gill slit was on the top anyway.

There are many asymmetries in amphioxus which can be explained in mitrate terms. For example, the velar mouth, when it first appears, penetrates the left mandibular somite. That almost certainly happened in the ontogeny of mitrates as well. The asymmetries of amphioxus have been known for a long time but they were ignored because their significance was unknown. Most of these asymmetries turn up again in mitrates.

Wood: What do you think are the selective advantages of bilateral symmetry, which according to your view was imposed on an asymmetrical animal?

Jefferies: In the first instance it has to do with locomotion. Bilateral symmetry began at the back end in the hind tail, then gradually worked into the fore tail, then spread into the head. It is related to the curious habit these animals had of moving backwards. There is strong evidence for backward motion, in particular in the mitrate *Mitrocystella*. This had cuesta-shaped ribs, with a steep slope and a gentle slope. The steep slope was always anterior. There are plenty of extant analogies to this, in crabs, for example, and in bivalve molluscs. Cuesta-shaped ribs are a sediment-gripping device that allow the animal to move easily in one direction and hinder movement in the opposite direction. The tail also fits in beautifully if you make the assumption that *Mitrocystella* did move backwards.

As far as the outside of organisms is concerned, locomotion was the driving force towards bilateral symmetry. As far as the inside is concerned, when you produce right gill slits, you presumably double the pumping rate through the pharynx, so you double the rate of extracting food particles from the water.

Stern: Why do you say that the solutes may have a notochord?

Jefferies: The tail of solutes is divided into fore, mid and hind portions. The hind portion is rigid and has a hook on the end. The fore portion has a large lumen with C-shaped skeletal elements. These overlap by the ends of the Cs in the dorsal and ventral midlines; they look as though they were designed to allow the fore tail to flex left and right, rather than to telescope, and this implies an anti-compressional organ, presumably a notochord. There is evidence of similar flexion in the fore tail of cornutes and in the fore tail of mitrates. I am pretty sure that in the origin of the mitrates the old mid and hind tail were lopped off by autotomy and the mitrate fore, mid and hind tail is a regionation of the old stump. The mitrate tail certainly acted by ventral flexion, with the motor in the fore tail. You can tell this by the way the various plates overlap each other.

Stern: Is that a sufficient criterion to call it a notochord?

Jefferies: It is an indication: there are others. There is a canal down the middle of the hind tail of solutes which presumably carried a notochord. There is a groove down the dorsal surface of certain ossicles in the hind tail of cornutes that presumably carried a notochord. In mitrates there is evidence of spinal ganglia and a dorsal nerve chord resting on a cylindrical organ which I assume to be the notochord.

This argument is like solving a crossword puzzle. If sufficient numbers of things fit with these very complicated fossils, I hope I am getting it right. Any particular argument will only be presumptive, but they all fit together so closely that I hope the ultimate result is correct.

Lawrence: In the origin of the central nervous system, the brain appears at the front end of the tail. Does that mean before there was a tail there was another brain somewhere else?

Jefferies: It means that the brain of solutes, cornutes and mitrates, and of chordates in general, can't be homologous to the brain of the primitive bilaterians. Nor to the brain of arthropods or annelids. It can't be homologous with the ganglion that *Cephalodiscus* has between its tentacles, which is in entirely the wrong place for such a homology. It is no real surprise that the brain should appear at the front end of the notochord.

Lawrence: The new brain is a different brain and arises at the front end of the tail because the tail is a motile organ. Not much happens in front of the tail that needs thought!

Jefferies: Essentially, you are right. Such sensory organs as mitrates had, with the exception of the olfactory system that was inside the buccal cavity, tended to be at the back end of the head, which was the leading end.

Stern: In comparing the vertebrate head to a part of the body plan in the solutes, cornutes and mitrates, you seem to put the limit of the homologous region at the anterior end of the tail. The viscera should then move down entirely into the region you call the tail. Is this correct?

Jefferies: It's a little more complicated than that. In the stem lineage of the craniates, evolving towards the crown group of the craniates after the mitrate skeleton was lost, lots of changes occurred. One was that the right and left pharynx grew back underneath the front part of the tail. That would have produced the notochordal part of the head and the trunk. That's two sentences for a very complicated operation. One likely result of this was that the anterior muscle blocks grew down the gill bars to give the classical alternation of muscle blocks and gill bars which people like van Wijhe and Goodrich were so keen on in the last century.

Wolpert: When is the earliest example of bilateral asymmetry in the fossil record?

Jefferies: In the early solutes, which occurred in the lower Cambrian. The *Dendrocystoides*-type of organism was thoroughly asymmetrical. But asymmetry existed before then.

Wolpert: So left–right asymmetry is a very primitive character.

Jefferies: So far as I am concerned it is primitive for the group Dexiothetica— the echinoderms plus the chordates.

Brown: Is it correct that in chordate evolution, after dexiothetism, bilateral symmetry was achieved once again in the mitrates, before the further development to the asymmetrical plan of current vertebrates?

Jefferies: Partial symmetry: the inside of mitrates is always thoroughly asymmetrical. Even the outside is never quite symmetrical.

Frankel: The organisms of the stem of your echinoderm–chordate lineage existed before the beginning of the Cambrian. Therefore the primitively bilateral ancestor of that stem lived before the beginning of the Cambrian. Is there any evidence from the earlier Ediacaran fauna that can tell you anything about that stem?

Jefferies: The earliest Metazoa are supposed to be one thousand million years old. Some of the evidence consists of worm tracks: I would guess those 'worms' were bilaterally symmetrical, they certainly moved in one direction. So the primitive bilaterality of the Bilateria is presumably about 1000 million years old.

Frankel: Are there bilateral organisms in the Ediacaran (late Precambrian)?

Jefferies: Yes. There is *Spriggina*, which is either an annelid or an arthropod. It looks like a trilobite that can wriggle sideways.

Frankel: I thought the Metazoa were supposed to have arisen rather suddenly at the beginning of the Ediacaran about 700 million years ago.

Jefferies: The age that I gave of over 1000 million years, I read in Glaessner (1984); he knew what he was talking about.

Morgan: After dexiothetism had occurred, there seem to be two importantly different ways in which bilateral symmetry could have been restored. Let's consider paired structures: if you start from a situation where the structure on the right has degenerated, one option is to develop two new paired structures from the left one. That is basically what echinoderms have done with all their body cavities.

Jefferies: No; one of the coeloms is the right metacoel; there is a right metacoel remaining. But with that exception, you are correct.

Morgan: The alternative is to resuscitate a right-hand structure without deriving it from the left. In general, which of these routes do you think has been followed?

Jefferies: I think mitrates found a way of replicating structures which existed on their left and producing them on the right. This is fairly clearly the case with the acusticolateralis system and with the pharynx. How would an organism know that its remote ancestor had lost a structure on the right? Also, as already mentioned, the right anterior coelom of mitrates represents an old right structure that had become new right secondarily by a sort of evolutionary twisting.

Galaburda: The internal structure would be different. It is like convergent evolution: the wing of a bird and the wing of a bat have evolved to do basically the same thing but internally they are quite different.

Wolpert: Internally they are not quite different.

Galaburda: At the level of what genes and what proteins are made to form a feather or a hair on a bat wing, they are different. If one derives the new left side from some portion of the right side which is now dorsal, would some of the internal structure, some of the detailed building blocks of this new structure, be different from the one opposite? This is a very important issue. If this is a mechanism by which we have derived some of our bilateral asymmetry, it leaves open the question as to whether my left and right hands are in some sense like a bat wing and a bird wing.

Wolpert: That they are intrinsically different? I don't believe that!

Galaburda: I don't have any evidence one way or another but it could be so if an ancient ancestor did 'fall over'.

References

Gehling JG 1987 Earliest known echinoderm—a new Ediacaran fossil from the Pound Subgroup of South Australia. Alcheringa 11:337–345

Glaessner MF 1984 The dawn of animal life. Cambridge University Press, Cambridge

Jefferies RPS 1986 The ancestry of the vertebrates. British Museum (Natural History), London

Jefferies RPS 1990 The solute *Dendrocystoides scoticus* from the Upper Ordovician of Scotland and the ancestry of chordates and echinoderms. Palaeontology (Lond) 33:631–679

Asymmetries during molluscan embryogenesis

Jo A. M. van den Biggelaar

Department of Experimental Zoology, University of Utrecht, Padualaan 8, 3584 CH Utrecht, The Netherlands

Abstract. In some molluscan species the unfertilized egg is symmetrical around its centre. The maturation divisions provide the egg with an axial symmetry with an animal–vegetal asymmetry. During the first two cleavages the egg loses its axial symmetry by the formation of unequal quadrants. The size differences may be very pronounced in species where the first two cleavages are accompanied by the formation of a polar lobe or where the first two cleavages are very unequal. There are some molluscan species in which at first glance the four quadrants appear equal. Exact measurements of the relative volumes have shown that the spiral character of the cleavages gives rise to minor differences between the quadrants. During further division this difference is limited to the vegetal macromeres; other corresponding blastomeres in the four quadrants are mutually equal. Therefore the absolute difference between the macromeres increases after each division. The size difference between the macromeres predisposes the biggest macromere to attain a central position and to become induced to develop the stem cell of the mesoderm. The bilateral symmetry is later lost by the counterclockwise rotation through 180° of the visceral mass in relation to the head and foot.

1991 Biological asymmetry and handedness. Wiley, Chichester (Ciba Foundation Symposium 162) p 128–142

Asymmetries in the architecture of the adult animal gradually arise during morphogenesis. Theoretically, the unripe egg may be regarded as a system symmetrical around its centre; each plane through the centre of the egg may divide it into symmetrical halves. During the maturation divisions, full rotational symmetry is maintained around one axis only, the animal–vegetal egg axis. This axial symmetry is lost by the development of dorsoventral asymmetries. From that time the embryo can be divided into two symmetrical halves by only one plane, the median plane. Finally, bilateral symmetry may be lost by the creation of additional asymmetries. In this paper the discussion will be limited to the development of asymmetries in molluscs. It is an attempt to reduce the causes of asymmetries in the molluscan body plan to two general phenomena: asymmetries in cell division and asymmetries in cell positions.

This paper was presented by Dr R. Dohmen in the absence of Professor van den Biggelaar through illness.

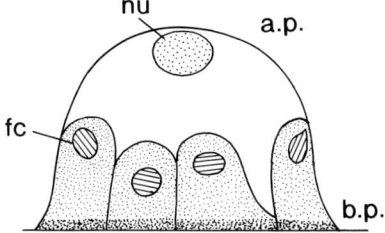

FIG. 1.   Schematic drawing of a molluscan oocyte attached with the basal pole (b.p.) to the ovarian wall and surrounded by follicle cells (fc). The nucleus (nu) marks the apical pole (a.p.), which faces the ovarian lumen.

Asymmetry of the ovarian oocyte

In the ovary one side of the oocyte rests on the ovarian wall and the opposite side faces the ovarian lumen (Fig. 1). This orientation provides the oocyte with an apical–basal asymmetry. The nucleus lies at the apical side. In some species the basal part is characterized by the localization of special structures, like the vegetal body in *Bithynia* (Dohmen & Verdonk 1974, 1979), or in Dentalium the accumulation of bacteria which remain attached to the egg after ovulation (Geilenkirchen et al 1971). Often a micropyle in the egg envelope marks where the basal part of the oocyte has been fixed to the ovarian wall (van den Biggelaar & Guerrier 1983).

Determination of the animal–vegetal asymmetry by the maturation divisions

Without any modification the apical–basal axis may normally coincide with the animal–vegetal egg axis. This does not imply that the two axes have to correspond. Experimental studies on the egg of *Limax* (Guerrier 1968) show that the direction of the second maturation spindle determines the animal–vegetal egg axis. The direction of the spindle may coincide with any arbitrary axis through the centre of the egg. Therefore, as assumed above, the immature egg can be regarded as a cell with a radial symmetry. However, as soon as the oocyte divides and forms the two polar bodies, the actual direction of the spindle and the plane of division provide the egg with an animal–vegetal asymmetry (Fig. 2). This asymmetry is also expressed in the differential segregation of ooplasmic substances along the animal–vegetal axis. At least in *Limax*, the unavoidable asymmetry associated with the second maturation division is sufficient to provide the egg with the animal–vegetal asymmetry. After maturation the animal–vegetal axis can no longer be changed, and it corresponds with the anteroposterior axis of the later embryo.

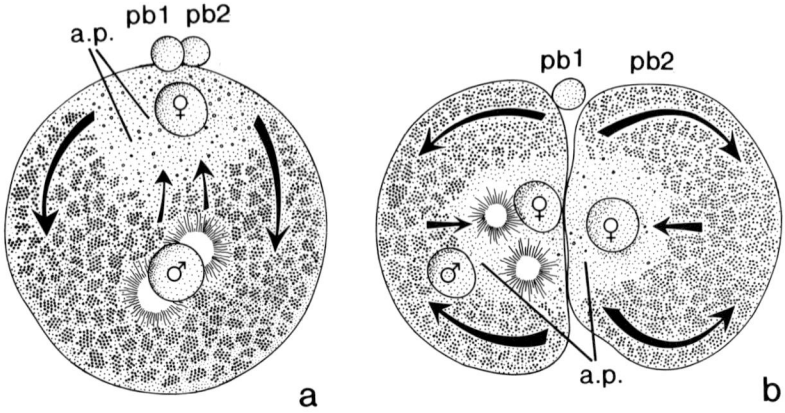

FIG. 2.   Direction of ooplasmic segregation (arrows) and the direction of the egg axis after the second maturation division. (a) Control egg with the two polar bodies (pb1, pb2), the female pronucleus surrounded by the animal pole plasm (a.p.) and the male pronucleus. (b) Position of the second polar body after displacement of the second maturation division at right angles to the original egg axis. Note the mirror image of the ooplasmic segregation in the giant second polar body with a female pronucleus only, and the mature egg with a female and a male pronucleus. The egg has been divided into two axially symmetrical halves with opposite directions of animal–vegetal polarity (after Guerrier 1968).

Determination of dorsoventral asymmetry by asymmetrical cleavages

Unequal cleavage

In molluscs, dorsoventral asymmetry can be traced back to the effects of cell division or to obligatory asymmetries in the configuration of cells produced by cell division. In species like *Spisula* and *Pholas*, axial symmetry is lost through the influence of sperm entry on the position of the spindles at the first two cleavages (Fig. 3). During the first mitotic division the spindle has a central position normal to the plane through the egg axis and the sperm entrance point. Before cytokinesis, the spindle moves randomly either to the right or to the left along the axis of the spindle (Guerrier 1970). This provokes an asymmetrical division into a smaller AB and a larger CD cell. At second cleavage the AB cell divides equally; in CD the spindle moves towards the sperm entrance point. This causes an asymmetrical division of CD into a smaller C and a larger D cell. The asymmetry of the 4-cell stage is directly related to the dorsoventral asymmetry of the embryo. The larger D quadrant becomes dorsal, B ventral, and A and C become lateral. Although this dorsoventral asymmetry refers back to the asymmetrical event of sperm entry, it appears that any unequal cleavage may produce an asymmetrical 4-cell stage and an embryo with a normal dorsoventral asymmetry. This demonstrates that it is essentially the asymmetrical

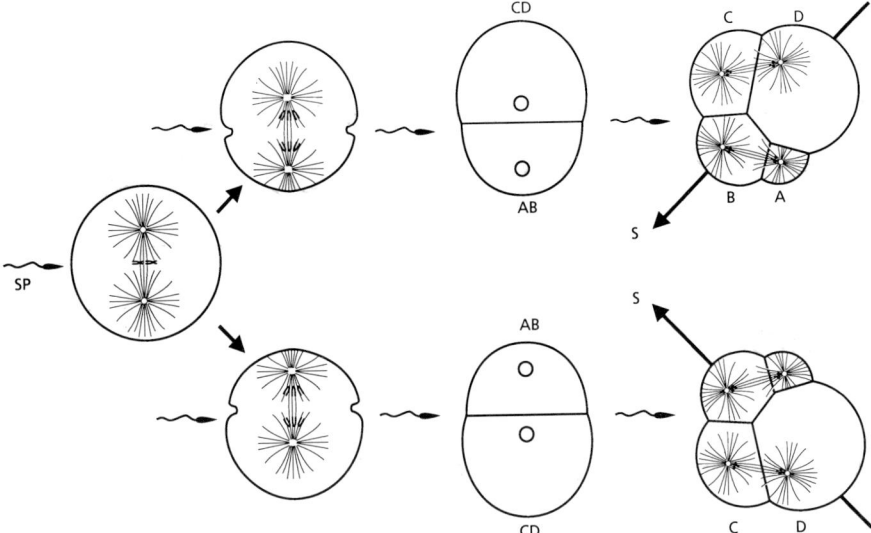

FIG. 3. Direct unequal division in *Pholas dactylus* in relation to sperm entry. Upper lane: displacement of the first mitotic spindle to the left with respect to the egg axis viewed from the sperm entry point; clockwise arrangement of the quadrants. Lower lane: displacement of the spindle to the right; counterclockwise arrangement of the quadrants. S, plane of bilateral symmetry; SP, sperm.

division itself which acts as the asymmetry-inducing mechanism, not the entry of the sperm cell.

Cleavage associated with polar lobe formation

In a number of other molluscan species an unequal 4-cell stage is reached indirectly (Fig. 4). At the first two cleavages the most vegetal domain is separated from the dividing egg in the form of a polar lobe. When the first cleavage plane reaches the narrow connection point between the egg and the polar lobe, it is forced to move laterally. Consequently, the polar lobe remains in touch with one blastomere, with which it re-fuses. The result is a small AB and a larger CD cell that contains the polar lobe. At the second cleavage AB divides equally and the CD cell forms another polar lobe, which, like the first, re-fuses with one of the two sister cells. This leads to an asymmetrical 4-cell stage with three equally sized quadrants A, B and C, and the larger, lobe-containing D quadrant. The D quadrant becomes dorsal, A left, C right, and B ventral. If the polar lobe is deleted at first cleavage, the dorsoventral asymmetry fails to appear and the embryo remains axially symmetrical.

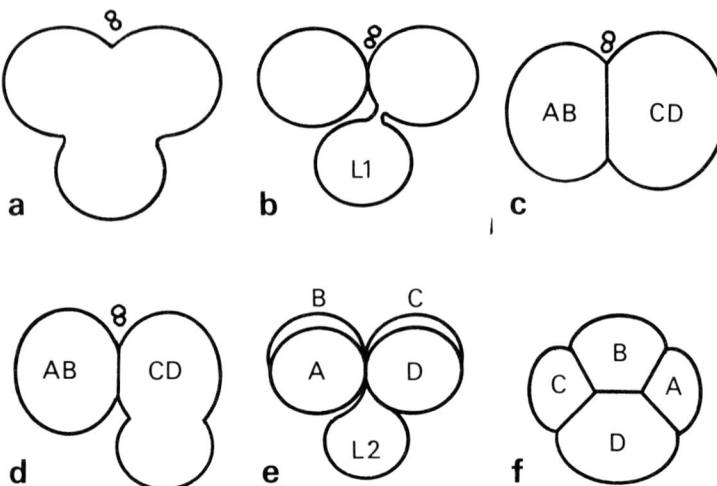

FIG. 4.   Creation of an unequal 4-cell stage by the formation of a polar lobe at the first two cleavages. (a) Formation of the first polar lobe. (b) Fusion of the first polar lobe with one of the two sister cells, which after fusion can be designated as CD. (c) 2-cell stage. (d) Formation of the second polar lobe by the CD cell. (e) Fusion of the second polar lobe with the sister cell (D) clockwise from C. (f) Unequal 4-cell stage.

Determination of dorsoventral asymmetry in embryos with an equal 4-cell stage

Spiral cleavage

By two parameridional divisions a 4-cell stage is reached in which, at first sight, the four quadrants seem to have the same size. A para-equatorial division leads to the 8-cell stage, which consists of an animal and a vegetal tier of four cells each. The third division is a spiral cleavage, so the animal cells are not placed directly above their vegetal sister cells, but are displaced clockwise or anti-clockwise (Fig. 5). The direction of this shift is species specific and appears to correspond with the direction of the coiling of the shell. Gastropods with a dextral third cleavage have dextrally coiled shells; gastropods with a sinistral third cleavage have sinistrally coiled shells.

The mutual shifting of sister cells observed at the third division reappears regularly during the successive cleavages. If shifting is clockwise at the third division, it will be clockwise during each odd-numbered division and anticlockwise in even-numbered divisions. The shifting of sister cells may be the result of two different mechanisms. The first stems from an oblique position of the spindles in relation to the egg axis. If the upper pole of an inclined spindle points clockwise, the animal daughter cell will be formed clockwise; if the upper pole points anticlockwise, the animal daughter cell will be formed anticlockwise.

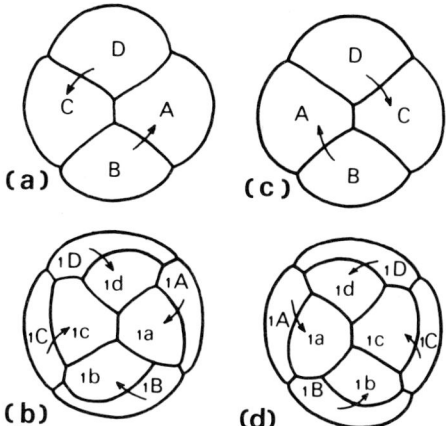

FIG. 5.   4- and 8-cell stages in a dextral (a,b) and a sinistral species (c,d). Arrows indicate the direction of the displacement of the cells.

A second mechanism demonstrated by Mescheryakov & Belusov (1975) is the mutual rotation of blastomeres during cleavage. The proposed explanation assumes that actin filaments are not arranged parallel to the circumference of the contractile ring, but deviate in one or the other direction. Consequently, the two sister cells rotate in opposite directions during cytokinesis. The observed enantiomorphic rotations in embryos of *Lymnaea* and *Physa* agree with this assumption (Mescheryakov & Belusov 1975). Both mechanisms, involving the oblique positioning of the spindle and blastomere rotation, are aspects of spiral cleavages.

Although the divisions are obviously spiral only from the third cleavage onwards, this does not imply that the first two are not spiral cleavages. Almost a century ago, Conklin (1897) observed that in *Crepidula* the first two cleavages can be regarded as spiral divisions, achieved by a rotation of the sister cells during cytokinesis preceded by a slight tilt of the spindles. Like all other odd cleavages, the first is clockwise, and like all even cleavages, the second is anticlockwise. Similar observations have been described by Mescheryakov & Belusov (1975) for eggs of *Lymnaea* and *Physa*. Torsion has even been observed in the formation of polar bodies during the maturation divisions.

Genetic aspects of dextrality and sinistrality

In *Lymnaea peregra* dextrality and sinistrality of the cleavages and the shell are determined maternally (Boycott & Diver 1923, Boycott et al 1930) and controlled by a single recessive locus called *sinistral* (Sturtevant 1923). Freeman & Lundelius (1982) have shown that after injection of cytoplasm from a wild-type (dextral) egg into an uncleaved, phenotypically sinistral egg, cleavage occurs with a dextral pattern. Cytoplasm of a sinistral form cannot reverse the cleavage pattern of a dextral egg. Reversal of the cleavage pattern is no longer possible after extrusion of the second polar body. It remains to be explained how a maternal gene product determines the cleavage pattern. The transplantation experiments of Freeman & Lundelius (1982) and the pressure experiments on the egg of *Limax* by Guerrier (1970) demonstrate that the animal–vegetal organization of the zygote and the cleavage pattern are irreversibly fixed after the second maturation division and before the first cleavage.

Dorsoventral asymmetry and spiral cleavage

When the first two cleavages are spiral, the spindles are not exactly normal to the egg axis; it may be assumed that neither the first nor the second cleavage is perfectly meridional, and henceforth, that they do not divide the egg into four equal quadrants. Bezem et al (1981) have measured the relative volumes of blastomeres during successive cleavage stages in the pond snail *L. stagnalis*, a species with an 'equal' 4-cell stage. Prior to the 24-cell stage only the lateral quadrants can be distinguished from the median quadrants. The former are in touch at the animal cross-furrow, the latter at the vegetal cross-furrow. At the 4-cell stage the lateral quadrants are already significantly smaller than the median quadrants. Apparently, the oblique position of the spindles at the 2-cell stage causes the formation of two slightly smaller blastomeres towards the animal pole and two slightly bigger cells towards the vegetal pole. In 8-cell embryos the difference is expressed only in macromeres 1A–1D; no differences are found in the micromeres 1a–1d. Similarly, at the 16-cell stage the mean volume of the lateral macromeres is significantly less than that of the median macromeres; no differences have been observed in any of the quartets of micromeres. Again, at the 24-cell stage there are significant differences in the relative volumes of only macromeres 3A–3D, not among the corresponding micromeres. The macromeres divide without an intervening growth phase: as the initial difference between the median and the lateral quadrants is attributed to progressively smaller macromeres, the relative difference between the median and the lateral macromeres increases from 5% at the 4-cell stage to 10% at the 8-cell stage and 13% at the 16-cell stage. In 24-cell embryos it is also possible to compare the four quadrants, as from that stage they can be distinguished unambiguously. There is then not only a significant difference between the pairs 3A/3C and 3B/3D, but also between 3B and 3D. The relative volumes of 3A and 3C are 5% and 5.1% of the whole egg, respectively; the volume of 3B is 6.4% and of 3D 7.4%. As the bigger of the two macromeres almost invariably becomes the central macromere 3D, it appears that a difference in size predisposes the quadrant with the biggest macromere to become central and to mark the dorsal quadrant. Comparable results have been observed during cleavage in embryos of

Physa fontinalis, *Patella vulgata* and *Haliotis tuberculata* (J. A. M. van den Biggelaar, unpublished observations 1990).

The morphogenetic significance of the central macromere 3D is that it forms the stem cell of the mesoderm and partially contributes to the development of the endoderm. Macromeres 3A, 3B and 3C produce only endoderm. This difference between 3D and the other macromeres is the result of an inductive interaction between 3D and the micromeres of the opposite animal pole. From cell deletion experiments it can be concluded that any of the macromeres may become central and behave as a normal 3D macromere (van den Biggelaar & Guerrier 1979, Arnolds et al 1983). The size difference between the two vegetal cross-furrow macromeres is normally sufficient to trigger the larger of these two cells to attain the central position and thus to become the stem cell of the mesoderm. Once this interaction has taken place, the direction of the dorsoventral asymmetry is determined.

Torsion of the visceral mass and coiling of the shell in gastropods

Initially, all molluscan embryos seem to be bilaterally symmetrical. The mouth is at the anterior end and the anus at the posterior end of the body axis. In gastropods the embryo loses its bilateral symmetry by counterclockwise rotation of the visceral mass through 180° in the anterodorsal direction. After this movement, the mouth and anus are placed anteriorly.

A second aspect of asymmetry in gastropods is the coiling of the shell. The visceral hump is coiled because of a difference between the growth rates of the two sides (Fig. 6). This is accompanied by the formation of the typical helicoid spiral of the shell. The relation between the chirality of the cleavages and the direction of the coiling of the visceral hump and the shell remains to be explained. In some gastropods, for instance in the genus *Odostomia* and allied species, the embryonic shell is sinistral and the embryonic body is dextral: as soon as the body has attained a certain size, the adult shell also becomes dextral (Lebour 1933).

At the moment, there is no general explanation for the relation between spiral cleavage, torsion and coiling of the shell. Verdonk (1979) has pointed out that in a number of species the second polar lobe is smaller than the first. This implies that some of the polar lobe substances are not incorporated into D, but are left in the C cell. This may be the first deviation from bilateral symmetry— quadrants A and B are free of polar lobe substances, C may contain a small amount and D the majority. Because the quadrants with polar lobe substances contribute more to the development of the adult structures and C is right of the median plane, this might foreshadow the torsion of the later embryo (Verdonk 1979).

In gastropods without a polar lobe, a different mechanism must be involved. As described above, spiral cleavage may necessarily lead to the formation of

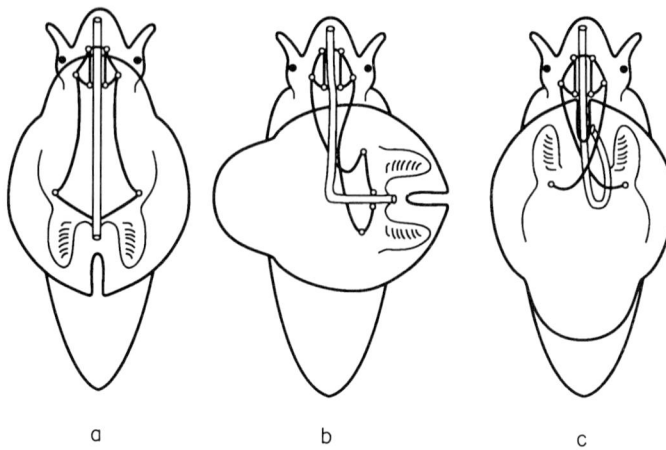

a   b   c

FIG. 6. Schematic drawing of the counterclockwise torsion of the visceral mass in gastropods. (A) Original situation with the mouth anterior and the anus posterior. (B) Counterclockwise torsion of the visceral hump. (C) Anterior location of the anus after a counterclockwise torsion of 180°.

an asymmetrical 4-cell stage. This asymmetry may persist during further development and influence the torsion. For instance, according to Robert (1902), in *Trochus* micromere 4a protrudes into the centre of the embryo during gastrulation more quickly than do micromeres 4b and 4c, and the blastoporus closes asymmetrically. Conklin (1897) has shown that in *Crepidula* there is an asynchrony in the division of macromeres 4A, 4B and 4C; 4C divides earlier and micromere 5c has an asymmetrical position, lying more dorsally than 5d. Probably as a consequence, the anlage of the hindgut is shifted towards the right.

Conclusion

Evidence has been given to support the assumption that before the maturation divisions the egg is a radially symmetrical system. Through the definitive position of the second meiotic spindle and the second meiotic division, radial symmetry is lost and the egg is transformed into an axially symmetrical system. The first mitotic spindle may be rotated in any direction at right angles to the egg axis. Its definite position, together with the cleavage plane, reduces the number of planes by which the egg may be divided into symmetrical halves to one. The position of the second mitotic spindle is determined by the position of the first, therefore the presumptive plane of bilateral symmetry, which passes through the egg axis at right angles to the vegetal cross-furrow, depends on the position of the first mitotic spindle. The spiral character of the first two divisions may produce an asymmetrical 4-cell stage, which adumbrates the direction of the torsion of the embryo. In this way the causes of animal–vegetal and dorsoventral asymmetries and of the torsion may be reduced to the disruption of symmetry by cell division.

References

Arnolds WA, van den Biggelaar JAM, Verdonk NH 1983 Spatial aspects of cell interactions involved in the determination of dorsoventral polarity in equally cleaving gastropods and regulative abilities of their embryos, as studied by micromere deletions in *Lymnaea* and *Patella*. Roux's Arch Dev Biol 192:75–85

Bezem JJ, Wagemaker HA, van den Biggelaar JAM 1981 Relative cell volumes of the blastomeres in embryos of *Lymnaea stagnalis* in relation to bilateral symmetry and dorsoventral asymmetry. Proc K Ned Akad Wet Ser C Biol Med Sci 84:9–20

Boycott AE, Diver C 1923 On the inheritance of sinistrality in *Limnaea peregra* (Mollusca, Pulmonata). Phil Trans Roy Soc Lond Ser B 95:207–213

Boycott AE, Diver C, Garstang SL, Turner FM 1930 The inheritance of sinistrality in *Limnaea peregra* (Mollusca, Pulmonata). Philos Trans R Soc Lond Ser B 219:51–131

Conklin EG 1897 Embryology of *Crepidula*. J Morphol 13:1–127

Dohmen MR, Verdonk NH 1974 The structure of a morphogenetic cytoplasm, present in the polar lobe of *Bithynia tentaculata* (Gastropoda, Prosobranchia). J Embryol Exp Morphol 31:423–433

Dohmen MR, Verdonk NH 1979 Cytoplasmic localization in mosaic eggs. In: Newth DH, Balls M (eds) Maternal effects in development. Cambridge University Press, London, p 127–145

Freeman G, Lundelius JW 1982 The developmental genetics of dextrality and sinistrality in the gastropod *Lymnaea peregra*. Wilhelm Roux's Arch Dev Biol 191:69–83

Geilenkirchen WLM, Timmermans LPM, van Dongen CAM, Arnolds WJA 1971 Symbiosis of bacteria with eggs of *Dentalium* at the vegetal pole. Exp Cell Res 67:477–479

Guerrier P 1968 Origine et stabilité de la polarité animale–végétative chez quelques Spiralia. Ann Embryol Morphol 1:119–139

Guerrier P 1970 Les caractères de la segmentation et la détermination de la polarité dorso-ventrale dans le développement de quelques Spiralia. III. *Pholas dactylus* et *Spisula subtruncata* (Mollusques, Lamellibranches). J Embryol Exp Morphol 23:667–692

Lebour M 1933 The eggs and early larvae of two commensal gastropods, *Stilifer stylifer* and *Odostomia eulimoides*. J Mar Biol Assoc UK 18:117–122

Mescheryakov VN, Belusov LV 1975 Asymmetrical rotations of blastomeres in early cleavage of Gastropoda. Wilhelm Roux's Arch Dev Biol 177:193–203

Robert A 1902 Recherches sur le développement des troques. Arch Zool Exp Gen 3rd Ser 10:269–359

Sturtevant AH 1923 Inheritance of direction of coiling in *Limnaea*. Science (Wash DC) 58:269–270

van den Biggelaar JAM, Guerrier P 1979 Dorsoventral polarity and mesentoblast determination as concomitant results of cellular interactions in the mollusk *Patella vulgata*. Dev Biol 68:462–471

van den Biggelaar JAM, Guerrier P 1983 Origin of spatial organization. In: Verdonk NH, van den Biggelaar JAM, Tompa AS (eds) The Mollusca, vol 3: Development. Academic Press, New York, p 179–213

Verdonk NH 1979 Symmetry and asymmetry in the embryonic development of molluscs. In: van der Spoel S, van Bruggen AC, Lever J (eds), Pathways in Malacology. Bohn, Scheltema and Holkema, Utrecht p 25–45

DISCUSSION

Dohmen: There are two phenomena observable before spiral cleavage which show a bias in laterality. These may be regulated by the same (unknown) mechanism that determines spiral cleavage. First, the surface of the first polar body extruded from the egg of *Lymnaea* shows a prominent right-handed spiral pattern of surface folds. The surface of the second polar body has not yet been studied in sufficient detail for conclusions to be drawn.

The second concerns the polar lobe, a peculiar equatorial constriction that temporarily separates the vegetal cytoplasm from the rest of the egg in a number of species (see Dohmen 1983). This constriction is accompanied by a spiral pattern of surface folds (Fig. 1; see also Fig. 9 in Dohmen 1983). A systematic

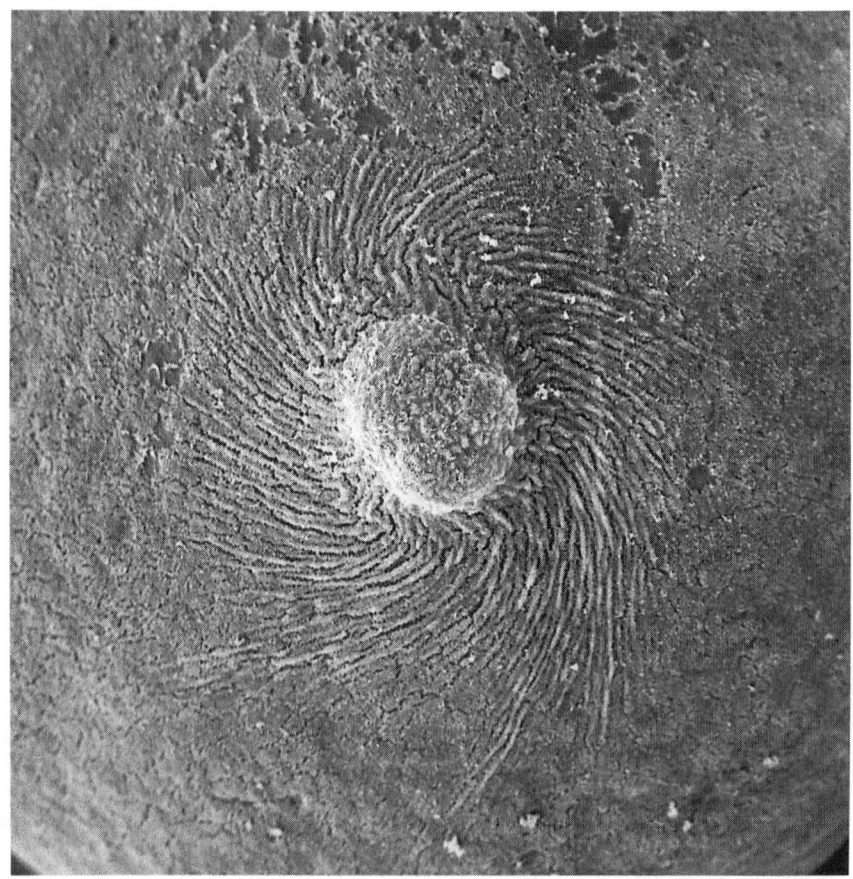

FIG. 1. *(Dohmen)* Scanning electron micrograph of the beginning of polar lobe formation in the egg of the mollusc *Buccinum*.

study of this phenomenon has not been done, so it is not known whether the handedness of this spiral pattern corresponds with that of the spiral cleavage.

Kondepudi: Is the handedness of the spiral on the surface the same in all eggs?

Dohmen: That has not been studied. Now this conference has convinced me of the importance of handedness, I shall continue these studies.

Wood: At what point in embryogenesis can you first identify the dorsoventral axis in molluscs?

Dohmen: That depends on the species. In *Lymnaea*, the dorsoventral axis is specified at the fifth cleavage when the blastomere that is the precursor of the mesoderm cell lineage is induced to become so by contact with another set of blastomeres.

Wood: So we can't talk about left and right until that point?

Dohmen: Not in *Lymnaea*: in the species that form a polar lobe, you can identify the dorsal side immediately after fusion of the polar lobe with one of the blastomeres, because this blastomere becomes the dorsal blastomere.

Wolpert: So all the axes are specified as soon as you know which blastomere is going to fuse with the polar lobe?

Dohmen: Yes, the animal–vegetal axis becomes the anteroposterior axis.

Wood: Can you see any asymmetry, besides that along the animal–vegetal axis, in the first cleavage? Is the spindle oriented exactly perpendicular to the animal–vegetal axis?

Dohmen: There are unpublished reports that it is not, but the obliqueness of the position of the spindle is so subtle that it is hard to be sure.

Yost: How large are these embryos? How was the polar lobe deleted at the first cleavage?

Dohmen: The eggs are about 200 μm in diameter. When the polar lobe is connected to the remainder of the egg by a thin stalk, it can be easily separated from the egg using a glass needle or a hair. This removes the morphogens present in the polar lobe, so in later development many things go wrong—many organs are completely absent, especially the mesoderm-derived organs.

Normally, during early cleavage stages there is a strictly determined asynchrony in the cleavage of the four quadrants of the embryo. Somewhat later there is also a typical size difference between certain cells in the D quadrant. All these asymmetries are abolished by removal of the polar lobe and you get a completely radialized embryo in which the four quadrants are equal.

Weber: Has the ultrastructure of the polar lobe been studied?

Dohmen: Ultrastructurally, the eggs of some species have been analysed. No difference between left and right has been found. There are differences between anterior and posterior.

The polar lobe contains special cytoplasmic components that are genetically important. However, if you centrifuge an uncleaved egg and displace the polar lobe cytoplasm into another region of the egg, development is normal.

The position of the cytoplasm is not crucial; if it is present anywhere in the egg, development is normal.

Kondepudi: Can you tell from the first two cleavages whether the animal will be dextral or sinistral?

Dohmen: After the second cleavage you cannot tell: after the third cleavage you can.

Burn: It was said that the maternal genome was responsible for the direction of coiling of the shell in *Lymnaea*. To be precise, that experiment says that the maternal phenotype is responsible, rather than the genome directly. When we talk about the maternal genome controlling something expressed in the offspring, we talk of the maternal copy of the gene being expressed in preference to the paternal copy; this is caused by genomic imprinting. I think what you are seeing is the phenotype in the sense of the cytoplasm having influence, which is an important distinction.

Dohmen: This product is made during oogenesis and laid down in the oocyte; it is not a result of zygotic transcription.

Wolpert: Embryologists may use 'maternal effect' in a different sense to geneticists. When we talk about maternally acting genes, we mean products that are laid down in the egg, or genes that are acting before the zygote has been formed.

Lewis: The wild-type gene gives right-handed snails. In the absence of this gene function, does the sinistral phenotype occur regularly or at random? Do phenotypically sinistral individuals breed true?

Dohmen: Freeman & Lundelius (1982) said that true-breeding sinistrals can be obtained but many sinistrals produce a small proportion of wild-type (dextral) progeny. When the gene function is absent, only sinistral phenotypes will occur.

Layton: I thought this aberration was not due to a genetic mutation, but rather to some sort of chromosome rearrangement.

Dohmen: That's a complication. According to Freeman, the production of wild-type offspring by purely sinistral parents is due to a crossing-over during meiosis or mitosis in the female germline. The cross-over either reconstitutes the wild-type gene from previously dissociated parts or activates a dextral gene through a position effect.

Lewis: What selection pressure keeps the dextral gene in existence?

Dohmen: One theory is that opposite phenotypes simply cannot mate.

Lewis: But sinistral could mate with sinistral.

Dohmen: I am not an evolutionary zoologist. Freeman suggested that the number of phenotypically sinistral individuals is always limited because of the dextral dominance. In a population, an individual with a left-handed coiled shell has difficulty in finding another left-handed individual with which to mate.

McManus: It is tempting to assume that because they look like mirror images of one another, they are mirror images. Gould et al (1985) analysed the morphology in detail of apparent mirror-image forms of the gastropod, *Cerion*. They found that mirror images are not exact, merely roughly approximate. They therefore argued that the selective advantage comes from the fact that the apparent enantiomorphs have different morphologies.

Dohmen: Similar observations have been made on other species with similar phenomena by Johnson (1982).

Morgan: The interpretation of the cytoplasmic injection experiments in *Lymnaea* was that there are functional gene products for both sinistral and dextral phenotypes, and that dextral is dominant over sinistral. An alternative interpretation is that there is no gene product that actually encodes a sinistral phenotype; rather, sinistrality is the autonomous state and the dextral gene product can interfere with it.

The sinistral phenotype may result from a gene dysfunction; there may be no functional product associated with it. The question is what determines the sinistrality in the first place? Is it the direction of the spindle right from the very first division, as was just suggested, and how can a gene product reverse that direction?

Frankel: Freeman & Lundelius (1982) made that specific point in the discussion of their paper, and argued further that the determinant, whatever it was, was not inherent for dextrality, it was for either one of the two forms of handedness, depending on the species of snail.

Morgan: It could just reverse whatever is around. It doesn't have to know anything about left and right to do that.

Stern: There is a problem with the interpretation Michael Morgan gave. It was mentioned earlier that in sinistral populations there are occasional natural revertants; that doesn't fit, does it?

Jefferies: There are some gastropods, mainly the Opisthobranchia, where the adult shell is dextral but the protoconch is sinistral. And there is one group of primitive pteropods, the Spiratellidae, where the adult shell is sinistral as a neotenous persistence of this condition (Cox 1960).

Yost: Is there any correlation between whole body torsion and asymmetries of the heart and viscera in different organisms? For example, in the chick the body axis always rotates and folds over to one side. In *Xenopus*, there is no correlation between the direction of body torsion and heart looping.

Wolpert: In the chick we have been singularly unsuccessful in getting the heart to go the wrong way. We have tried embryological grafts from left to right and it is very difficult to affect the direction of heart looping. It is easy to make the head go the wrong way, just by a little UV irradiation. It may be mechanical; I don't know anything about the mechanism. You can certainly uncouple the turning of the head from that of the heart.

Brown: They can also be uncoupled in the rat and in the mouse.

References

Cox LR 1960 Gastropoda. General characteristics of Gastropoda. In: Moore RC (ed) Treatise on invertebrate paleontology. Part 1. Mollusca 1. Geological Society of America & University of Kansas Press, Lawrence, KS, p I84–I169

Dohmen MR 1983 The polar lobe in eggs of molluscs and annelids: structure, composition, and function. In: Jeffery WR, Raff RA (eds) Time, space, and pattern in embryonic development. Liss, New York p 197–220

Freeman G, Lundelius JW 1982 The developmental genetics of dextrality and sinistrality in the gastropod *Lymnaea peregra*. Roux's Arch Dev Biol 191:69–83

Gould SJ, Young ND, Kasson B 1985 The consequences of being different: sinistral coiling in *Cerion*. Evolution 39:1364–1379

Johnson MS 1982 Polymorphism for direction of coil in *Partula suturalis*: behavioural isolation and positive frequency dependent selection. Heredity 49:145–151

Handed asymmetry, handedness reversal and mechanisms of cell fate determination in nematode embryos

William B. Wood* and Douglas Kershaw†

*Department of Molecular, Cellular, and Developmental Biology, University of Colorado, Box 347, Boulder CO 80309, USA and †MRC Laboratory of Molecular Biology, Cambridge CB2 2QH, UK

Abstract. Embryos of the nematode *Caenorhabditis elegans* exhibit left–right asymmetry with an invariant handedness. The embryonic cell lineage is asymmetrical: although the animal is generally bilaterally symmetrical with only a few left–right asymmetries, many of its contralaterally analogous cells arise via different lineages on the two sides of the embryo. Larvae and adults also exhibit left–right asymmetries with a handedness that is normally invariant. The frequency of animals with opposite handedness was increased among the progeny of adults exposed to the mutagen ethyl methanesulphonate and among animals that developed from embryos treated in early cleavage with chitinase to destroy the egg shell. Reversal of embryonic handedness was accomplished directly by micromanipulation at the 6-cell stage, resulting in mirror-image but otherwise normal development into healthy, fertile animals with all the usual left–right asymmetries reversed. This demonstrates that (1) the handedness of cell positions in the 6-cell embryo dictates handedness throughout development; (2) at this stage the pair of anterior blastomeres on the right is equivalent to the pair on the left; and (3) the extensive differences in fates of lineally homologous cells on the two sides of the animal must be dictated by cellular interactions, most of which are likely to occur early in embryogenesis and appear to have been conserved in widely diverged nematode species.

1991 Biological asymmetry and handedness. Wiley, Chichester (Ciba Foundation Symposium 162) p 143–164

Many animals with overall bilateral symmetry also exhibit some consistent left–right (*l-r*) asymmetries. Familiar examples in mammals are the asymmetrical placement and morphology of the heart, colon and other visceral organs. The generally invariant handedness of such asymmetries implies that the *l-r* embryonic axis, like the anteroposterior and dorsoventral axes, has a consistent polarity.

143

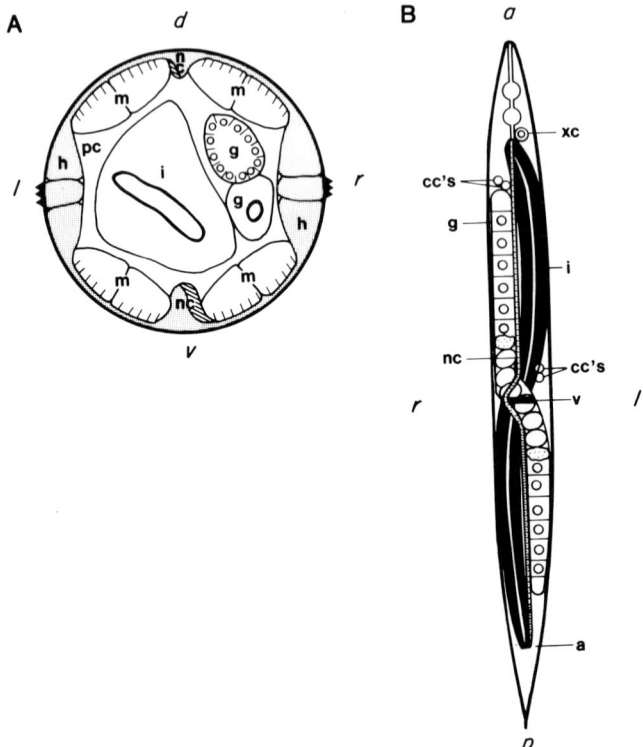

FIG. 1. Bilateral symmetry and left–right (*l-r*) handed asymmetries in the adult
C. elegans hermaphrodite. A) Cross section through the anterior, between pharynx and
uterus, viewed from the posterior. B) Ventral view of interior organs. Anterior, posterior,
dorsal, ventral, left and right are indicated by *a*, *p*, *d*, *v*, *l* and *r*, respectively. Hypodermis
(h) and muscle (m) are arranged symmetrically. The bilobed gonad (g) and the intestine
(i) lie in the pseudocoelom (pc) with the anterior gonad lobe to the right and the posterior
lobe to the left of the intestine, which terminates at the anus (a). The arm of the
gonad distal to the vulva (v), showing syncytial germ cells, is dorsal to the proximal
arm, which in the section shown (A) contains an oocyte nucleus. The excretory cell
nucleus (xc) lies to the left of the midline; the anterior pair of coelomocytes (cc's) is
located on the right and the posterior pair on the left of the pseudocoelomic cavity (B).
The ventral nerve cord (nc) lies on the midline except that it passes around the
vulva (v) on the right side (B). In the ventral nerve cord the cell bodies all lie to
the right of the neuronal processes, which run along the left side of the cord (A; see
also Fig. 5f); the dorsal nerve cord shows the opposite asymmetry. The gonad and intestine
asymmetries can be clearly seen with a dissecting microscope if animals are rolled 90°
from their normal orientation (on one side or the other) to ventral or dorsal side
up (e.g. Fig. 4e); the remaining asymmetries are as observed under a Zeiss compound
microscope equipped with Nomarski optics, in animals mounted ventral side up.
(Based on Sulston & Horvitz 1977, White et al 1986, White 1988 and observations
of the authors.)

This polarity is obvious in some invertebrate embryos, which are markedly asymmetrical from early stages onward, and less evident in vertebrates until later in development. The origin of *l-r* polarity and the mechanism by which it dictates the development of handed asymmetries pose fundamental and intriguing embryological questions, some with clinical significance for humans (reviewed by Brown & Wolpert 1990).

Symmetry, asymmetry and handedness in *Caenorhabditis elegans*

Nematodes, such as the free-living, hermaphroditic soil nematode *C. elegans* are similar to vertebrates in having a generally bilaterally symmetrical body plan with several handed asymmetries. In the adult *C. elegans* hermaphrodite, for example, the bilobed gonad and the intestine are normally arranged within the pseudocoelom such that the anterior gonad lobe lies to the right and the posterior lobe lies to the left of the intestine as shown in Fig. 1. More subtle asymmetries are seen in the placement of a few specific cells, some of which are also shown in the figure. Outside of the gonad and intestine, however, most of the remaining 782 somatic cells in the animal, comprising the nervous system, musculature, hypodermis and a few other tissues, are arranged in a bilaterally symmetrical fashion; that is, functionally analogous cells and associated structures are found in equivalent contralateral positions along the anteroposterior axis of the animal (Sulston & Horvitz 1977, Sulston et al 1983, White et al 1986).

The departures from bilateral symmetry presumably originate in the early embryo, which shows marked handed asymmetry, beginning at the 6-cell stage, and becomes gradually more symmetrical as embryogenesis progresses. Priess & Thomson (1987) showed that *l-r* polarity cannot be fixed until after dorsoventral polarity is established between the 2- and 3-cell stages, since switching the positions of the two AB cells in the 3-cell embryo by micromanipulation reversed the dorsoventral axis but did not affect handedness. In the 4-cell embryo the ABa (anterior), ABp (dorsal), EMS (ventral) and P_2 (posterior) cells lie in a plane, and no *l-r* asymmetry is apparent (Fig. 2A,B). In the next round of cell division, however, skewing of the *l-r* cleavages of ABa and ABp results in positioning of the ABal and ABpl daughters anterior to ABar and ABpr, respectively; ABal is more ventral than ABar (Figs. 2C, 4d). The subsequent, primarily anteroposterior cleavages of the EMS and P_2 cells are also skewed, so that the 8-cell embryo is markedly asymmetrical (Fig. 2D). The eight cells present at this stage give rise to specific tissues and asymmetrically placed cells in the first-stage larva (L1) as shown in Fig. 2E.

After the 8-cell stage, there are substantial differences between the two sides of the embryo. AB descendants on the left continue to be positioned anteriorly to their lineal homologues on the right; E and C descendants lie initially to the left of the midline and MS descendants to the right. Moreover, there are substantial differences in the cell lineages that produce AB progeny cells of similar fates on either side of the animal (contralaterally analogous cells). As shown in Fig. 3, these lineages are particularly asymmetrical for the ABa daughters, which give rise to much of the anterior nervous system as well as muscles of the pharynx. Although ABal and ABar generate similar sets of contralateral analogues, they do so by quite different lineal patterns (Sulston et al 1983, Sulston 1983).

This asymmetry of the AB lineages poses a paradox. Lineal programming by cell-autonomous determinants has traditionally been assumed as the major mechanism of cell fate determination in *C. elegans* embryos, yet many

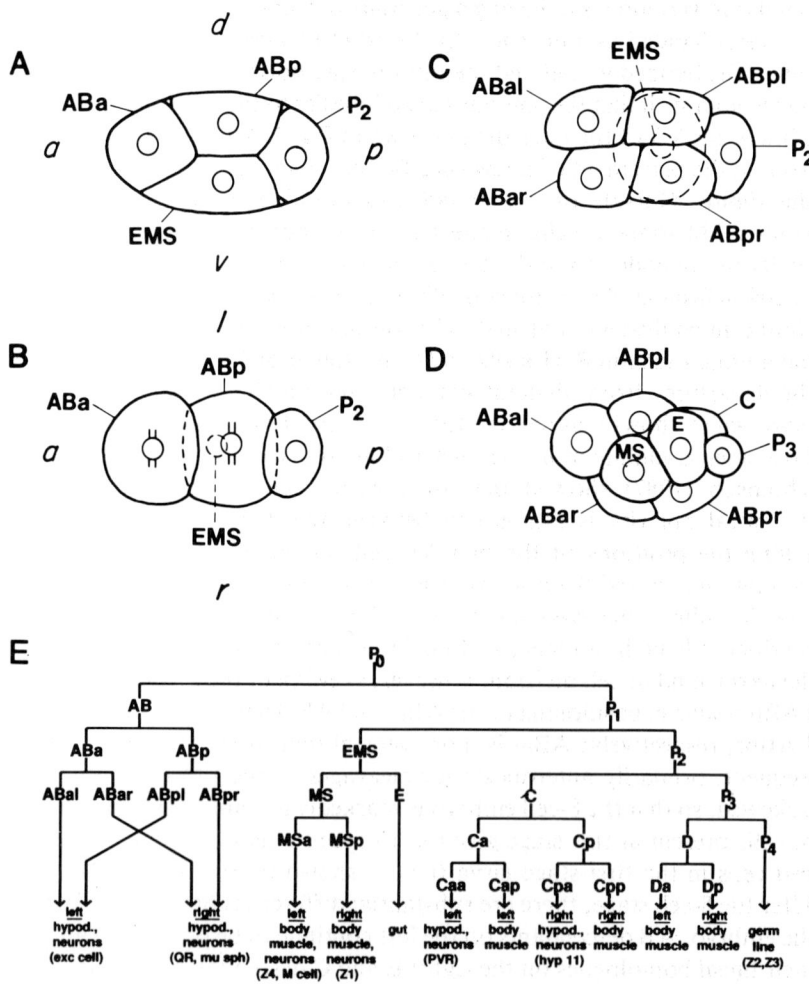

FIG. 2.   (*see opposite*)

contralaterally analogous cells are not lineally homologous. How is bilateral symmetry generated by asymmetrical lineages? Two extreme possibilities can be imagined. Lineal programming could simply have evolved differently for AB descendants receiving 'left' and 'right' determinants, such that similar patterns of cell fates are determined cell-autonomously on the two sides of the animal via different lineal routes. Alternatively, contralateral analogues could be determined not cell-autonomously, but rather by cellular interactions that dictate similar cell fates in contralaterally equivalent positions.

These two possibilities might be distinguishable experimentally if one could reverse the relative positions of the ABal-pl and ABar-pr cell pairs at the 6-cell stage, thereby reversing the normal handedness of the embryo. If the *l-r* differences in cell fates of lineal homologues are programmed cell-autonomously, then grossly abnormal development should result, as the characteristic left and right lineages would be executed to produce their respective patterns of cell fates in the wrong relative anteroposterior positions. However, if the *l-r* differences are dictated by cellular interactions, then a mirror-image but otherwise normal lineage might result, possibly giving rise to a functional animal with reversed handedness of all *l-r* asymmetries. This paper describes the results of experiments to test these predictions, some of which have been reported previously (Wood 1991).

Invariance of handedness in normal development

The handed asymmetry most easily scored in *C. elegans* is position of the gonad in adult hermaphrodites (Fig. 1). Because the animals always lie or move on one side or the other on a solid surface such as an agar plate, presenting a lateral aspect to the observer, this asymmetry is not apparent. However, rolling the animal with a platinum worm pick to either dorsal or ventral side up allows

---

FIG. 2.   Cell orientations in 4-cell to 8-cell embryos. A) 4-cell embryo, left lateral view [see (E) below for explanation of cell nomenclature]. B) 4-cell embryo, ventral view focused on lower plane. Mitotic spindles in ABa and ABp cells are forming perpendicular to the anteroposterior axis and will become skewed in a counterclockwise direction as cleavage proceeds. Dashed lines show positions of EMS cell and its nucleus in upper focal plane. C) 6-cell embryo, ventral view focused on lower plane. AB cells are arranged asymmetrically as a result of the normal skewing with ABal and pl daughters somewhat anterior to ABar and pr, respectively. ABal is also somewhat ventral to ABar. D) 8-cell embryo, ventral view. EMS and P_2 have divided, also asymmetrically, so that MS and P_3 are to the right and E and C to the left of the midline. E) Lineage diagram of early cleavages showing origins of the six founder cells AB, MS, E, C, D and P_4 and major developmental fates of their progeny. Relative timing of divisions, on the vertical axis, are not accurately depicted. Progeny of founder cells are named according to their positions following cleavage; for example, ABa is the anterior daughter of AB; ABal is the left daughter of ABa, and so on.

anterior

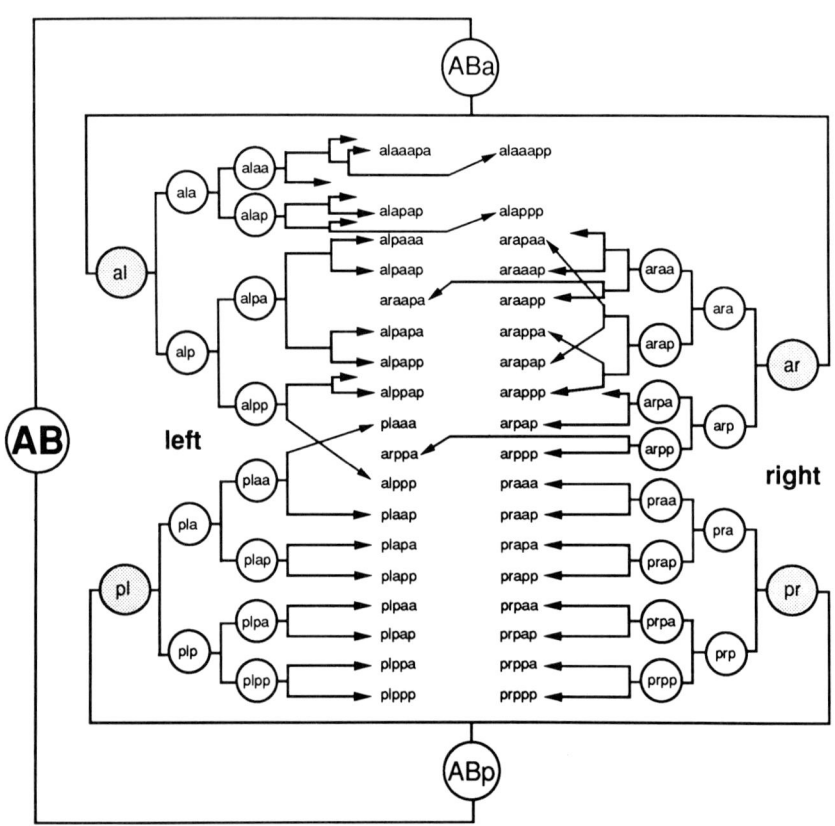

posterior

FIG. 3. Lineal relationships among 18 pairs of contralaterally analogous cells in the AB lineage of *C. elegans*. Pairs of cells shown opposite each other in the centre of the diagram are analogous in both position and fate. The members of each pair occupy approximately equivalent positions on the left and right sides of the embryo at the time they are born (32-AB-cell to 128-AB-cell stages); they give rise to nearly or completely identical numbers and types of progeny cells during subsequent development. Their ancestors, back to the 4-AB-cell stage, are represented in their approximate relative anteroposterior positions, with cells on the left more anterior than their lineal homologues on the right. Analogues among ABp descendants in the seven most posterior pairs are also lineal homologues, whereas the remaining more anterior analogues have different lineal ancestries on the two sides of the embryo (as can be seen from their cell names as well as from the lineage diagram). Four pairs of analogues are generated by divisions that cross the midline. (Based on Sulston et al 1983, Sulston 1983.)

convenient scoring of gonad handedness under a dissecting microscope (Fig. 4e). Examination of more than 2500 adult hermaphrodites (var. Bristol, N2 strain, cultivated at 16°C; Brenner 1974) revealed only one animal with a partially reversed and somewhat abnormal appearing gonad; the rest all showed the anterior-right, posterior-left gonad placement diagrammed in Fig. 1.

The other asymmetries shown in Fig. 1 must be scored in animals mounted for observation at higher magnification with Nomarski optics. As controls for this study, several normal adult hermaphrodites were examined. All exhibited the asymmetries shown in Fig. 1 and described in the figure legend. Embryos with reversed handedness have not been reported and, to the authors' knowledge, have never been observed during extensive studies of *C. elegans* embryogenesis over the past 20 years in several laboratories (W. B. Wood, unpublished observations; J. Sulston, E. Schierenberg, personal communications 1990). J. Sulston (personal communication 1990) encountered only two animals with at least some asymmetry reversals in large-scale screening of fixed adults. It can be concluded that the handedness of the asymmetries in normal development scored here is essentially invariant for *C. elegans*, although this is not true for some other nematodes (see below).

Treatments leading to increased frequency of handedness reversal

The frequency of gonad handedness reversal was increased to about 0.2% among the F3 progeny of animals that had been treated with the mutagen ethyl methanesulphonate in an attempt to isolate mutations affecting asymmetry. The reversed animals could be grouped into two classes: (1) no reversal of most other asymmetries, and (2) reversal of all asymmetries scored. With the exception of three mutant lines that gave a heritable low frequency (10–20%) of class 1 animals (to be described elsewhere), the reversed animals of both classes produced no reversed progeny, as if they had resulted from developmental abnormalities not caused by mutation. Nevertheless, the finding of apparently healthy fertile animals of class 2 indicated that complete asymmetry reversal could occur without adverse developmental effects.

The frequency of handedness reversal was also increased by treatment of early embryos with chitinase and chymotrypsin (Edgar & McGhee 1988) to remove the egg shell that normally gives the embryo its oblate shape and appears to constrain cell positions during the early cleavages. Chitinase-treated embryos become spherical and the cellular configuration at the early 4-cell stage is more T-shaped than rhomboid, quite similar to that seen normally in embryos of the parasitic nematode *Ascaris* (zur Strassen 1896). Some chitinase-treated embryos appear highly abnormal after a few cleavages and do not survive, but if the vitelline membrane that normally surrounds the embryo inside the shell remains intact, then about 60% proceed through embryogenesis, hatching and larval development to produce healthy, fertile adults. Among 91 animals developing from chitinase-treated early embryos scored in several experiments, five showed reversed gonad handedness; all of these when examined by Nomarski microscopy were of class 2, with reversal of all asymmetries scored. This observation suggests that relaxation of spatial constraints in the embryo occasionally permits early asymmetry reversal, leading to reversed development.

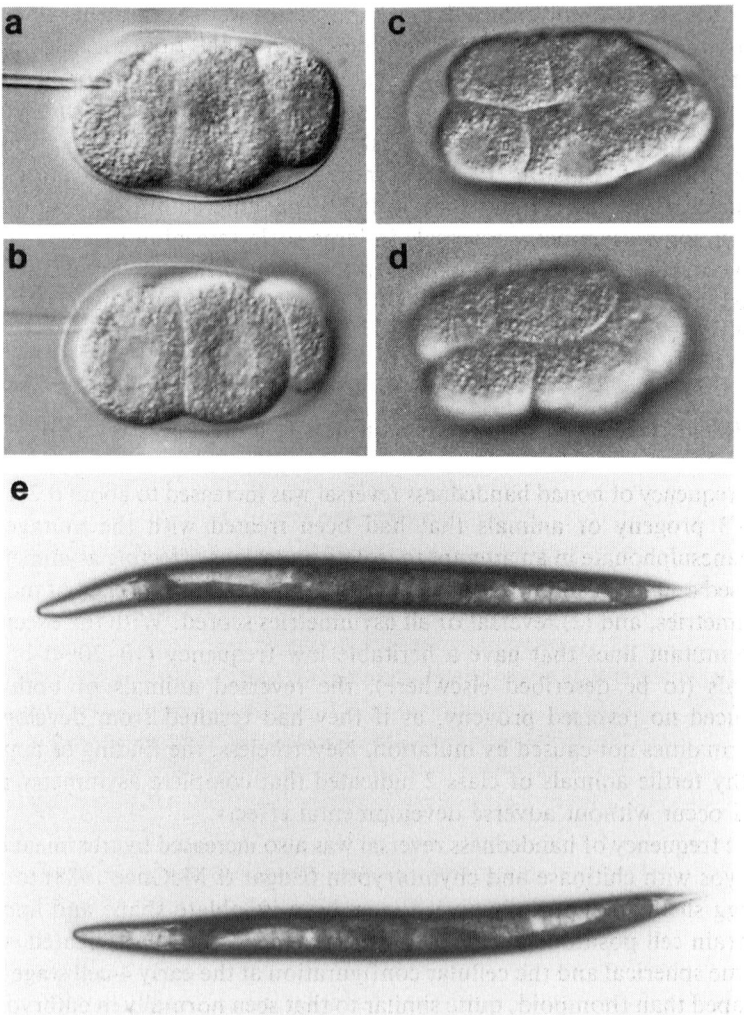

Reversal of embryonic handedness by micromanipulation

Priess & Thomson (1987) showed that the positions of ABa and ABp could be reversed in the *C. elegans* embryo by applying pressure on one pole of the AB spindle with a microneedle during cytokinesis of the AB cell division between the 2- and 3-cell stages. Essentially the same technique was used to accomplish handedness reversal of otherwise untreated N2 strain embryos, by rolling them to ventral side up at the 4-cell stage (as in Fig. 2B), then exerting downward and rearward pressure on the left ventral surface of the ABa cell during spindle formation and subsequent cleavage of ABa and ABp (Fig. 4a,b). When successful, this procedure reversed the normal skewing of both AB spindles and resulted in ABar and ABpr being anterior to ABal and ABpl, respectively, at completion of the cleavage, producing a reversed 6-cell embryo (compare Fig. 4, c and d). The subsequent divisions of EMS and P_2 (see Fig. 2D) gave a mirror-image 8-cell embryo (not shown) in which the normal asymmetrical positions of non-AB cells were also reversed, with MS to the left and C to the right of the midline. Reversed embryos underwent the next few cleavages on schedule and hatched as viable larvae that grew into fertile adults (Fig. 4e).

FIG. 4.   Handedness reversal: operation and results in embryos and adults. (a) Nomarski photomicrograph of 4-cell embryo, ventral side up as in Fig. 2B, with microneedle in position for operation (see below). (b) Same embryo as in (a), lower focal plane showing ABa nucleus at onset of spindle formation. (c) Reversed 6-cell embryo following successful operation, focused near lower surface to show cleavage planes optimally. (d) Normal 6-cell embryo, similar stage and focal plane to (c). (e) The reversed (above) and normal (below) adult hermaphrodites that developed from the operated and control embryos shown in (c) and (d), respectively, viewed ventral side up. Embryonic handedness reversal, judged by configuration at the 6-cell stage (c,d), was achieved with a success rate of about 25%, probably depending most critically on positioning of the microneedle, timing and orientation of the embryo. All reversed embryos (6/6) developed into reversed but otherwise normal, fertile adults. About a dozen unsuccessfully operated embryos, which appeared as in (d) at the 6-cell stage, were incubated as controls; all developed into fertile adults with normal handedness. *C. elegans* was cultivated and embryos obtained by standard procedures (Sulston & Hodgkin 1988). Microneedles were prepared and embryos mounted for micromanipulation according to Priess & Thomson (1987). Embryos were rolled to ventral side up at the 4-cell stage, then downward and rearward pressure was exerted on the left ventral surface of the ABa cell as in (a), during spindle formation and subsequent cleavage of ABa and ABp. The two adult hermaphrodites (e) were lightly anaesthetized in 0.7% phenoxypropanol (Sulston & Hodgkin 1988), rolled to ventral side up on an agar surface, and photographed together using a Wild M400 microscope. Embryos are about 70 μm long, adult worms about 1 mm.

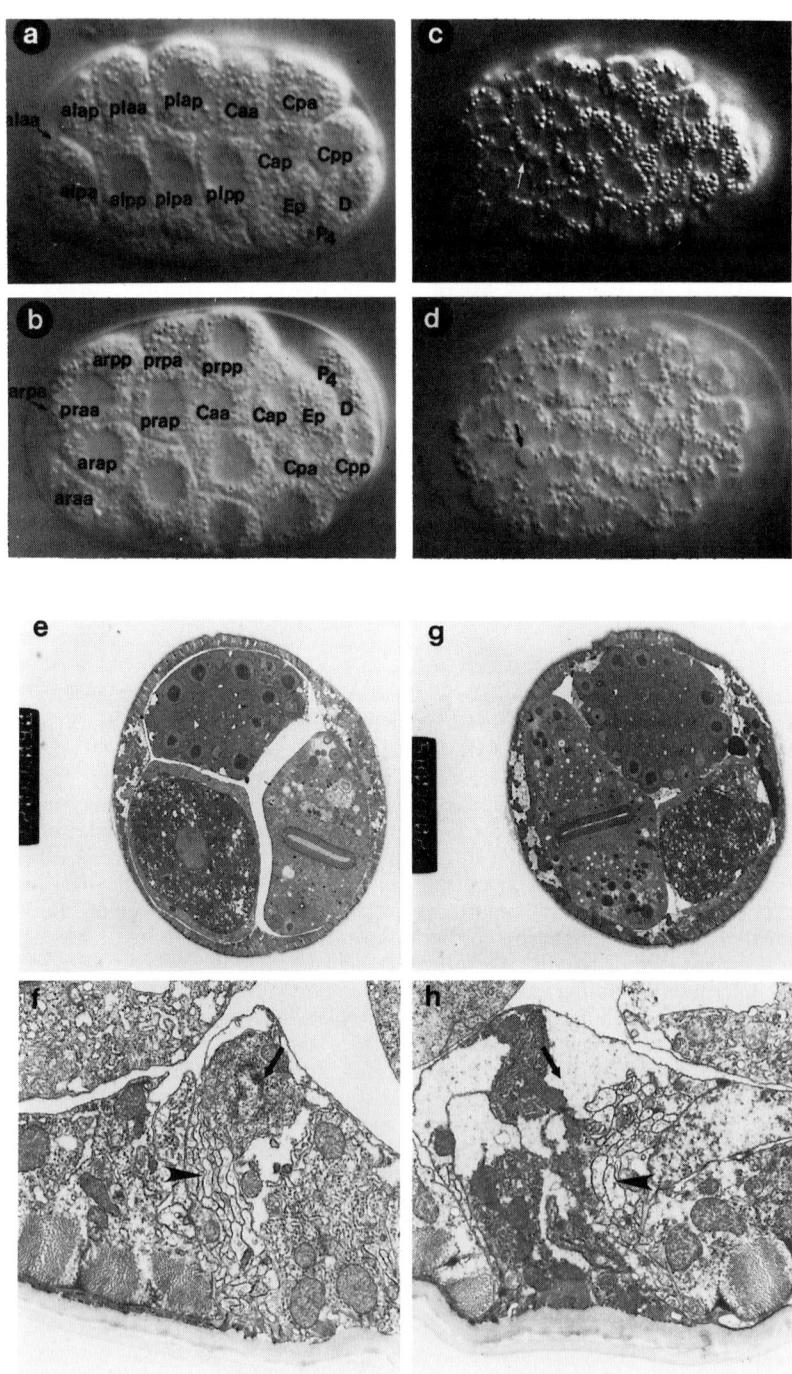

The complete embryonic development of a reversed embryo was observed and recorded for subsequent lineage determination using a programmable video disk system (J. White, personal communication 1990; see legend to Fig. 5). The timing of divisions and positioning of specific cells in the reversed embryo were consistent with an enantiomorphic but otherwise normal pattern of development, as illustrated by examples from two stages in Fig. 5. For example, at the 28-cell stage the MS descendants, normally on the right (Panel a), were instead located on the left side of the reversed embryo (Panel b); likewise the C descendants, normally on the right, were located on the left. MSa and Ca descendants,

---

FIG. 5. Comparisons of normal and reversed embryos and adults. Panels (a)–(d) are Nomarski photomicrographs of normal and reversed embryos at two stages, with selected cells labelled. (a,b) Normal 28-cell embryo, superficial left lateral view, and reversed 28-cell embryo, superficial right lateral view, respectively. Labelled cells with no founder cell designation are AB descendants. C cell progeny, normally on the left side as in (a), are on the right side in (b); conversely, MS cell progeny, that would normally be seen on the right side are not visible (in a lower focal plane) in either (a) or (b). (c,d) Superficial dorsal views of normal and reversed embryos, respectively, at about the 360-cell stage. White arrow on the normal embryo points to ABarpaaapp undergoing programmed cell death (reproduced, with permission, from Sulston et al 1983); black arrow on the reversed embryo points to the analogous dying cell, which was shown by lineage determination to be ABalpaaapp (see text). Nuclei scored for position in the L1 larva that hatched from the reversed embryo (see text) were as follows. AB descendants: QR (previously known as Q2; Sulston & Horvitz 1977), the excretory cell, and the rectal sphincter muscle cell mu sph. MS descendants: Z1 and Z4 in the gonad primordium, the mesoblast M, and the four coelomocytes. C descendants: hyp11 and PVR in the tail. P_4 descendants: Z2 and Z3 in the gonad primordium (see Fig. 2E for cell ancestries). Embryos were mounted as described by Sulston & Hodgkin (1988) and their development was recorded (see text) using the '4-D' video recording system developed by J. White and M. Thompson, which consists of a Nomarski microscope (Zeiss Axioplan) equipped with a video camera and connected to a computer-controlled optical disk recorder (Sony) and focusing drive motor (Biorad). The system automatically records complete sets of serial optical sections at predetermined intervals and allows subsequent playback of consecutive images from any desired section to facilitate lineage mapping. Photographs were taken from the video monitor. Panels (e)–(h) are electron micrographs of transverse sections through the posterior portions of adult normal (e,f) and reversed (g,h) hermaphrodites, viewed from the posterior. Panels (e) and (g) are relatively low magnification images showing that the two animals have mirror-image arrangements of internal organs (identifiable by comparison to Fig. 1A, which, however, represents an anterior section). Panels (f) and (h) are higher magnification views of the same two sections, showing mirror-image arrangement of cell body region (arrowheads) and processes (arrows) in the ventral nerve cord. To obtain comparable sections, one reversed and two normal animals were fixed for electron microscopy (Sulston & Hodgkin 1988), arranged side-by-side in the same orientation, and embedded in the same block, from which 80 nm serial sections were cut, stained and photographed in a Phillips CM10 electron microscope. The reversed animal was identified by comparison at low magnification with the two normal animals in each section.

normally to the left of MSp and Cp descendants, respectively, were instead to the right of these cells, and so on. Embryos at about the 360-cell stage are shown in Panels (c) and (d) in superficial dorsal views. Panel (c) shows the programmed death of ABarpaaapp (white arrow) in a normal embryo. Panel (d) shows a similar view of the reversed embryo; the analogous dying cell (black arrow) was identified by lineage mapping as ABalpaaapp, which in a normal embryo would give rise to a pharyngeal neuron (Sulston et al 1983).

In the L1 larva that hatched from the reversed embryo, 14 normally asymmetrically placed nuclei (Sulston & Horvitz 1977) that represented descendants of founder cells AB, MS, C and P_4 (see legend, Fig. 5) were scored. All were found in their usual anteroposterior positions, but on the opposite side of the animal than normally (not shown). The adults that developed from reversed embryos, when examined by Nomarski (not shown) and electron microscopy (Fig. 5,e–h), exhibited reversed handedness for all the cells and tissues shown in Fig. 1. These animals were healthy, moved normally and produced normal numbers of self progeny, all of which showed normal *l-r* asymmetry.

Conclusions and discussion

These experiments show that reversing the relative positions of the ABal-pl and ABar-pr cell pairs by micromanipulation results in a mirror-image but otherwise normal process of development. Four significant conclusions can be drawn from our results. First, not surprisingly, handedness in *C. elegans* must be determined genetically, as in the snail *Lymnaea peregra* (van den Biggelaar, this volume), rather than epigenetically, as in protozoa (Frankel, this volume), since artificially reversed animals do not produce reversed progeny. Second, handedness of the embryo at the 6-cell stage determines the handedness of subsequent asymmetries throughout development and in the adult. Third, at this stage ABal must be developmentally equivalent to ABar, and ABpl to ABpr. The differences in the fates of these pairs of cells cannot be attributed to intrinsic 'left' and 'right' determinants segregated to left and right daughter cells, respectively, in the ABa and ABp cleavages. Fourth, therefore, the extensive differences in fates of AB-derived lineal homologues on the two sides of the embryo (Fig. 3) must be determined by cellular interactions, which differ on the left and right because of the asymmetrical positioning of these cells relative to each other and to other cells in the embryo.

Whether the last conclusion also applies to the smaller number of *l-r* differences in fates of non-AB lineages, for example those giving rise to the asymmetrically placed MS and C derivatives in the L1 described above and shown in Fig. 2E, cannot be determined from our results. This is because MSa and Ca probably give rise to the same cells in reversed embryos (not all lineages were followed) as in normal embryos, although in mirror-image *l-r* positions;

for example, MSa in reversed embryos gives rise to *right* body muscle and neurons (opposite to normal embryos), and also to Z4 and M [as in normal embryos but on the opposite (right) side; see Fig. 2E]. Similar statements apply to MSp and Cp. Therefore, the reversal experiments cannot distinguish whether the *l-r* differences in these lineages are determined by lineal programming or cellular interactions.

The determinative interactions in the AB lineage probably take place early in embryogenesis. The contacts of AB homologues at the 6-cell stage are still equivalent (Fig. 2C), but by the 8-cell stage they are different on the two sides of the embryo (Fig. 2D). In later embryos of 51 cells and more, the laser ablation experiments of Sulston et al (1983) provided convincing evidence for cell autonomy of many AB-cell fates. Therefore, most of the determinative interactions are likely to occur between the 8-cell and 51-cell stages.

The conclusions listed above relate to a large body of previous classical as well as more recent work on nematode embryogenesis. The conclusion that reversed embryos develop into reversed animals confirms a conjecture put forward nearly a century ago by zur Strassen (1896), who analysed embryonic development in *Ascaris*, a large nematode parasite of mammals. Using fixed preparations of *A. megalocephala* embryos, he was able to describe the cell lineage up to 182 cells. The early embryos are also *l-r* asymmetrical, though less markedly so than those of *C. elegans*. After the *Ascaris* 4-cell stage resolves from the 'T' form mentioned above to the 'rhomboid' configuration (as in Fig. 2A), the patterning of divisions and cell placement in the two species is extremely similar. In *Ascaris*, however, zur Strassen (1896) observed that about 2.5% of early embryos had reversed *l-r* asymmetry; he postulated that these 'inverse' embryos, as he called them, would give rise to inverse adults. Unable to confirm this directly, since *Ascaris* is an obligate parasite, he noted that the frequency of adults with reversed asymmetry of the excretory system was also approximately 2.5%.

In later studies, zur Strassen (1951, 1959) examined embryos from several phylogenetically diverse free-living and parasitic nematodes and found characteristic but very different frequencies of inverse embryos, ranging from <0.1% for some species to 50% for the dung beetle parasite *Bradynema rigidum*, with several species showing intermediate frequencies. However, he found the same handedness preference for all species showing a bias: that is, early AB derivatives on the left are always anterior to their lineal homologues on the right, as in *C. elegans*. Thus many, if not all, nematodes have the same apparent embryonic *l-r* polarity, resulting in the same handedness bias, which is more stringently reinforced in some species than in others. The results described above with chitinase-treated *C. elegans* embryos suggest that constraints imposed by the egg shell might contribute to the interspecific differences in the frequency of spontaneous handedness reversal.

The conclusion that Abal-pl and ABar-pr cell pairs are equivalent relates to an old controversy (see Ludwig 1932 for review) between, on the one hand, zur Strassen and his students, for example Dunschen (1929), and, on the other, Schleip (1929) and his students, for example Bonfig (1925). The zur Strassen school maintained that *Ascaris* embryonic development was strictly mosaic in character and that asymmetry and *l-r* differences, like all other differences in cell fates, must be dictated by determinants present in the zygote. Schleip (1929) defended the alternative possibility that the early AB cells are equivalent, becoming different only later as a result of cellular interactions. Bonfig (1925) had reported, for example, that the entire quadrant of AB cells at the 6-cell stage could occasionally spontaneously rotate 90° (reversing handedness) or even 180° (not reversing handedness) without impairing viability of the embryo. The results presented here add to previous evidence from Priess & Thomson (1987) in clearly supporting the view of Schleip and Bonfig.

However, the handedness bias at the 6-cell stage in both *Ascaris* and *C. elegans* suggests that there is either a pre-existing *l-r* polarity in the apparently symmetrical 4-cell embryo (but not earlier for *C. elegans* as shown by Priess & Thomson 1987) or some intrinsic property of the ABa and ABp spindles that causes them to skew during cleavage in the characteristic counterclockwise direction as viewed from the ventral side. The basis for this bias remains unclear. It may be similar to the handedness bias in *Lymnaea*, in which the function of a recessive gene called *sinistral* is required maternally for the normal right-handed coiling of the shell during development (Sturtevant 1923). Absence of this function causes a change in spindle orientation at the second embryonic cleavage, resulting in left-handed coiling (Crampton 1894, van den Biggelaar, this volume). Freeman & Lundelius (1982) showed that injection of wild-type ooplasm into zygotes from homozygous *sinistral* animals could induce the normal right-handed coiling, but the nature and action of the active factor were not investigated.

The final conclusion, based on our evidence that the extensive embryonic *l-r* differences in fates of lineally homologous cells must be dictated by cellular interactions, represents a fundamental change from previous assumptions about *C. elegans* early embryogenesis. Nematode embryos have been viewed as classic examples of 'mosaic' development, in which most, if not all, cell fates were thought to be dictated by autonomous determinants, distributed by unknown mechanisms to appropriate cells via the invariant cell lineage. Some lineages in both early (Laufer et al 1980, Wood et al 1984, Cowan & McIntosh 1985, Edgar & McGhee 1988) and later (Sulston et al 1983) embryos do appear to be autonomously determined in this manner. However, it is becoming clear that the fates of many cells in embryonic as well as larval development (reviewed by Greenwald 1989, Han & Sternberg 1991) are determined by cellular interactions, and that *C. elegans* embryos in this regard are much more similar to those of vertebrates than previously assumed. Earlier evidence on this point came from Schierenberg (1987) and in particular from the experiments of Priess

& Thomson (1987) referred to above, which demonstrated equivalence of ABa and ABp at the 4-cell stage and implicated interactions of AB descendants with EMS descendants in the induction of AB-derived pharyngeal muscle cells. Further evidence was provided by the observation (Priess et al 1987) that mutations in the *glp-1* gene, known to code for a transmembrane protein with homologies to epidermal growth factor and other components of intercellular signalling pathways in other organisms (Yochem & Greenwald 1989), caused defects in AB-derived pharyngeal muscle induction. The results reported here indicate that extensive cellular interactions must be required for determination of the many AB-derived lineal homologues that show *l-r* differences in developmental fate.

Similar cell interactions are probably required for embryogenesis in *Ascaris* and other nematodes. Remarkably, comparison of the embryonic lineages for *Ascaris* (zur Strassen 1896) and *C. elegans* (Sulston et al 1983) shows that the *l-r* differences in lineal ancestries of cells in contralaterally analogous positions in Ascaris are *identical* to the *l-r* differences in lineal ancestries of cells with contralaterally analogous fates shown in Fig. 3 for *C. elegans* (R. P. Woods, personal communication 1990). Since *Ascaris* and *C. elegans* are only distantly related, belonging to different orders of the class *Secernentea*, this finding suggests that both the pattern of cell divisions and the cellular interactions responsible for *l-r* differences have been conserved in the evolution of nematode development.

Acknowledgements

W.B.W. is grateful for discussion, helpful suggestions and hospitality to J. White, J. Sulston, J. Rothman, A. Chisolm and others in the Cell Biology Unit of the MRC Laboratory of Molecular Biology, Cambridge, UK, where most of this work was done during a sabbatical leave from University of Colorado, Boulder. He also thanks J. White for use of the '4-D' video recording system, J. Rothman for providing optical disk recordings of normal embryonic development, and R. P. Woods for communicating his unpublished comparative analysis of *Ascaris* and *C. elegans* lineages. This research was supported by funds from the Medical Research Council in the UK and by grants from the National Institutes of Health (HD-11762 and HD14958) in the US.

References

Bonfig R 1925 Die Determination der Hauptrichtungen des Embryo von *Ascaris megalocephala*. Z Wiss Zool 124:407–456

Brenner S 1974 The genetics of *Caenorhabditis elegans*. Genetics 77:71–94

Brown NA, Wolpert L 1990 The development of handedness in left/right asymmetry. Development 109:1–9

Cowan AE, McIntosh JR 1985 Mapping the distribution of differentiation potential for intestine, muscle, and hypodermis during early development in *Caenorhabditis elegans*. Cell 41:923–932

Crampton HE 1894 Reversal of cleavage in a sinistral gastropod. Ann NY Acad Sci 8:167–170

Dunschen F 1929 Inversentwicklung und Mosaikfrage bei *Ascaris megalocephala*. Arch Entwicklungsmech Org (Wilhelm Roux) 115:237–335

Edgar LG, McGhee JD 1988 DNA synthesis and the control of embryonic gene expression in *C. elegans*. Cell 53:589–599

Frankel J 1991 Intracellular handedness in ciliates. In: Biological asymmetry and handedness. Wiley, Chichester (Ciba Found Symp 162) p 73–93

Freeman G, Lundelius JW 1982 The developmental genetics of dextrality and sinistrality in the gastropod *Lymnaea peregra*. Wilhelm Roux's Arch Entwicklungsmech Org 191:69–83

Laufer JS, Bazzicalupo P, Wood WB 1980 Segregation of developmental potential in early embryos of *Caenorhabditis elegans*. Cell 19:569–577

Ludwig W 1932 Das rechts-links Problem im Tierreich und beim Menschen. Springer Verlag, Berlin

Priess J, Thomson JN 1987 Cellular interactions in early *Caenorhabditis elegans* embryos. Cell 48:241–250

Greenwald I 1989 Cell–cell interactions that specify certain cell fates in *C. elegans* development. Trends Genet 5:237–241

Han M, Sternberg PW 1991 Pattern formation in *C. elegans*. JAI Press, Greenwich CT, USA. Adv Dev Biol, in press

Priess J, Schnabel H, Schnabel R 1987 The *glp-1* locus and cellular interactions in early *Caenorhabditis elegans* embryos. Cell 51:601–611

Schierenberg E 1987 Reversal of cellular polarity and early cell–cell interactions in the embryo of *Caenorhabditis elegans*. Dev Biol 122:452–463

Schleip W 1929 Die Determination der Primitiventwicklung. Akademische Verlagsanstalt, Leipzig

Sturtevant AH 1923 Inheritance of direction of coiling in Lymnaea. Science (Wash DC) 58:269–270

Sulston JE 1983 Neuronal cell lineages in the nematode *C. elegans*. Cold Spring Harbor Symp Quant Biol 48:443–452

Sulston JE, Horvitz HR 1977 Post-embryonic cell lineages of the nematode *Caenorhabditis elegans*. Dev Biol 56:110–156

Sulston J, Hodgkin J 1988 Methods. In: Wood WB (ed) The nematode *Caenorhabditis elegans*. Cold Spring Harbor Laboratory Press, Cold Spring Harbor, NY p 587–606

Sulston JE, Schierenberg E, White JG, Thomson JN 1983 The embryonic cell lineage of the nematode *Caenorhabditis elegans*. Dev Biol 100:64–119

van den Biggelaar JAM 1991 Asymmetries during molluscan embryogenesis. In: Biological asymmetry and handedness. Wiley, Chichester (Ciba Found Symp 162) p 128–142

White J 1988 The anatomy. In: Wood WB (ed) The nematode *Caenorhabditis elegans*. Cold Spring Harbor Laboratory Press, Cold Spring Harbor, NY p 81–122

White JG, Southgate E, Thomson JN, Brenner S 1986 The structure of the nervous system of *Caenorhabditis elegans*. Philos Trans R Soc Lond B Biol Sci 314:1–340

Wood WB 1991 Evidence from reversal of handedness in *C. elegans* embryos for early cell interactions determining cell fates. Nature (Lond) 349:536–538

Wood WB, Schierenberg E, Strome S 1984 Localization and determination in early embryos of *Caenorhabditis elegans*. In: Davidson EH, Firtel R (eds) Molecular biology of development. Alan R. Liss, New York, p 37–49

Yochem J, Greenwald I 1989 *glp-1* and *lin-12*, genes implicated in distinct cell–cell interactions in *C. elegans* encode similar transmembrane proteins. Cell 58:553–563 zur Strassen O 1896 Embryonalentwicklung der *Ascaris megalocephala*. Arch Entwicklungsmech Org (Wilhelm Roux) 3:27–105

zur Strassen O 1951 Der Erbgang der Nematoden-Asymmetrie. Verh Dtsch Zool Ges p 77–81

zur Strassen O 1959 Neue Beiträge zur Entwicklungsmechanik der Nematoden. Zoologica (Stuttgart) 107:1–142

DISCUSSION

Galloway: Is it true that different adults of *C. elegans* all have the same number of cells?

Wood: Somatic cells, yes: 959 for the hermaphrodite, 1031 for the male. The number of cells in the germline is indeterminate.

Peters: There is a relatively high number of nerve cells in *C. elegans*, about 300. What does the brain of *C. elegans* look like? Is there a distinct structure or is it as in *Aplysia*, where ganglia are found in various places?

Wood: There is a major area of neuropil called the nerve ring, which encircles the base of the pharynx and includes the ring ganglion and the anterior ganglion containing over half the animal's 302 neurons. There are also several smaller ganglia located laterally and in the tail.

Peters: Is cell death a factor in the final sorting out of cell destinies, or do lineages just come to a stop?

Wood: Cell death is often involved in the lineages that give rise to nerve cells.

Lawrence: Perhaps the dogma that lineage is the predominant determinant of cell fate in the nematode should be consigned to the dustbin of history? Your evidence that equivalent cells on the left and right have different lineages is a powerful argument against the idea that precise cell fate depends on ancestry of non-interacting groups in the nematode. That idea has died in most other systems.

Wood: My results provide strong new evidence against lineage being the major determinant of cell fate in the early embryo, but it is not the first such evidence. Bonfig's observations in 1925 were similarly interpreted, though not widely accepted at the time. Priess & Thomson's experiment (1987) and other work indicated that there must be some cell interactions in the early *C. elegans* embryo; my results show that there must be many more. On the other hand, some lineages do seem to be determined by lineally inherited determinants that segregate: the germline is probably a cell-autonomously determined lineage, the gut lineage is probably another. The embryo clearly uses both mechanisms. The idea we should discard is the notion that nematodes are very different from other embryos in this respect. Localization phenomena and cell interactions are both going to be important.

McManus: What happens if one of these cells gets damaged very early on, let's say at the 4-cell stage? Does the animal die or does an adult develop that is missing parts or what?

Wood: The simple answer is that such embryos die about half way through embryogenesis and the cells that would normally come from the damaged blastomere are missing. But more interesting and detailed information on this is just becoming available. When Sulston and co-workers tried such experiments in their 1983 study of lineage, they found the results difficult to interpret because the laser they used destroyed the embryos when they irradiated the large cells at early stages enough to kill them. Now, however, there are lasers that can be more finely controlled. Ralph Schnabel (1991) has shown that if he ablates certain cells at the 4-cell or 8-cell stages, there are effects on other lineages, as you would expect from the results I have presented.

Wolpert: Could you speculate on the mechanism of the handedness bias in nematode embryos?

Wood: Let me first speculate about what the egg shell is doing. Nobody has done any genetics, as far as I know, on any of the worms that zur Strassen looked at that show intermediate levels of spontaneous reversal (1951, 1959), so we don't know whether that's a genetic or an epigenetic phenomenon. The experiments involving chitinase treatment that I described would support the view that there is a similar bias in all nematode embryos at the second AB cell cleavage, and that it is more or less rigorously enforced, depending on the constraints of the egg shell. In *C. elegans* the egg shell is very tight and this cleavage always skews the same way. If you loosen the constraints with chitinase treatment, you get a few reversed animals. *Ascaris* seems to have a looser egg shell and 2.5% of the embryos reverse spontaneously. So, the extent of the bias may be determined by the egg shell.

That doesn't address where the bias comes from in the first place. I don't know the answer. We are going to keep looking for mutants; the screen I described was not exhaustive. There is no obvious left–right asymmetry before the AB cleavage at the 4-cell stage but there may be a subtle one and we are going to look for that. Mounted embryos, as we usually observe them, lie on their side so minor left–right differences would be difficult to see. We need to look at embryos mounted dorsal or ventral side up, which is more difficult. We will also try reconstructing such views from serial sections rotated on a computer.

Wolpert: But it can't have anything to do with the shell—Priess' experiments showed that if one inverted the dorsoventral axis, development was still normal.

Wood: That's right, there can't be a pre-existing bias in the form of a fixed polarity along the future left–right axis, but there could be a pre-existing left–right distinction based on the handedness of some internal component, like the spindle. The centrioles probably give the spindles an invariant handedness, as seen in other organisms; in the AB cells this might somehow lead to the skewing that we normally see. I was able to override that mechanically in *C. elegans*, but otherwise it gives the normal handedness at the 6-cell stage.

Lewis: Joe Frankel described how ciliates can reverse their global structure, just as your embryos do, without reversing the handedness of their individual cilia. This has bad consequences for the ciliate because it upsets the relationship between the direction of the feeding currents and the position of the mouth (Frankel, this volume). In *C. elegans* could there be a selection pressure operating in favour of a particular handedness, on the basis of that same set of principles? In other words, does *C. elegans* depend on ciliary beating?

Wood: No, there are no cilia in *C. elegans*, only ciliary microtubules in a few sensory cells. The reversed animals that have been made artificially move normally, look perfectly healthy, give the normal number of progeny and so on. They have no clear disadvantage as seen in ciliates.

Jefferies: Spiral cleavage is probably as old as the bilaterians, but the sinistrality of the shell can't be older than the gastropods, which are a group within the Bilateria. The handedness of the shell is determined by something that is evolutionarily much older, which must have existed before shells did. The same is true for these nematodes. It is as if new asymmetries are determined by an old mechanism.

Frankel: In this early developmental event that determines handedness is there something analogous to the *lin-12*-dependent interaction between cells to determine alternative fates (Greenwald et al 1983)? There, one cell becomes A and the other becomes B. In lack-of-function mutants of *lin-12* both cells become A; in gain-of-function mutants they both become B. Could the situation with handedness be analogous, so the embryo could develop either two left halves or two right halves and become mirror symmetrical?

Wood: I doubt it; mostly because of the constraints of the egg shell. When the AB cells undergo left–right division at the 4-cell stage, they simply can't fit inside the egg shell unless they skew one way or the other. Once they have skewed, the contacts on the two sides are different.

Frankel: But if you remove the egg shell and leave only the vitelline membrane, is handedness less rigidly determined?

Wood: Yes, however, when I was able to watch the second AB cell division after removal of the egg shell, the ABa and ABp spindles seemed to skew in the same sense, i.e. counterclockwise as viewed from the ventral side.

Frankel: In amphibian embryos, Gerhart et al (1981) prevented rotation by embedding the eggs in a Ficoll gel. Could you do the same with *C. elegans*? Hold the two cells right next to each other to prevent either from getting in front of the other?

Wood: I don't think so. I found it very difficult even to do the reversal. In many cases, the spindles would resist the pressure from the microneedle and come around to give a normally handed embryo. The skewing mechanism, whatever it is, seems to be quite robust.

Galaburda: Has anybody put these animals in magnetic fields to test the effect on handedness?

Wood: A few such experiments were tried looking for general effects on embryos. I doubt that handedness was looked at; although it seems straightforward when it's pointed out, the distinction between normal and reversed embryos and animals is easy to miss if you're not looking for it.

Morgan: May I report the silliest experiments I have ever done? I was interested some years ago in the habenular nucleus asymmetry of the tadpole. There are two habenular nuclei on the left and one on the right. I was continually pestered by people who asked about coriolis forces and what would happen in the Southern hemisphere! I fertilized some frog eggs in a very rapidly rotating glass jar in which they remained until they metamorphosed into little frogs: whether the jar was rotated clockwise or counterclockwise had no effect on the direction of habenular asymmetry!

Wolpert: Bill, is it not true that your system requires some molecular mechanism whereby left–right is specified with respect to both the anteroposterior axis and the dorsoventral axis?

Wood: I think so. I have not thought enough about the kind of mechanism seen in *Lymnaea* where the bias may be established as early as the end of meiosis.

Dohmen: The bias is not seen until the fifth cleavage.

Wood: But from that little spiral you showed as the second polar body is coming up (p 138), if that correlates with the handedness of cleavage, conceivably the meiotic apparatus is imparting handedness to the whole organism.

Dohmen: That is possible, but I have no idea how it might operate.

Wolpert: I would argue very strongly that you cannot specify left and right—in fact left and right have no meaning—until you have the anteroposterior and also the dorsoventral axis.

Jefferies: I think left–right came before dorsoventral in the evolution of chordates, nonetheless.

Wood: It's semantics: you specify a second axis after you specify the first axis. The second one may end up being the final left–right axis or the final dorsoventral axis. The third axis is dependent on the first two.

Brown: It depends whether you are talking about an animal which is bilateral or not.

Morgan: Euclidean space has three orthogonal dimensions and thus two degrees of freedom: I think that says everything.

Wolpert: No, it doesn't. From the point of view of left and right, the axes are quite different. I would argue that left and right have no *meaning* until you specify the other two axes. That is why you get mirror-image reversal. Usually, when you look in a mirror, you reverse the dorsoventral axis, or if you lie on your tummy, you reverse the anteroposterior axis, and that's why you reverse left and right.

Wood: You can specify the axis; you just can't define the polarity of the axis, you can't say which is left or which is right.

Wolpert: I agree. So if you have an animal that can do this, it doesn't exist!

Wood: Could I comment about mirror symmetry in general? It is not just the makers of engravings who don't care about left–right reversal: it is also the people who make microscopes! The electron microscope is set up with a potential for left–right reversal at several stages. You have to be very careful. Many light microscopes invert images optically. I was shocked when I found that one of the fancy microscopes in my lab not only inverts the image as you observe it, it also takes photographs that are left–right reversed. This is obviously something that most people don't need to worry about, but in this type of research we have to be careful.

Frankel: Could you tell us more about the mutant lines that gave a heritable low frequency of class 1 animals?

Wood: I treated wild-type animals with the mutagen ethyl methanesulphonate, then looked in the F3 generation for reversed adult animals by rolling them over and observing their morphology. The frequency went up from essentially zero to about 0.2% after the mutagenesis treatment. From such animals I isolated three true-breeding lines. All three showed gonad and intestine reversal at a consistent low frequency of about 10%, but nothing else was reversed.

The genes affected in these mutants seem to influence the placement of the gonad primordium, which lies near the ventral surface of an L1 animal and consists of a linear arrangement of two somatic and two germline cells that migrate together during embryogenesis. The primordium is normally positioned on the midline but already skewed in the same sense as the adult gonad (anterior right, posterior left), which develops by elongation of the primordium. In the mutants, the initial placement of the primordium seems to be sloppy: sometimes it's reversed in the handedness of its orientation, other times it is dorsal instead of ventral and so on. These mutants may be interesting, but they are not what I was looking for, because they never reverse the asymmetry of the entire embryo.

Kondepudi: At what stage is the mutagen applied?

Wood: The standard procedure is to treat young adult animals so that the gametes, both sperm and oocytes, are mutagenized as they are being produced.

Kondepudi: Can you subject a fertilized egg to the same treatment?

Wood: Not with ethyl methanesulphonate because it doesn't get into the egg. You can mutagenize at other stages, but this is the most efficient way.

References

Frankel J 1991 Intracellular handedness in ciliates. In: Biological asymmetry and handedness. Wiley, Chichester (Ciba Found Symp 162) p 73–93

Gerhart J, Ubbels G, Black S, Hara K, Kirschner M 1981 A reinvestigation of the role of the grey crescent in axis formation in *Xenopus laevis*. Nature (Lond) 292:511–516

Greenwald IS, Sternberg PW, Horvitz HR 1983 The *lin-12* locus specifies cell fates in *Caenorhabditis elegans*. Cell 34:435–444

Priess J, Thomson JN 1987 Cellular interactions in early *Caenorhabditis elegans*. Cell
    48:241–250
Schnabel R 1991 Cellular interactions involved in the determination of the early *C. elegans*
    embryo. Mechanisms of Development, in press
zur Strassen O 1951 Der Erbgang der Nematoden-Asymmetrie. Verh Dtsch Zool Ges
    p 77–81
zur Strassen O 1959 Neue Beiträge zur Entwicklungsmechanik der Nematoden. Zoologica
    (Stuttgart) 107:1–142

Development of the left-right axis in amphibians

H. Joseph Yost

Department of Molecular and Cell Biology, Life Science Addition, Box 301, University of California, Berkeley, CA 94720, USA

Abstract. The heart and viscera of vertebrates are formed from primordia that are apparently bilaterally symmetrical. This symmetry is broken during development, yielding organs that develop characteristic asymmetries along the left–right axis. Results from three lines of experimentation on embryos of the amphibian *Xenopus laevis* indicate that left–right asymmetries are established early in development and that cellular interactions transmit left–right information from one primordium to another. First, a cytoplasmic rearrangement that occurs during the first cell cycle after fertilization may establish left–right asymmetry in some regions of the embryo. Second, a variety of experimental results indicate that embryonic ectoderm or its basal extracellular matrix may transmit left–right axial information to cardiac mesoderm and visceral endoderm. Third, inhibition of proteoglycan synthesis during a narrow period of development, concurrent with the migration of the cardiac primordia to the ventral midline, prevents asymmetrical development of the heart.

1991 Biological asymmetry and handedness. Wiley, Chichester (Ciba Foundation Symposium 162) p 165–181

Two distinct questions must be addressed when considering left–right pattern formation. First, how is asymmetry generated from bilaterally symmetrical tissue? Second, how are left–right asymmetries consistently aligned with respect to the dorsoventral and anteroposterior axes? To understand the mechanisms that establish left–right asymmetries during embryogenesis, we must first discover when left–right asymmetry is embryologically determined in specific tissues. The experiments reviewed here focus on the asymmetrical development of the heart and viscera in the amphibian *Xenopus laevis*; they indicate that the generation of asymmetry can be separated from the alignment of left–right asymmetry with the other axes.

Present address: Department of Cell Biology and Neuroanatomy, University of Minnesota, 4-135 Jackson Hall, 321 Church Street S.E., Minneapolis, MN 55455, USA.

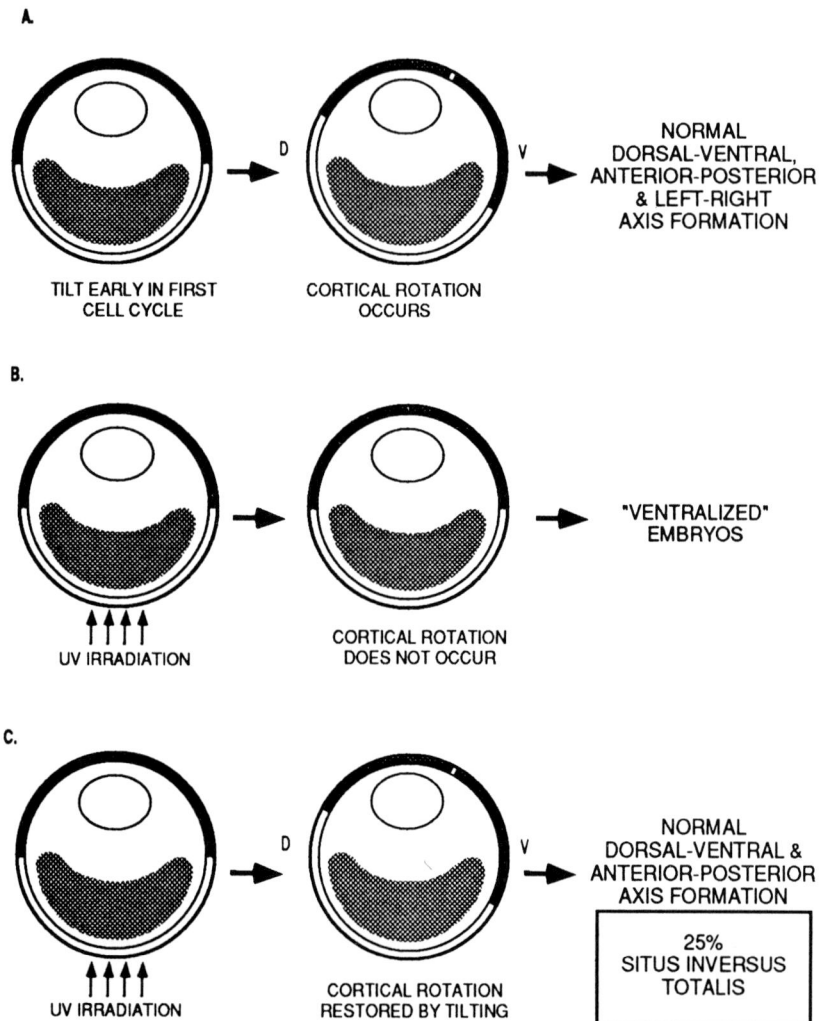

A.

TILT EARLY IN FIRST CELL CYCLE → CORTICAL ROTATION OCCURS → NORMAL DORSAL-VENTRAL, ANTERIOR-POSTERIOR & LEFT-RIGHT AXIS FORMATION

B.

UV IRRADIATION → CORTICAL ROTATION DOES NOT OCCUR → "VENTRALIZED" EMBRYOS

C.

UV IRRADIATION → CORTICAL ROTATION RESTORED BY TILTING → NORMAL DORSAL-VENTRAL & ANTERIOR-POSTERIOR AXIS FORMATION

25% SITUS INVERSUS TOTALIS

FIG. 1. Concomitant establishment of the dorsoventral and left–right axes during subcortical cytoplasmic rotation in the first cell cycle. (A) The dorsal midline is established at the egg meridian along which the cytoplasm moves the greatest extent toward the vegetal pole. This meridian can be selected by manually tilting the fertilized egg before the period of natural subcortical rotation. (B) UV-irradiation of the vegetal hemisphere blocks formation of the parallel array of microtubules, subcortical rotation does not occur and normal dorsoanterior structures do not develop (for review, see Gerhart et al 1989). (C) Subcortical rotation, dorsoventral and anteroposterior axis formation can be rescued in UV-treated eggs by tilting the egg. However, approximately 25% of the embryos exhibit situs inversus of both the heart and viscera (H. J. Yost, manuscript in preparation).

Concomitant establishment of the left–right and dorsoventral axes

Before fertilization, the amphibian egg is cylindrically symmetrical from the animal pole to the vegetal pole. This symmetry is broken in the first cell cycle by a rotation of the subcortical cytoplasm with respect to the cortex (Fig. 1). The dorsal midline is established at the meridian ($\pm15°$) along which the subcortical cytoplasm moves to the greatest extent vegetally. This rotation is thought to be driven by parallel arrays of microtubules near the surface of the vegetal hemisphere. A variety of treatments that block formation of the microtubule arrays, including UV-irradiation of the vegetal hemisphere, also block the subcortical rotation. Embryos in which rotation has been blocked do not develop dorsoanterior structures. Subcortical rotation and the formation of dorsoanterior structures can be rescued in UV-treated eggs by manually tilting the egg during the first cell cycle. Thus, it is the subcortical cytoplasmic rotation *per se*, not the microtubule array, that determines the meridian of the egg at which the dorsoventral axis is formed (for review, see Gerhart et al 1989).

In normal animals, left–right asymmetries are not random with respect to the dorsoventral and anteroposterior axes, suggesting that the mechanisms determining all three body axes must somehow be linked. Do the events of the first cell cycle that determine the dorsoventral axis also affect left–right development? In approximately 25% of the embryos that are UV-treated and manually tilted to rescue the subcortical rotation, the left–right axis is fully reversed (Fig. 1C). Thus, the presence of the microtubule array during cortical rotation is not necessary to generate left–right asymmetry, but is required to align left–right asymmetry consistently with the other body axes (H. J. Yost, manuscript in preparation). What mechanism might bring the left–right axis into register with the dorsoventral axis? Polymerized microtubules are helical in structure. In theory, a parallel array of helical molecules could provide a handedness to the subcortical rotation. This would cause an unequal distribution of cytoplasm from left to right across the concomitantly established dorsoventral midline. Interestingly, dynein defects in humans may lead to situs inversus (for review, see Brown & Wolpert 1990). Since situs inversus occurs in less than 50% of the embryos in these experiments, other cellular structures or events during the first cell cycle that may also contribute to a left–right asymmetrical rotation of the subcortical cytoplasm are currently being sought.

Is left–right axial information established during the first cell cycle distributed to all cells during cleavage of the fertilized egg or passed from one subset of cells to another during development? I shall address this question by describing the results of experimental manipulations carried out on two early stages of development (Fig. 2). Manipulations during one stage lead to randomization of the left–right axis but do not alter the generation of asymmetry. A treatment during a subsequent stage blocks the generation of asymmetry.

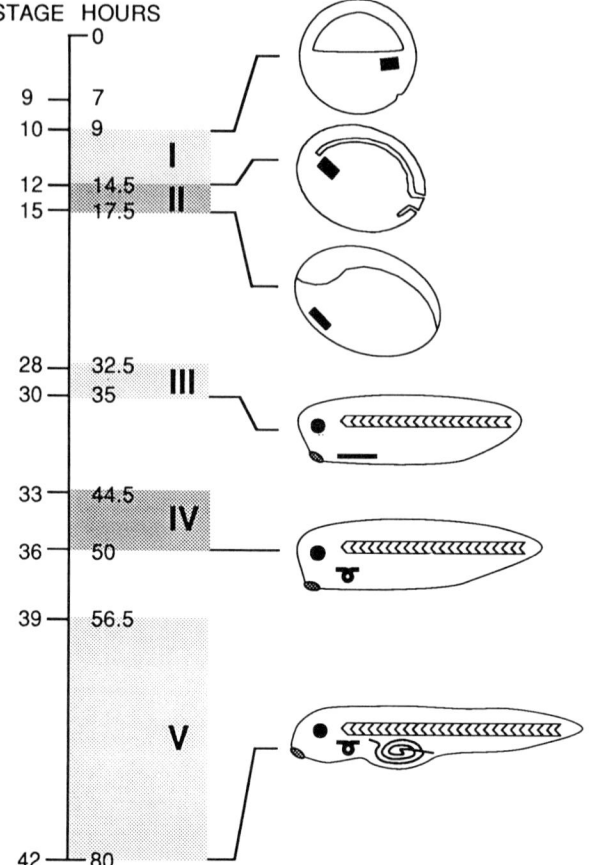

FIG. 2. Cardiac and visceral development in *Xenopus laevis*. Embryonic stages (Nieuwkoop & Faber 1967), time in hours at 25°C, and positions of cardiac primordia (black patches) are indicated. Five periods are highlighted: (I) Gastrulation. Extracellular matrix is deposited on the roof of the blastocoel. A pair of cardiac primordia arise in the mesodermal mantle on either side of the dorsal lip at the early gastrula stage. Induction by the dorsal lip cells (the 'Spemann Organizer' region) probably directs non-differentiated mesoderm to become cardiac mesoderm (Sater & Jacobson 1990). The cardiac primordia cells are swept up as part of the leading edge of the mesodermal mantle, along with presumptive head mesoderm, in the movements of gastrulation. Near the end of gastrulation the cardiac primordia lie just lateral to the dorsal midline, at approximately the hindbrain level along the anteroposterior axis. (II) Ventral migration. At the end of gastrulation, the cardiac primordia migrate toward the ventral midline. When they reach it they slowly fuse to form a single sheet. (III) Cardiac tube formation. The cardiac sheet is rolled into a tube, running along the ventral midline from anterior to posterior. (IV) Heart looping. The bilateral symmetry of the cardiac tube is broken by twisting and looping to the right of the embryo, to form an S-shaped tube. (V) Gut coiling. The viscera of the embryo is also initially bilaterally symmetrical, and does not begin asymmetrical coiling until after cardiac looping is complete. Adapted from Keller (1975, 1976), Nieuwkoop & Faber (1967).

Experimental manipulations of ectoderm and extracellular matrix result in situs inversus

Early in development, cardiac primordia migrate across an extracellular matrix (ECM) between an epithelium and anterior endoderm (Figs 2 and 3). The anterior endoderm is brought into the pathway of cardiac migration by the movements of gastrulation, and is fated to become anterior pharyngeal endoderm. The epithelium is derived from the ectoderm at the animal pole of the blastula. An ECM is deposited on the basal side of the ectoderm (i.e. the roof of the blastocoel) in blastula and gastrula stage embryos (Karfunkel 1977, Keller 1975, 1976, Nieuwkoop & Faber 1967). In *Xenopus*, the predominant known component of this ECM is fibronectin (Lee et al 1984). Laminin is not synthesized and deposited until the end of gastrulation (Fey & Hausen 1990), when there is an overall increase in proteoglycan synthesis (Kosher & Searls 1973, Brickman 1990).

Are any of these components involved in establishing the left-right axis in cardiac mesoderm? One classical approach is to assess the effects of tissue extirpation on cardiac development. Removal of endoderm from urodele neurulae prevents heart formation (Jacobson & Sater 1988). In contrast, removal of the prospective superficial pharyngeal endoderm at the beginning of gastrulation in the anuran *X. laevis* does not alter the ability of cardiac primordia to form asymmetrically looped hearts (Sater & Jacobson 1989), although the

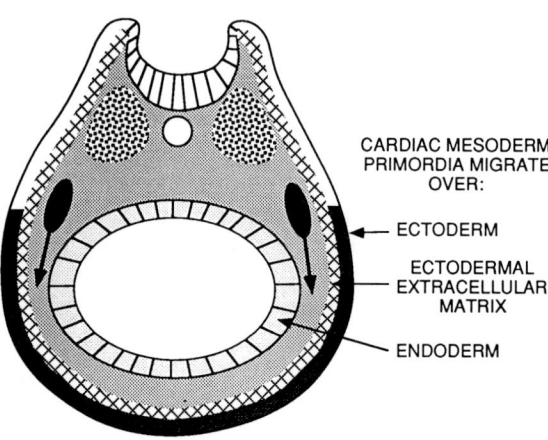

FIG. 3. Schematic cross-section of early neurula stage embryo at the prospective midbrain level, indicating the migratory pathway of cardiac primordia (black) to the ventral midline. The cardiac mesoderm contacts ectoderm, ectodermal extracellular matrix and endoderm. The endoderm also contacts the ectoderm and its extracellular matrix.

orientation of looping has not been reported. It is possible that in some vertebrates anterior endoderm is involved in the initial induction of mesoderm to become cardiac primordia, but it may not be involved in establishing the left–right axis of those primordia.

Throughout this century, many microsurgical manipulations of amphibian embryos have been shown to result in cardiac and visceral situs inversus (for reviews in English, see Wehrmaker 1969, Oppenheimer 1974, Brown & Wolpert 1990). Problems arose in the interpretation of these experiments when it was found that manipulations of parts of the embryo thought to be located a great distance from the heart and visceral primordia would result in situs inversus. However, precise fate maps for many of the species used were not available, and the relative positions of various tissues are greatly altered during early development by epiboly, gastrulation, migration of primordia and elongation of the embryo.

To begin addressing these problems, I have expanded some of the classical transplant experiments by fate mapping transplanted tissues. First, wounding the blastocoel roof of late blastulae or early gastrulae results in some cardiac situs inversus and some visceral situs inversus. To fate map the wounding sites precisely, small pieces of ectoderm in pigmented gastrulae were explanted and replaced with similar pieces of ectoderm from albino gastrulae. To eliminate anteroposterior and dorsoventral biases, these axes had been eliminated in the albino (ectoderm-donor) embryos by UV-irradiation of the vegetal hemisphere during the first cell cycle (for review, see Gerhart et al 1989). After culturing to the tailbud stage, the position of the transplanted albino ectoderm was mapped in individual pigmented embryos. On further development, the periodic albinism disappears and the orientation of the heart and viscera of each individual can be assessed without bias.

Before consideration of the fate maps of the transplanted ectoderm, the results resemble those of earlier transplantation or wounding experiments: situs inversus occurs in 50% or fewer of the embryos and the left–right orientation of one organ is not linked to the orientation of the other organs. However, fate mapping of the transplanted ectoderm reveals a strong correlation between the site of ectoderm transplantation and the orientation of the underlying heart and viscera. In cases in which the transplanted ectoderm does not appose an organ, the organ always develops in the normal left–right orientation. When the transplanted ectoderm directly apposes an organ, orientation of the left–right axis occurs randomly, that is, in half the cases the organ develops in the normal situs and in half the cases the organ is situs inversus. For example, when the transplanted ectoderm apposes only the midgut region, the pancreas, gall bladder and liver are on the left side and in opposite orientation in 50% of the embryos, whereas the heart and lower viscera (intestinal coiling) are always normal. Similarly, when the transplanted ectoderm apposes only the heart, the direction of heart coiling is random but development of the more posterior viscera is normal (Fig. 4). Thus, wounding or transplantation of ectoderm causes random orientation of the left–right axis in the cardiac mesoderm or visceral endoderm that directly opposes the treated ectoderm (Yost 1990a).

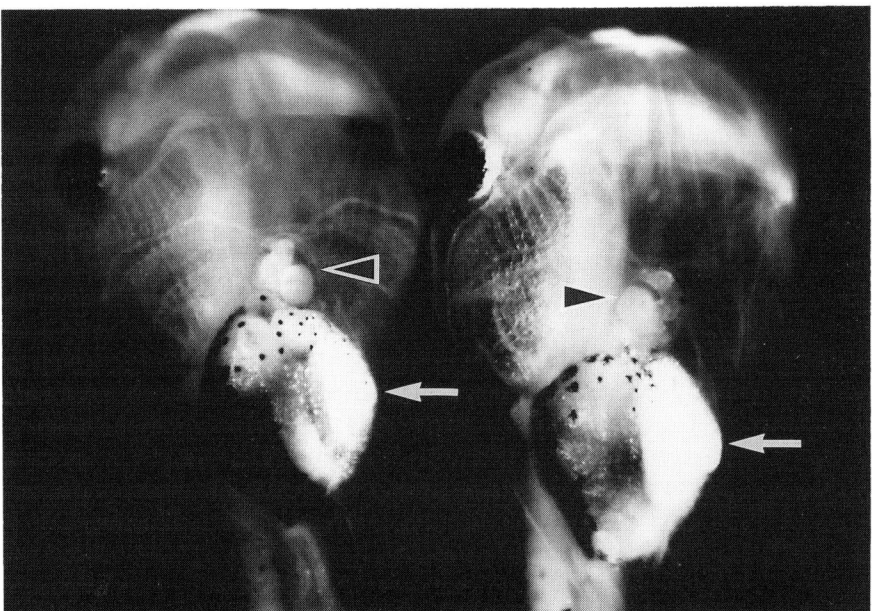

FIG. 4.   Perturbing the ectoderm at the animal pole in the early gastrula stage affects the left–right axis of the apposing heart and viscera. Ventral view of an unoperated tadpole (left) and a tadpole (right) in which transplanted ectoderm apposed the heart but not the lower viscera. The embryo on the right developed cardiac situs inversus and normal viscera. Small pieces of pigmented animal caps were replaced with corresponding pieces from albino embryos. The fate map of the albino ectoderm was scored at stage 28; the orientations of the heart (arrowhead) and viscera (white arrow) were scored in stage 46 tadpoles. The left–right axes of the heart and viscera were altered only when apposed to transplanted ectoderm.

These results suggest that by early gastrula stages left–right information is distributed in the ectoderm of the animal pole and/or its ECM. This information might be passed to the animal pole ectodermal cells directly after cytoplasmic events that occur during the first cell cycle, or might be the result of inductive events during subsequent development. Left–right information might be destroyed or perturbed by the manipulations described above. Why is the transplanted area affected, but not the surrounding tissue? Considering the short period in which transplantations are most effective in causing situs inversus, one possibility is that a signal arising from the vegetal part of the embryo crosses the plane of the ectoderm toward the animal pole. This signal might direct the modelling of left–right axial information in the ectoderm and its ECM. Transplantation or wounding during a critical phase would disrupt the continuity of this signal on the animal pole side of the wound but not on the vegetal side. Researchers studying neural induction have suggested that in blastulae and early gastrulae signals pass from the dorsal lip region through the plane of the animal cap ectoderm to participate in neural tissue formation (Spemann 1938, Slack

1983, Jacobson & Sater 1988, Dixon & Kintner 1989). Signalling through the ectoderm up to the animal pole is a possible mechanism for assuring that the dorsoventral and left–right axes are in register throughout the embryo.

Transplantation of ectoderm appears to cause situs inversus most effectively when performed on early gastrulae, coincident with the period of active ECM deposition. A second interpretation of these results is that wounding or transplantation of ectoderm perturbs the deposition of ECM in the transplanted ectoderm, but not in surrounding ectoderm of the transplant recipient. Wounding in other systems has been shown to induce local remodelling of ECM, possibly by the release of enzymes that digest the matrix (Baird & Ling 1987). Random orientation of left–right axial information in the ECM would also explain why 180° rotation and replacement of ectoderm results in only 50% situs inversus, not more (Wehrmaker 1969, H. J. Yost, unpublished 1990).

To assess whether ectodermal ECM provides left–right asymmetrical cues to cardiac mesoderm and visceral endoderm, I used methods that specifically perturb the ECM. Polypeptides containing the amino acid sequence arginine-glycine-aspartic acid (RGD) interfere with the attachment of ECM molecules (e.g. fibronectin, vitronectin, fibrinogen, collagen) to cells. This interference is mediated by competition of the peptide for binding sites on cell surface receptors (Ruoslahti & Pierschbacher 1987). Polypeptides containing the amino acid sequence arginine-glycine-glutamic acid (RGE) are inactive and serve as controls. RGD treatment disrupts gastrulation in urodeles, presumably by interfering with cell migration along the blastocoel roof. It does not affect gastrulation in the anuran *X. laevis*, probably because gastrulation is driven predominantly by convergent extension movements of the mesoderm (for review see Keller & Winklbauer 1990).

X. laevis embryos that are microinjected during blastula or early gastrula stages with RGD-containing polypeptides form normal anteroposterior and dorsoventral axes. However, approximately 50% of the embryos exhibit situs inversus of the heart and viscera. Thus, asymmetry is generated but with random left–right orientation. Microinjection of either control peptide (RGE) or buffer does not affect left–right axis formation. These results suggest that interaction of cells with extracellular material is an important step in establishing the left–right orientation of both the heart and the viscera (Yost 1990a).

There are two points at which RGD-containing polypeptides might have interfered with cell–ECM interactions. The first is during the deposition of ECM on the roof of the blastocoel. For normal deposition to occur, cell surface receptors must be present and functional. Alternatively, RGD peptides may perturb the interaction of cell receptors on the cardiac and endodermal primordia with either the ECM or cell surface molecules in the ectoderm. Microinjection during the developmental period in which ECM deposition occurs (Fig. 2, period I) leads to random orientation of the left–right axis. Microinjection of

RGD-containing polypeptides after the ECM has been deposited, but during the period of cardiac primordia migration (Fig. 2, period II), has no effect (Yost 1990a).

Another method of altering ECM composition is by treatment with heparinase, an enzyme that specifically digests heparan sulphate glycosaminoglycans on heparan sulphate proteoglycans. These proteoglycans are likely candidates for mediators of cell–matrix interactions. They bind growth factors in the ECM and serve as one class of cell surface receptors for growth factors (Cheifetz et al 1988, Roberts et al 1988, Saksela et al 1988, Segarini & Seyedin 1988). They also interact with fibronectin and facilitate the binding of at least one cell–cell adhesion molecule (Cole et al 1986, Saunders & Bernfield 1988). Microinjection of heparinase into the blastocoel of blastula or early gastrula stage embryos (Fig. 2, period I) leads to random orientation of the left–right axis. Microinjection of buffer alone or heat-inactivated heparinase does not. Similar to the results with RGD, heparinase injected into late gastrulae (Fig. 2, period I) has no effect (Yost 1990a). Specific digestion of heparan sulphate proteoglycans may cause situs inversus either by indirectly disorganizing the ECM or by directly disrupting an asymmetrical distribution of one proteoglycan or of a factor bound to the proteoglycans.

These results suggest that left–right axial information is sequestered on the basal surface of animal pole ectoderm and/or in its ECM during the period in which the matrix is initially deposited, and that this information can be transmitted to underlying presumptive cardiac mesoderm and visceral endoderm. Two intriguing possibilities currently being explored are that left–right axial information is derived from an asymmetrical distribution of a specific molecule embedded in the ECM, or from symmetrical alignment of a chiral or helical molecule that would then confer asymmetry on cardiac and visceral primordia.

Eliminating the transition from bilateral symmetry to bilateral asymmetry

Proteoglycans are major components of cell surfaces and extracellular matrices; they are thought to be involved in cell–cell and cell–matrix interactions, including tissue remodelling, cell migration and signal reception (for reviews, see Gallagher 1989, Gallagher et al 1986, Ruoslahti 1988, 1989). Treatment of *Xenopus* embryos with p-nitrophenyl-β-D-xylopyranoside (β-xyloside), an inhibitor of proteoglycan synthesis, prevents cardiac looping. Treatment of embryos with p-nitrophenyl-α-D-xylopyranoside (α-xyloside), an analogue that does not inhibit proteoglycan synthesis, has no effect on cardiac morphogenesis (Yost 1990b).

It is not surprising that proteoglycan synthesis is required for a major morphogenetic event. What is of interest is that the effective developmental period of β-xyloside treatment (stages 12 to 15) is well before cardiac tube formation occurs. Strikingly, the effective period is coincident with the migration of the paired cardiac primordia to the ventral midline after gastrulation (Fig. 2, period II). Results from a variety of tissue explant and tissue co-culture experiments indicate that the effects of the β-xyloside treatment are cell autonomous. In cardiac primordia treated with β-xyloside during the effective period, cardiac looping cannot be restored by proximity to untreated cardiac primordia or adjacent ectoderm. Proximity to β-xyloside-treated tissue could not inhibit looping in untreated cardiac primordia. Biochemical analysis demonstrates that most of the proteoglycan synthesis inhibited by β-xyloside during the effective period is heparin sulphate proteoglycan synthesis. The results indicate that during migration to the ventral midline cardiac primordia acquire the capacity to loop that will not be expressed until 1.5 days later (Yost 1990b).

One model for asymmetrical cardiac looping is that the left primordium and the right primordium attempt to loop in opposite directions. This is reasonable, given the mirror symmetry of the primordia. The consistent looping to one side may then be explained by a consistent dominance of one side. The acquisition of this dominance is currently inexplicable and is the focus of much discourse (Brown & Wolpert 1990). Results of earlier transplantation and extirpation experiments suggest that situs inversus reflects loss of dominance on the left side (Wehrmaker 1969, Oppenheimer 1974). Treatment of a cardiac primordium with β-xyloside neutralizes its ability to loop, providing an opportunity to test this model. By combining lateral halves of embryos just before the cardiac primordia reach the ventral midline, hearts can be made of a β-xyloside-treated primordium and an untreated primordium. The model outlined above predicts that if the left primordium is normally dominant, an untreated left primordium and a β-xyloside-treated right primordium should form a normal heart. Conversely, an untreated right primordium and a β-xyloside-treated left primordium should form a situs inversus heart. All the hearts looped in the normal orientation, regardless of whether the untreated primordium was on the left or the right (Yost 1990b), which suggests that by the time the cardiac primordia reach the ventral midline, each primordium is individually determined to loop and has gained left–right axial information. Thus, looping may not be a contest between the two primordia, but an agreement when they meet that right is right and left is left.

Concluding remarks

Although morphological evidence of left–right asymmetries does not arise until relatively late in development, the distinction between the left and right sides of an embryo appears to be established during the first cell cycle. Later in development, the acquisition of the ability to generate asymmetry (by coiling) is separate from the acquisition of the ability to distinguish left from right. These two processes can be experimentally perturbed at non-overlapping developmental stages. During the blastula to early gastrula stages (Fig. 2, period I), three separate treatments (microsurgical wounding, injection of RGD peptides or injection of heparinase) result in random orientation of the left–right axis but do not interfere with the generation of asymmetry. During the late gastrula to early neurula stages (Fig. 2, period II), β-xyloside treatment inhibits generation of asymmetry.

Presumably, left–right axial information is not initially contained in the cardiac mesoderm and visceral endoderm, but is imparted by interaction with the animal pole ectoderm or its ECM. Investigations of left–right autonomy will require technical improvements in our ability to isolate and culture primordia in ways that permit reliable detection of their expression of left–right asymmetry, either morphologically or through some molecular marker.

Acknowledgements

I gratefully acknowledge Drs M. L. Condic, J. Gerhart and R. Keller for discussions and advice during the course of this work. I also thank Dr M. L. Condic for encouraging a re-examination of first cell cycle experiments, critical reading of the manuscript and assistance with drawing the figures. H.J.Y. was supported by an NIH postdoctoral fellowship. This work was supported by USPHS GM19363 to Dr John C. Gerhart.

References

Baird A, Ling N 1987 Fibroblast growth factors are present in the extracellular matrix produced by endothelial cells in vitro: implications for a role of heparinase-like enzymes in the neovascular response. Biochem Biophys Res Commun 142:428–35

Brickman M 1990 Isolation and characterization of glycosaminoglycans from *Xenopus laevis* gastrula stage embryos. J Cell Biol 111:484(abstr)

Brown NA, Wolpert L 1990 The development of handedness in left/right asymmetry. Development 109:1–9

Cheifetz S, Andres JL, Massagué J 1988 The transforming growth factor β-receptor type III is a membrane proteoglycan. J Biol Chem 263:16984–16991

Cole GJ, Loewy A, Glaser L 1986 Neuronal cell–cell adhesion depends on interactions of N-CAM with heparin-like molecules. Nature (Lond) 320:445–449

Dixon JE, Kintner CR 1989 Cellular contacts required for neural induction in *Xenopus* embryos: evidence for two signals. Development 106:749–757

Fey J, Hausen P 1990 Appearance and distribution of laminin during development of *Xenopus laevis*. Differentiation 42:144–152

Gallagher JT 1989 The extended family of proteoglycans: social residents of the pericellular zone. Curr Opinion Cell Biol 1:1201–1218

Gallagher JT, Lyon M, Steward WP 1986 Structure and function of heparan sulphate proteoglycans. Biochem J 236:313–325

Gerhart J, Danilchik M, Doniach T, Roberts S, Rowning B, Stewart R 1989 Cortical rotation of the *Xenopus* egg: consequences for the anteroposterior pattern of embryonic dorsal development. Development (Suppl):37–51

Jacobson AG, Sater AK 1988 Features of embryonic induction. Development 104:341–359

Karfunkel P 1977 SEM analysis of amphibian mesodermal migration. Roux's Arch Dev Biol 181:31–40

Keller RE 1975 Vital dye mapping of the gastrula and neurula of *Xenopus laevis*. I. Prospective areas and morphogenetic movements of the superficial layer. Dev Biol 42:222–241

Keller RE 1976 Vital dye mapping of the gastrula and neurula of *Xenopus laevis*. II. Prospective areas and morphogenetic movements of the deep layer. Dev Biol 51:118–137

Keller R, Winklbauer R 1990 The role of the extracellular matrix in amphibian gastrulation. Semin Dev Biol 1:25–33

Kosher RA, Searls RL 1973 Sulfated mucopolysaccharide synthesis during the development of *Rana pipiens*. Dev Biol 32:50–68

Lee G, Hynes R, Kirschner M 1984 Temporal and spatial regulation of fibronectin in early *Xenopus* development. Cell 36:729–740

Nieuwkoop PD, Faber J 1967 Normal table of *Xenopus laevis* (Daudin). North-Holland, Amsterdam.

Oppenheimer JM 1974 Asymmetry revisited. Am Zool 15:867–879

Roberts R, Gallagher J, Spooncer E, Allen TD, Bloomfield F, Dexter TM 1988 Heparan sulphate bound growth factors: a mechanism for stromal cell mediated haemopoiesis. Nature (Lond) 332:376–378

Ruoslahti E 1988 Structure and biology of proteoglycans. Annu Rev Cell Biol 4:229–255

Ruoslahti E 1989 Proteoglycans in cell regulation. J Biol Chem 264:13369–13372

Ruoslahti E, Pierschbacher MD 1987 New perspectives in cell adhesion: RGD and integrins. Science (Wash DC) 238:491–497

Saksela O, Moscatelli D, Sommer A, Rifkin DB 1988 Endothelial cell derived heparan sulfate binds basic fibroblast growth factor and protects it from proteolytic degradation. J Cell Biol 107:743–751

Sater AK, Jacobson AG 1989 The specification of heart mesoderm occurs during gastrulation in *Xenopus laevis*. Development 105:821–830

Sater AK, Jacobson AG 1990 The role of the dorsal lip in the induction of heart mesoderm in *Xenopus laevis*. Development 108:461–470

Saunders S, Bernfield M 1988 Cell surface proteoglycan binds mouse mammary epithelial cells to fibronectin and behaves as a receptor for interstitial matrix. J Cell Biol 106:423–430

Segarini PR, Seyedin SM 1988 The high molecular weight receptor to transforming growth factor-β contains glycosaminoglycan chains. J Biol Chem 263:8366–8370

Slack JMW 1983 From egg to embryo. Determinative events in early development. Cambridge University Press, Cambridge

Spemann H 1938 Embryonic development and induction. Hafner, New York

Wehrmaker A 1969 Right–left asymmetry and *situs inversus* in *Triturus alpestris*. Roux's Arch Dev Biol 163:1–32

Yost HJ 1990a The involvement of extracellular matrix in the establishment of the left–right axis in *Xenopus laevis*. J Cell Biol 111:483(abstr)

Yost HJ 1990b Inhibition of proteoglycan synthesis eliminates left–right asymmetry in *Xenopus laevis* cardiac looping. Development 110:865–874

DISCUSSION

Kondepudi: You said that the single egg cell is cylindrically symmetrical, but it contains microtubules and things. In what sense is it cylindrically symmetrical?

Yost: Only in the sense that we can put a dorsoventral midline at any longitude.

Wolpert: I thought the sperm entry point determined the anterior side.

Yost: That's not true! Each longitude of the embryo has the capacity to be the dorsal midline.

Wolpert: But what determines that? Not sperm entry any more?

Yost: It is cortical rotation *per se* that's most important for breaking the cylindrical symmetry. The sperm entry point may bias the direction of the rotation initially. That bias may then be picked up by the deposition of parallel tracks of microtubules, which complete the rotation. We can eliminate the sperm entry point experimentally and as long as the rotation occurs so does normal anteroposterior development (for review, see Gerhart et al 1989).

Thwaites: Surely the fact that it rotates presupposes that there is an axis about which it rotates, which somehow has been determined already?

Yost: It can rotate in many directions, so there isn't any axis. The microtubule track is laid down right at the beginning of the rotation. You can just give a very slight bias around this cylinder that will start the rotation going in one direction.

Thwaites: When you say in one direction, do you mean about a particular axis?

Yost: About a given meridian; any meridian has an equal chance of becoming the dorsoventral axis.

Kondepudi: The less we see, the more symmetrical an object seems. Have you seen no topology inside the early *Xenopus* embryo that could confer any particular direction or any kind of handedness?

Yost: In the early embryo, as a result of the cortical rotation, you can see some differences; for example, the pigment is concentrated more on one side of the embryo than the other.

Wolpert: The egg is symmetrical in the sense that you can decide where the asymmetry can be. You can put it anywhere you like by the appropriate experimental manipulation.

Berg: What about the anteroposterior axis? Is that not determined until after the dorsoventral axis?

Yost: Transplantation experiments show that the anteroposterior axis is not set up until the gastrula stage.

Berg: Does it not have anything to do with animal and vegetal poles?

Yost: It's probably an interplay between the dorsoventral axis and the animal and vegetal poles. During gastrulation, cells move towards the animal pole. The cells that are dorsal-most and also closest to the animal pole become the anterior-most cells in the embryo.

Wolpert: There is one axis present at the beginning, the animal–vegetal axis; even though it doesn't correspond precisely at this stage with the future anteroposterior axis, it does have polarity.

Lewis: The animal–vegetal axis roughly corresponds to external versus internal, because of the movements of gastrulation.

Is the ectoderm ciliated at the relevant stage of development, when the handedness of the embryo is being established?

Yost: Not this early in development, not at the gastrula stage. Cilia appear sometime in the tailbud stages, I'm not sure exactly when.

Frankel: If I recollect correctly, in the development of the anteroposterior axis the only thing that matters is the *amount* of cortical rotation before the first cleavage. There is a strict correlation between the rotation and completeness of the anteroposterior axis (Gerhart et al 1989). The way in which the rotation is generated is totally irrelevant.

But for the right–left axis, you have shown that the manner of rotation counts. Can one then conclude that there is something about the normal manner of rotation, perhaps microtubule-directed rather than gravity-directed, that is important for the right–left axis?

Yost: Yes, it is important that parallel microtubules are present during the rotation to get normal left–right axis formation. If you remove the microtubules, you randomize asymmetry.

It is an interesting point. We have talked about how evolution might select consistent left–right asymmetries. I believe evolution might be selecting not for left–right asymmetry but for the mechanism that gives normal dorsoventral and anteroposterior axis formation. Because that mechanism is driven by the microtubule tracks, the presence of these tracks, by coincidence, might give consistent left–right asymmetry.

Wood: Is there something special about UV treatment, or does anything that disrupts the microtubule network give the same results?

Yost: I have blocked microtubules so far only with UV treatment. If rotation is allowed to occur before the eggs are treated with UV, there is no effect on left–right axis formation.

Weber: What does the UV-irradiation do? How does it work?

Yost: There are some unpublished experiments by Susan Roberts that address this. The thinking is that there are microtubule subunits in the vegetal region that are ready to be polymerized. UV-irradiation causes cross-linking of a GTP molecule to the microtubule subunits which prevents them being polymerized (S.J. Roberts, J.C. Gerhart, unpublished paper, UCLA Symposium on Developmental Biology 1989).

Morgan: What is the mechanism of the cardiac looping? Is it differential growth?

Yost: The mechanism is not really understood. As far as we can tell from other experiments, it probably is not differential cell division.

Layton: Cardiac looping seems to involve changes in the shape of myocardial cells (Manasek et al 1984a). The cells on the convex surface of the heart are thinner and have a greater surface area than those on the concave side. It appears that neither differential cell division nor blood flow pattern is important in loop formation. As might be expected, the cytoskeleton seems to be responsible for the changes in cell shape, and cytochalasin does block cardiac looping.

Frankel: It seems to be something about the normal mode of rotation and not rotation *per se* that is necessary for cardiac looping. One could imagine that cytoskeletal intracellular processes are the primary events of determination and the events that one sees involving extracellular matrix are secondary. In *C. elegans*, Hill & Strome (1988, 1990) did an elegant series of experiments to study the segregation of polar granules. They applied pulses of cytochalasin B to 1-cell embryos and found that actin is necessary at a particular stage for normal polar granule migration and cleavage. Could one do similar experiments on an amphibian embryo? Would disrupting microtubules and other cytoskeletal elements at an early stage affect handedness?

Yost: That should be possible. I have talked only about the extracellular matrix. There are many examples where this matrix and intracellular structures interact. The experiments would be technically difficult: one would probably do them in explants of primordia to avoid growth perturbation of the whole embryo.

So there is an extracellular event occurring early in embryogenesis, but I have no evidence on how that interaction is translated into something the cells do a day and a half later.

Lawrence: What's known about the anatomy of the extracellular fibres that are laid down when the heart is still a single uncoiled tube? Is there any helical arrangement of those fibres and is it disturbed by your experiments? If so, there might be a link between the molecular structure and the asymmetry.

Yost: I haven't looked carefully at the cardiac tube at the time it should be folding in embryos that have been treated so that the folding is blocked.

Lawrence: Has it been looked at in untreated animals with fluorescent antibodies or with any other method of seeing the fibres?

Layton: Manasek et al (1984b) examined scanning electron micrographs of the cytoskeleton of the developing chick heart after removal of plasma membranes using Triton X-100. The pattern of cytoskeletal elements was what would be expected in cardiac myocytes and showed some regional variation. However, there was no apparent overall alignment of fibres that could explain the shape of the cardiac loop.

Wood: It is surprising that you can totally disrupt the extracellular matrix and see an effect only on handedness. Is that true or are there other defects?

Yost: I have found no defects in the other two axes. I was surprised; I thought we would see inhibition of migration of cells down the ventral midline, for example. The tadpoles develop normal-sized hearts that are normally looped; it is just that the orientation is randomized.

Layton: If you knocked out the fibronectin on which these cells are supposed to migrate, wouldn't that interfere with neural crest migration?

Yost: It is a different area of the embryo where the fibronectin is being knocked out. Extracellular matrix is being deposited at the roof of the blastocoel; I inject RGD into the blastocoel to interfere with this early deposition. Neural crest migration occurs on an extracellular matrix that is deposited later among somitic mesoderm.

Wolpert: There's a real puzzle in these very interesting experiments. Embryologists always thought that the mesoderm instructed the ectoderm. You have shown that the ectoderm seems to control the mesoderm as regards handedness. And why should simply wounding the ectoderm interfere with the extracellular matrix anyway?

Yost: In other systems, wounding leads to a reorganization of the extracellular matrix. My interest is whether there is a left–right asymmetry in the matrix or in the ectoderm. One could then ask, is it asymmetrical distribution of a molecule or is something else going on? The question then, embryologically, is: When is that information set up? Is it inherited from the first cell cycle or is it set up by interactions perhaps with the Spemann organizer cells, the dorsal cells?

Galaburda: Does the production of growth or trophic factors related to injury play a role? In other systems, injury does up-regulate production of trophic factors.

Wolpert: You must understand the peculiarity of this system. Joe Yost wounds the embryo at a very early stage. The wounded area does not meet the region it is going to affect until several hours later. Joe, have you ever tried wounding locally, in the region of the future heart? In other words, not wounding early but at the time of cardiac development.

Yost: I haven't tried that. This raises the question of exactly when the interactions between the ectoderm and the heart primordium happen. Fate mapping gives a general localization of the patches overlying the heart region. It would be interesting to get transplants in which a path is cut across where the heart primordia are migrating.

Jefferies: How phylogenetically old are these asymmetries? The tunicate heart is usually on the right side of the stomach, is that the same thing?

Yost: As far as I know, all vertebrates have the heart looping in the embryonic rightward direction.

Morgan: I am slightly confused about the relationship between your experiments and those of von Woellwarth (1950) on producing situs inversus. He cut out pieces of neural plate then replaced them with or without a rotation. He got situs inversus totalis; everything always went the same way, including the habenular nuclei. What's the difference between the two sets of experiments?

Yost: von Woellwarth manipulated much larger pieces of embryos, with or without rotation of the pieces. His experiments (reviewed in English by Wehrmaker 1969) were on embryos that had already gastrulated. Gastrulation places all three germ layers (ectoderm, mesoderm and endoderm) near the site of wounding, so one cannot determine which germ layer is crucial in his experiments. In my experiments, embryos were wounded before gastrulation had moved the mesoderm and endoderm into position, so only the ectoderm was wounded. Also, he used *Triturus*, a urodele, whereas I used *Xenopus*, an anuran. There are important differences between urodeles and anurans in how they gastrulate and in their fate maps, which makes comparisons between the two difficult. I would like to repeat some of my experiments in urodeles to allow a better comparison.

Lewis: Does *Xenopus* gut have spiral smooth muscle coats like those of mammals? If so, is their handedness reversed in your experimental animals?

Yost: I don't know the answer to that.

References

Gerhart J, Danilchik M, Doniach T, Roberts S, Rowning B, Stewart R 1989 Cortical rotation of the *Xenopus* egg: consequences for the anteroposterior pattern of embryonic dorsal development. Development (Suppl):37–51

Hill DP, Strome S 1988 An analysis of the role of microfilaments in the establishment of asymmetry in *Caenorhabditis elegans*. Dev Biol 125:75– 84

Hill DP, Strome S 1990 Brief cytochalasin-induced disruption of microfilaments during a critical interval in 1-cell *C. elegans* embryos alters the partitioning of developmental instructions to the 2-cell embryo. Development 108:159–172

Manasek FJ, Kulikowski RR, Nakamura A, Nguyenphuc Q, Lacktis JW 1984a Zak R (ed) Growth of the heart in health and disease. Raven Press, New York p 105–130

Manasek FJ, Isobe Y, Shimada Y, Hopkins W 1984b The embryonic myocardial cytoskeleton, interstitial pressure, and the control of morphogenesis. In: Nora JJ, Takao A (eds) Congenital heart disease: causes and processes. Futura Publishing Co, Mount Kisco, New York p 359–376

von Woellwarth C 1950 Experimentelle Untersuchungen über den Situs Inversus der Eingeweide und der Habenula des Zwishenhirns bei Amphibien. Wilhelm Roux's Archiv Entwicklungsmech Org 144:178–256

Wehrmaker A 1969 Right–left asymmetry and *situs inversus* in *Triturus alpestris*. Roux's Arch Dev Biol 163:1–32

Development of handed body asymmetry in mammals

Nigel A. Brown*, Afshan McCarthy* and Lewis Wolpert†

*MRC Experimental Embryology and Teratology Unit, St George's Hospital Medical School, Cranmer Terrace, London SW17 0RE and †Department of Anatomy and Developmental Biology, University College & Middlesex Hospital School of Medicine, Windeyer Building, Cleveland Street, London W1P 6DB, UK

Abstract. We have proposed a three step model for the specification of left–right in mammalian embryos. The fundamental assumption is that handedness is imparted by an asymmetrical molecule. *Conversion* of molecular asymmetry to the cellular level gives a property to one side of the embryo to bias an otherwise *random generation* of an asymmetrical gradient which can be *interpreted* by developing organs. Rat embryos, treated at discrete stages, show a window of sensitivity for disruption of handedness, which may reflect the time of *conversion*/biasing. Heat shock and several chemicals cause left–right inversion in up to 50% of embryos exposed during neural groove formation. Earlier stages are less sensitive; no treatment begun after foregut pocket formation influences asymmetry. Evidence for cellular interactions in left–right specification comes from the apparent rescue of *iv/iv* mutant embryos in chimeras. We are looking for molecular left–right disparity before morphological asymmetry but detect no differences in two-dimensional protein profiles. Using an indirect measure, we find a right–left gradient of tissue oxygen in embryos at the 20–30 somite stage. This may reflect asymmetrical vasculature, as we have suggested to explain drug-induced asymmetrical limb malformations.

1991 Biological asymmetry and handedness. Wiley, Chichester (Ciba Foundation Symposium 162) p 182–201

The mammalian body is overtly symmetrical about the midline that divides left and right. However, virtually all internal structures are asymmetrical in position, shape, or both. This asymmetry is reproducibly handed, not random, and is remarkably consistent across species. The aorta loops to the left, the right lung has more lobes than the left, the stomach and spleen lie on the left, the liver has a single left lobe, and so on. How do these handed asymmetries develop? And how does the embryo tell its left from right?

A body plan can be considered in terms of a coordinate system with three axes: anteroposterior, dorsoventral and left–right. Left–right is fundamentally different in that it must be specified in relation to the other axes.

Although there are theoretically other possibilities, it seems highly likely that left-right is specified in the embryo after the other two axes have been established. The anteroposterior axis is first apparent with the appearance of the primitive streak at about 6½ days in the mouse (see Theiler 1972; 8½ days in the rat, two weeks in humans). Dorsoventral polarity is already apparent by this stage in the arrangement of the germ layers. The embryo remains evidently symmetrical about the neural groove until the heart loops to the right, at about 8½ days in mouse (Fig. 1A). This is followed by the rotation (flexion or turning) of the embryo, again to the embryo's right. Thus, some, if not all, left-right specification has taken place by 8½ days in mouse (Fig. 1).

We have proposed a model for this breaking of symmetry that produces the distinct left and right sides of mammals (Brown & Wolpert 1990). Here, we will summarize the model, review supporting observations, describe initial experimental testing, suggest further studies, and examine the relevance to abnormalities in left-right asymmetry, both clinical and experimental.

A model of left-right specification

The fundamental assumption is that handedness is signalled by a molecule (or larger structure) which is itself handed and can be fixed in a particular orientation in relation to the anteroposterior and dorsoventral axes. We make this assumption in the absence of plausible alternatives. Magnetic fields induced by an electric current down the midline have been proposed (Huxley & de Beer 1934), but this hypothesis does not fit well with known cellular properties. Maternal positional cues seem to be ruled out by the facts that cultured embryos can develop normally (but see below), mouse embryos implant at two opposed uterine orientations (Smith 1980) and *iv* mutant embryos develop abnormally in a normal uterus (Brown et al 1990a, see below).

The model has three components: *conversion*, in which the molecular asymmetry is transmitted to the cellular level; *random generation of asymmetry*, which can be biased by *conversion* to produce handed asymmetry; and *interpretation* in which individual organs use left-right information.

We have proposed a specific mechanism for *conversion*, starting with an embryo in which the anteroposterior axis, dorsoventral axis and plane of bilateral symmetry are established (Fig. 2). Cells become polarized with respect to the midline; for example, a cell component becomes more concentrated at the side of the cell opposite to the midline, perhaps in response to a diffusible substance produced by midline cells. There is ample precedent for unequal distribution of cellular constituents in relation to polarity (e.g. Schierenberg 1989). We depict the handed molecule as an 'F', which is held in a specific orientation with respect to both the anteroposterior and dorsoventral axes. This breaks the mirror symmetry across the midline. An appealing example of such an orientation of a handed structure is seen in the cilia of tadpole tails. These

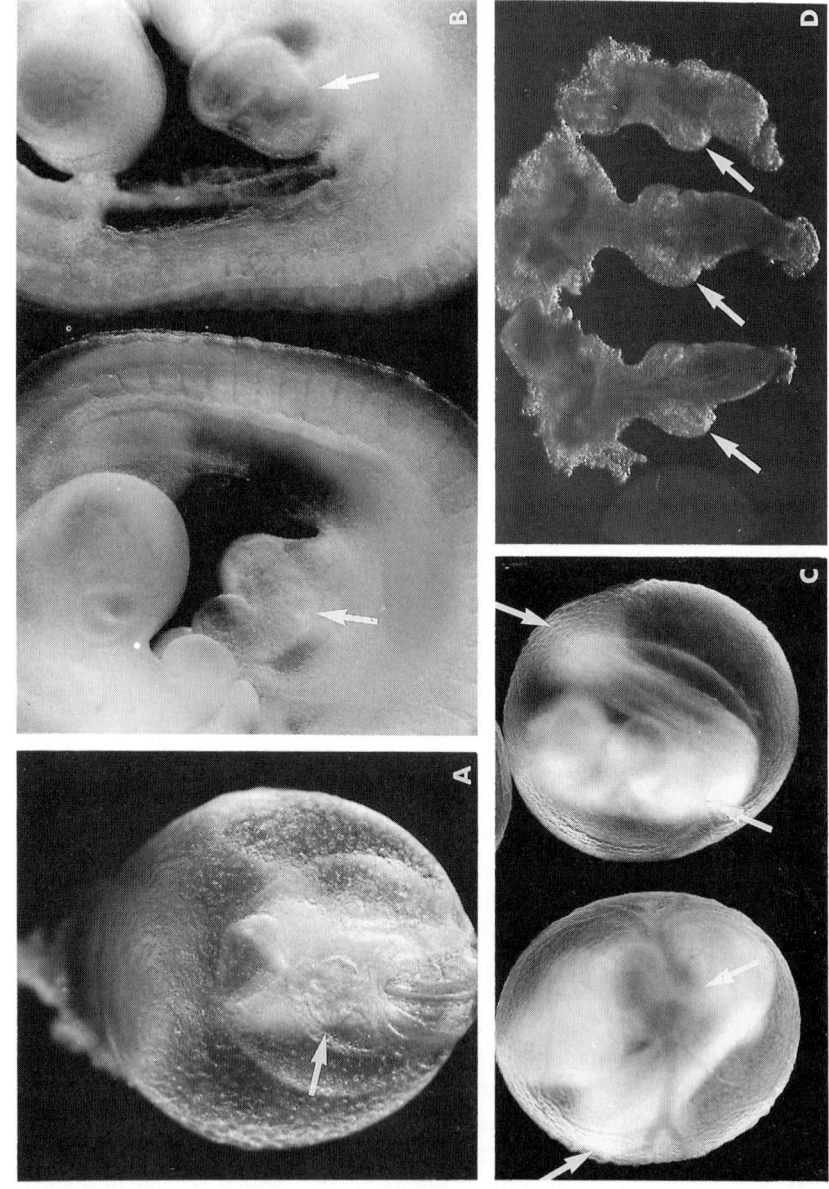

are fixed in the same orientation on left and right sides, so they all beat caudally.

To establish a left–right difference, 'F' interacts with the cell polarity. For example, if F caused transport of a molecule down its 'arms', then the molecules would accumulate towards the midline on the left side and away from it on the right side (Fig. 2). This would produce a clear difference between the cells on the left and those on the right. The transport would be in the same direction as the increased polarity concentration on the right side, but in the opposite direction on the left (Fig. 2). We suggest that a stable property is acquired by the right side, but not the left, as the result of the interaction of the two molecules.

This would be sufficient to account for handed asymmetry were it not for the patterns of development in situations where there appears to be a loss of function in *conversion*. In the human Kartagener's syndrome and in *iv/iv* mutant mice, loss of handed asymmetry results not in symmetry but in random asymmetry. Thus, there must be a separate mechanism for the *generation of random asymmetry*. We favour a simple mechanism provided by a gradient in a morphogen produced by reaction–diffusion (Kauffman et al 1978). This could occur across the midline and left–right asymmetry would be established, the concentration rising on one side and falling on the other. Most importantly, the mechanism is, in principle, random so that a slight initial asymmetry can bias the reaction to one side rather than the other (Almirantes & Nicolis 1987). In our model, the initial asymmetry is provided by the left–right difference arising from *conversion*.

Interpretation is what cells do with the information that they are on the left or right of the embryo. Interpretation is, in principle, even handed; there is no reason to suggest 'dominance' of one side over the other.

We propose that *conversion* is an early event, perhaps affecting the whole embryo. Support comes from classical experimental manipulations of amphibians (reviewed by Oppenheimer 1974) and observations of human conjoined twins. In conjoined twins of the newt, *Triturus*, caused by mechanical constriction, viscera are normal in the left twin but often reversed in the right. Spemann & Falkenberg (1919) found that 50% of right twins showed situs inversus and Wilhelmi (1921) suggested that when the influence that biased curvature on the left side was removed, curvature would be random. A similar phenomenon in human conjoined twins has been reported; normal situs in left

---

FIG. 1. Early manifestations of asymmetry in rodent embryos. (A) 8½ day mouse embryo showing bulging of the heart tube to the right. (B,C,D) 11½ day rat embryos. (B) Heart loops. Left, normal (dextral); right, inverted (sinistral). (C) Embryo turning. Left, normal, caudal trunk to right side of head, vitelline vessels emerging from left side of body; right, inverted. (D) Lung buds. Note the more caudal orientation of the right bud (arrows).

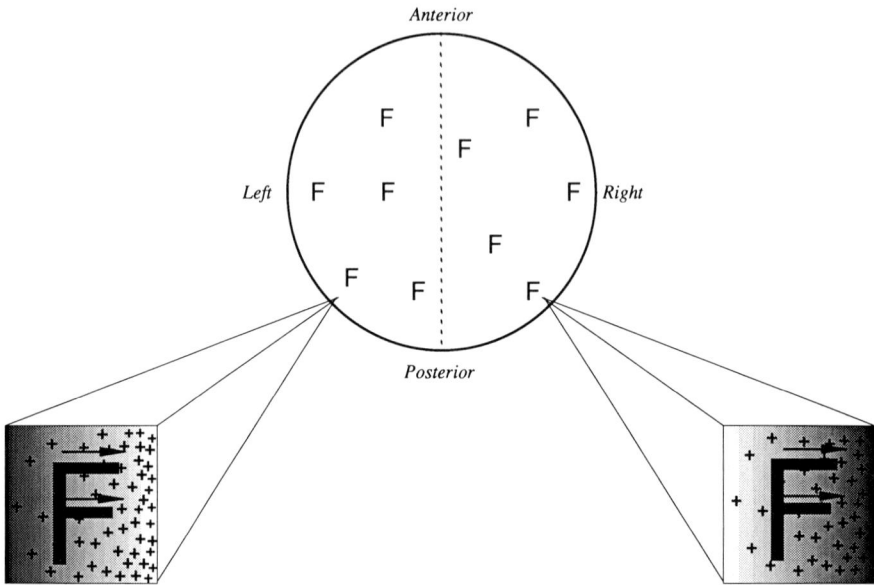

FIG. 2.   Proposed mechanism for *conversion*. The embryo has anteroposterior and dorsoventral axes established and cells are polarized with respect to the midline, which results in the distribution of some cellular constituent (shading) away from that side of each cell closest to the midline. A handed molecule 'F' is inserted into the ventral membrane and is aligned by extracellular fibres along the anteroposterior axis. The F molecule causes transport of another cellular constituent (crosses) down its 'arms'. This results in a distinct difference: on the left side, the two molecules (shading and crosses) concentrate at opposite sides of the cell; on the right side, they are together on one side of the cell.

twins, frequently abnormal situs in right (e.g. Siebert et al 1989), but the data are not conclusive and they are complicated by the various planes of twinning (Layton 1989).

Our model provides a simple explanation for defects of asymmetry in right half-embryos (Fig. 3). If *conversion* results in a stable property on the right, but not the left, then an embryo bisected after this process is complete has a left half that is still labile. Thus, *conversion* can occur again, new left and right sides can be specified, and development is normal. In contrast, the right half is already fixed and cannot be reprogrammed. The two sides that develop from the right half do not differ, there is nothing to bias the *generation of random asymmetry*, so asymmetry develops randomly. It follows that an embryo bisected before *conversion* should develop normally. Separation of *Triturus* at the two-cell stage (Mangold 1921) results in both twins being normal; mammalian embryos split at early cleavage stages develop normally, as do the vast majority of human monozygotic twins, where presumably twinning takes place earlier than in those embryos that develop into conjoined twins.

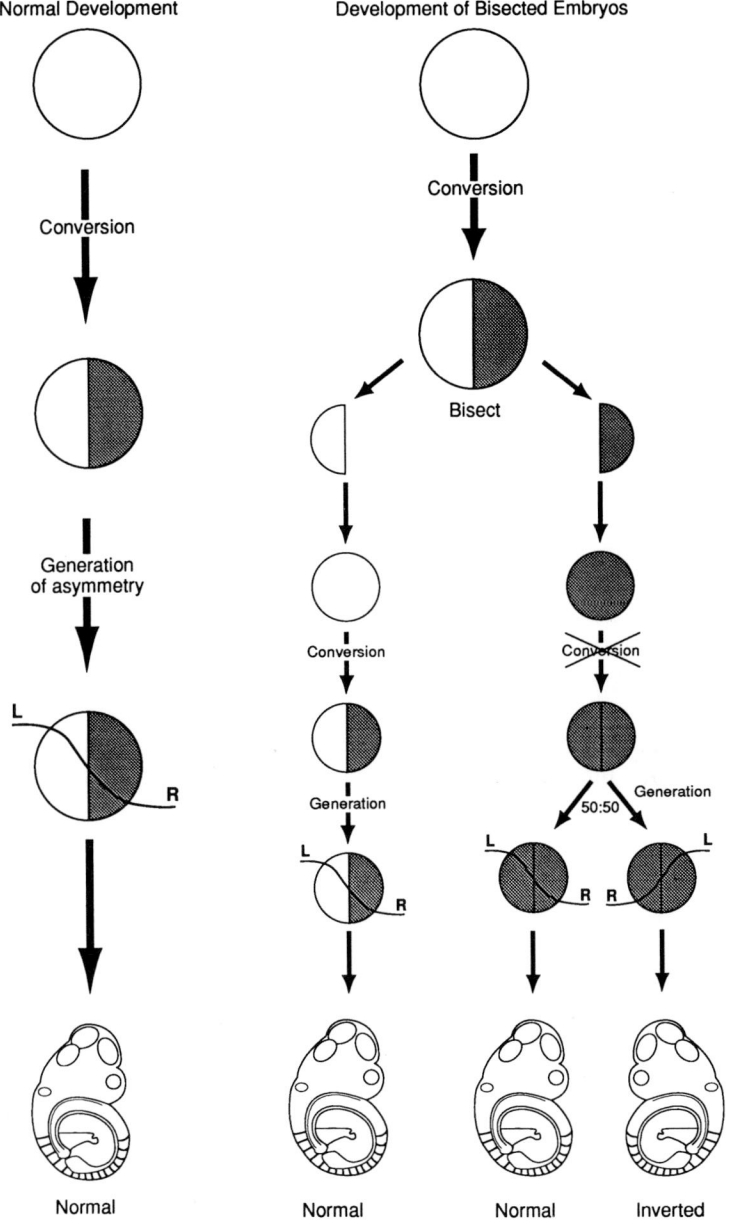

FIG. 3. The development of asymmetry in bisected embryos. We suggest that *conversion* results in a stable property in right, but not left, cells. The left half of an embryo bisected after *conversion* is labile and able to go through *conversion* again and develop normally. The right half cannot be respecified so asymmetry is random and half the embryos develop with situs inversus.

The *generation* of the gradient of asymmetry seems to occur in individual organ fields, independently and sequentially. This is clear from the organ relationships in *iv/iv* mice and humans with asymmetry defects. *iv/iv* mice show heterotaxia, that is, organs of normal and inverted asymmetry in one individual (Hummel & Chapman 1959, Layton 1976) in about 40% of fetuses (Brown et al 1989). Thus, in the absence of correct left–right specification, asymmetry can develop independently in different fields. Heterotaxia can even occur within a single organ, for example, human hearts with discordant ventricular and atrial situs (transpositions) are well known. Initial heart looping specifies ventricular situs, but atrial asymmetry can be specified separately and seems to be closely related to that of the lung, since atrial and lung asymmetry are rarely discordant (Macartney et al 1980).

Experimental testing of the model

Chimeras

Chimeras of normal and developmental mutant mice have often been used to examine the role of cellular interactions in development. In our chimeras of *iv/iv* and *+/+* mice (Brown et al 1990a), neither cell type was phenotypically dominant (Table 1). Defective asymmetry was rare, suggesting a rescue of the mutant phenotype. This implies a role for cellular interactions in the process of left–right specification, as required by our model. From these data we cannot exclude, statistically, the alternative interpretation that cells specify left–right autonomously. Interestingly, the chimeras that developed abnormalities of asymmetry were distinguished by having a greater than 67% contribution of *iv/iv* cells to both the heart and the visceral yolk sac.

TABLE 1   Tissue genotype and embryo phenotype in *iv/iv* plus *+/+* mouse aggregation chimeras examined at 9½ days

Tissue genotype *iv/iv* contribution to heart tissue	Embryo phenotype Number of embryos	
	Normal	*Inverted*
All	9	5
>50%	7	2[a]
<50%	12	0
None	47	0

Embryos were aggregated at the 8-cell stage and examined on Day 10. Genotype was determined by glucose phosphate isomerase isoenzyme assay in tissue homogenates.
[a]One embryo with situs inversus, 71% and 73% *iv/iv* in heart and visceral yolk sac; one with sinistral heart loop, 68% and 71% *iv/iv* contributions. Data from Brown et al (1990a).

Reciprocal transfers of *iv/iv* and + / + 8-cell embryos failed to show any influence of the uterine environment on the development of asymmetry (Brown et al 1990a). *iv/iv* embryos developed with random asymmetry in + / + mothers, and + / + embryos developed normally in *iv/iv* mothers.

Timing

To examine when early events in left–right specification take place, we have treated embryos at various stages (Fig. 4) and observed the development of the first two signs of handed asymmetry: heart looping and embryo turning (Fig. 1B, C). We have used rat embryos because their postimplantation development in culture is better than that of mouse. Following the observations of Fujinaga & Baden (1991) that α_1-adrenergic agonists can disrupt the development of asymmetry, we used methoxamine treatment. We identified a window of sensitivity beginning when the neural groove deepens, just before head fold formation, and ending by the stage of formation of the first somite (McCarthy et al 1990, Table 2). The coincidence of this sensitivity with a distinct stage in midline development may fit with the establishment of polarity relative to the midline proposed in our model.

To evaluate if this stage specificity was restricted to α_1-adrenergic agonists and thus due, for example, to transient receptor expression, we have used several other treatments at the same stages (Brown et al 1991). Serotonin, ionomycin

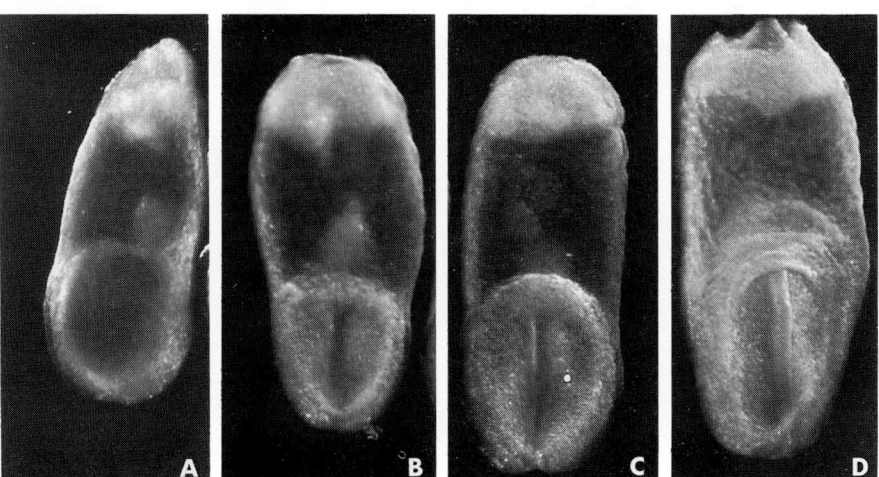

FIG. 4.   Frontal views of the treatment stages of 9–9½ day rat embryos. (A) Stage I, late primitive streak, 'pre-midline'. (B) Stage II, distinct neural groove, 'midline'. (C) Stage III, neural groove and 'head fold'. (D) Stage IV, immediately pre-somite with 'foregut' pocket.

TABLE 2 Development of asymmetry in rat embryos treated at different stages with methoxamine

Stage of treatment	Control					Methoxamine				
	Total embryos	% abnormal asymmetry	Situs inversus	Sinistral heart	Inverted turning	Total embryos	% abnormal asymmetry	Situs inversus	Sinistral heart	Inverted turning
I 'Pre-midline'	31	6	2	—	—	38	3	1	—	—
II 'Midline'	102	11	6	4	1	110	38	27	10	5
III 'Head fold'	89	12	6	3	2	97	33	20	9	3
IV 'Foregut'	76	7	4	1	—	78	5	1	2	1

Embryos were explanted at 9–9½ days, grouped into the defined stages shown in Fig. 4, cultured for four hours in the presence of 100 μM methoxamine, washed, cultured for 48 hours, then examined.

TABLE 3 Development of asymmetry in rat embryos treated at different stages with lithium chloride or colcemid

Stage of treatment	Control					Treated				
	Total embryos	% abnormal asymmetry	Situs inversus	Sinistral heart	Inverted turning	Total embryos	% abnormal asymmetry	Situs inversus	Sinistral heart	Inverted turning
Lithium chloride										
I	9	22	—	2	—	12	58	5	1	1
II	88	8	5	2	—	33	55	11	1	6
III	82	11	3	5	1	25	32	4	2	2
IV	20	5	—	—	1	20	15	1	—	2
Colcemid										
I	8	12	1	—	—	6	83	3	1	1
II	33	12	2	2	—	29	31	8	1	—
III	38	10	1	3	1	41	41	9	6	2
IV	12	0	—	—	—	12	0	—	—	—

Embryos were explanted at 9–9½ days, grouped into the defined stages shown in Fig. 4, cultured for four hours in the presence of 16 mM (32 mM for stage IV) lithium chloride or for 75 minutes in 0.4 µg/ml colcemid, washed, cultured for 48 hours, then examined.

and phorbol dibutyrate were ineffective; heat shock and retinoic acid were each moderately effective; lithium chloride and colchicine were each distinctly effective (Table 3). There were two significant differences between these agents and the α_1-adrenergic agonists: all these treatments induced other defects in addition to their effects on asymmetry; all were also effective at the earliest stage tested (late primitive streak).

The specificity of the α_1-adrenergic agonists makes their mechanism worth investigating. It is unlikely that an endogenous α_1-adrenergic agonist is involved in left–right specification, because the α_1-adrenergic antagonist prasozin has no effect on the development of asymmetry, although it does block the effects of methoxamine treatment (McCarthy et al 1990). More plausible is an interaction between some step in α_1-adrenergic signal transduction and the normal process of left–right specification. Since there are multiple α_1-adrenergic receptors with different transduction pathways (Schwinn et al 1990), identifying the receptor subtype expressed at these embryonic stages is an important next step.

The lability of left–right specification is also illustrated by the development of a significant number of reversals of asymmetry in untreated, cultured rat embryos (Tables 2, 3). This incidence is higher if embryos are handled during the sensitive period (for example, in controls for washing out short-term treatments). Similarly, the background rate of reversal in mouse embryos cultured over the sensitive phase is about 20% in our hands, presumably because optimal culture conditions have not been devised for this species. One possible interpretation of the effectiveness of the non-α_1-adrenergic agents at the late primitive streak stage is that they induced non-specific disruptions that persisted and affected specification of asymmetry at later stages.

The hunt for molecular left–right differences

Protein/mRNA profiles

We have analysed the profiles of proteins synthesized in left and right hemi-embryos by two-dimensional gel electrophoresis (A. McCarthy, N. A. Brown, unpublished). Several replicate pairs of individual rat embryo halves, from neural plate to 30 somite stages, were analysed using several techniques: [^{3}H]leucine and [^{35}S]methionine labelling; isoelectric focusing and non-equilibrium first dimensions; and linear and gradient second dimensions. Although some 600 proteins are resolvable, in no case have we detected a consistent difference between left and right halves.

We are moving on to subtractive hybridization as an approach to identify components of the left–right system. mRNA will be prepared from left and right lung buds. A cDNA library will be made from one set of mRNAs and hybridized with mRNA from the other side of the lung. Any cDNA or mRNA that does not hybridize is from a gene differentially expressed in right and left lung buds. We would expect very few differences since the fate of cells is essentially identical in the two buds. However, the pattern of lobation and even the pattern of the primary buds are quite distinct in the two sides (Fig. 1D). This distinction is presumably driven by differential gene expression as a result of *interpretation*.

Tissue oxygen

Misonidazole is a nitroheterocyclic compound that is reductively metabolized at a rate that is inversely proportional to the concentration of tissue oxygen. The production of metabolites, which bind irreversibly to cell components, has been widely used to map, indirectly, areas of hypoxia. We have shown that there is a left–right (low to high) gradient of misonidazole metabolites in rat embryos at the 20–30 somite stage (Brown & Coakley 1987). Recently, we found that this gradient is inverted in *iv/iv* mouse embryos with situs inversus (Fig. 5, Brown et al 1990b). This may reflect some metabolic left–right difference, as suggested by Fantel et al (1989). However, by the 20–30 somite stage the embryo is profoundly asymmetrical, and we suggest that these left–right differences may be a consequence of asymmetrical development of the vasculature (Brown et

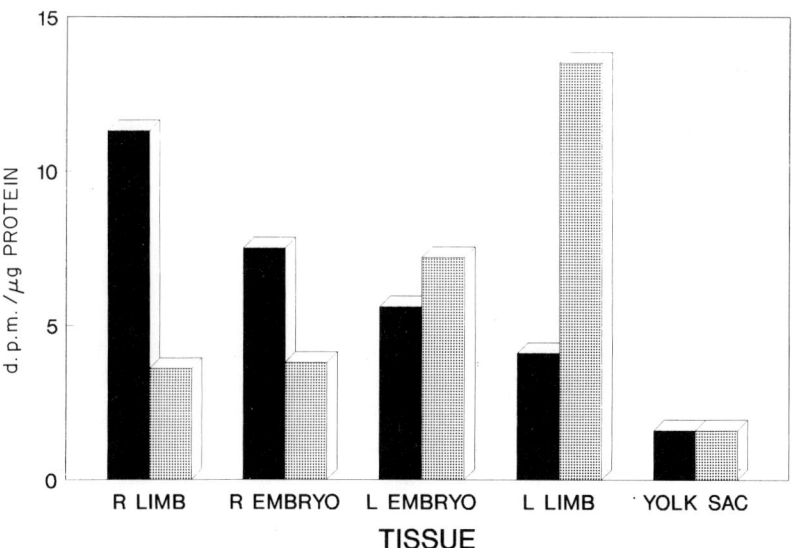

FIG. 5.   The distribution of tissue-bound misonidazole metabolites in normal and situs inversus *iv/iv* mouse embryos. Embryos were explanted at 8½ days and cultured for 48 hours in the presence of [¹⁴C]misonidazole. Tissues were dissected and TCA-insoluble counts extracted. Data are the mean of 31 normal (■) and 21 situs inversus (▦) embryos.

al 1989). In support of this we have found that the left–right gradient of misonidazole distribution is destroyed if rat embryos are cultured outside but still attached to their extraembryonic membranes, in which case oxygen may be delivered directly from the medium rather than via the vascular supply.

Gene defects

The search for the murine *iv* gene is described elsewhere in this volume (Brueckner et al 1991). Another genetic lead is provided by the murine insertional mutant *legless* (McNeish et al 1990), which has multiple developmental effects, including 50% situs inversus. In a preliminary study we located the insert, by *in situ* hybridization, to the distal end of chromosome 12 (J. Crolla, N. A. Brown unpublished), and this has been confirmed by P. Jenkins & P. Copeland (unpublished). This is close to the location of *iv* (Brueckner et al 1989) and allelism is being tested (W. Scott, personal communication 1991). Analysis of flanking sequences shows a strong CG-rich island, a 1–2 kb deletion, and potential coding sequences homologous from *Drosophila* to humans, as shown by moderate stringency Southern blotting (W. Scott, personal communication 1991).

Kartagener's syndrome and dynein

Kartagener's syndrome is an autosomal recessive defect leading to dextrocardia, bronchiectasis and sinusitis, in which partial or complete situs inversus is common. The respiratory disease is considered to be due to ciliary dyskinesia caused by defective dynein. Dyneins are chemomechanical ATPases that react with cellular filaments to generate force. Afzelius (1976) has proposed that the situs inversus in Kartagener's syndrome is caused by ciliary dysfunction, citing the finding of single cilia on non-mitotic cells in the chick epiblast and hypoblast. It is not clear how cilia could affect the asymmetry of, for example, the lung buds or the liver.

We suggest that the situs inversus in Kartagener's syndrome may be associated with a defect in cytoplasmic, rather than ciliary, dynein. Ciliary and cytoplasmic dyneins are distinct, although some of the constituent proteins of these multimeric complexes are shared by both forms. Defects of chemotaxis and other cell motilities have been described in patients with Kartagener's syndrome (Walter et al 1990), but it is not known if cytoplasmic dynein is involved in such movement.

Dynein has one property that makes it an attractive candidate for a role in the specification of left–right: it is capable of asymmetrical transport. With few exceptions, dynein moves specifically towards the minus end of microtubules (McIntosh & Porter 1989). This polarity of movement is exactly the property attributed to the 'F' molecule in the *conversion* component of our model.

Conclusions

The development of the handed asymmetry of the mammalian body plan has intrigued scientists for more than a century. Our model for the processes involved provides a framework for evaluating a large number of natural and experimental observations. Several aspects of the model are directly testable; so far, studies are consistent with its predictions. Progress in the genetics of mouse mutations affecting asymmetry and understanding the mechanism of some treatments that cause random asymmetry seem most likely to provide clues to the molecular components of left–right specification.

References

Afzelius BA 1976 A human syndrome caused by immotile cilia. Science (Wash DC) 193:317–319

Almirantes Y, Nicolis G 1987 Morphogenesis in an asymmetric medium. Bull Math Biol 49:519–538

Brown NA, Coakley ME 1987 Tissue oxygen as a determinant of axially asymmetric teratologic responses; misonidazole as a marker for hypoxic cells. Teratology 35:73(abstr)

Brown NA, Wolpert L 1990 The development of handedness in left/right asymmetry. Development 109:1–9

Brown NA, Hoyle CI, McCarthy A, Wolpert L 1989 The development of asymmetry: the sidedness of drug-induced limb abnormalities is reversed in *situs inversus* mice. Development 107:637–642

Brown NA, McCarthy A, Wolpert L 1990a The development of handedness in aggregation chimeras of situs inversus mutant and wild-type embryos. Development 109:949–954

Brown NA, McCarthy A, Wolpert L 1990b *iv/iv* Mice and lateral asymmetry: the development of visceral situs in *iv/iv* plus +/+ chimeric embryos and the inversion of a left–right tissue oxygen gradient in *situs inversus*. Teratology 41:541(abstr)

Brown NA, McCarthy A, Wolpert L 1991 Further studies on the embryonic stage at which left/right asymmetry is specified in rat embryos. Teratology 43:446(abstr)

Brueckner A, D'Eustacio P, Horwich AL 1989 Linkage mapping of a mouse gene, *iv*, that controls left–right asymmetry of the heart and viscera. Proc Natl Acad Sci USA 86:5035–5038

Brueckner M, McGrath J, D'Eustachio P, Horwich AL 1991 Establishment of left–right asymmetry in vertebrates: genetically distinct steps are involved. In: Biological asymmetry and handedness. Wiley, Chichester (Ciba Found Symp 162) p 202–218

Fantel AG, Mackler B, Juchau MR, Person RE, Burroughs-Gleim C 1989 Biochemical basis of the dysmorphogenic effects of niridazole. Teratology 39:451(abstr)

Fujinaga M, Baden JM 1991 Evidence for an adrenergic mechanism in the control of body asymmetry. Dev Biol 143:203–205

Hummel KP, Chapman DB 1959 Visceral inversion and associated anomalies in the mouse. J Hered 50:10–13

Huxley JS, de Beer GR 1934 The elements of experimental embryology. Hafner, New York

Kauffman SA, Shymko R, Trabert K 1978 Control of sequential compartment formation in *Drosophila*. Science (Wash DC) 199:259–270

Layton WM 1976 Random determination of a developmental process. J Hered 67:336–338

Layton WM 1989 *Situs inversus* in conjoined twins. Am J Med Genet 34:297

Mangold O 1921 Situs inversus bei *Triton*. Arch Entwicklungsmech Org (Wilhelm Roux) 48:505–516

Macartney FJ, Zuberbuhler JR, Anderson RH 1980 Morphological considerations pertaining to recognition of atrial isomerism. Consequences for sequential chamber localization. Br Heart J 44:657–667

McCarthy A, Wolpert L, Brown NA 1990 The development of the left/right axis in neural plate phase rat embryos is disrupted by α-adrenergic agonists in an apparent stage-specific and receptor-mediated manner. Teratology 42:33(abstr)

McIntosh JR, Porter ME 1989 Enzymes for microtubule-dependent motility. J Biol Chem 264:6001–6004

McNeish JD, Thayer J, Walling K, Sulik KK, Potter S, Scott WJ 1990 Phenotypic characterization of the transgenic mouse insertional mutation, legless. J Exp Zool 253:151–162

Oppenheimer JM 1974 Asymmetry revisited. Am Zool 14:867–879

Schierenberg E 1989 Cytoplasmic determination and distribution of developmental potential in the embryo of *Caenorhabditis elegans*. BioEssays 10:99–104

Schwinn DA, Lomasney JW, Lorenz W et al 1990 Molecular cloning and expression of the cDNA for a novel α_1-adrenergic receptor subtype. J Biol Chem 265:8183–8189

Siebert JR, Machin GA, Sperber CH 1989 Anatomic findings in dicephalic conjoined twins: implications for morphogenesis. Teratology 40:305–310

Smith LJ 1980 Embryonic axis orientation in the mouse and its correlation with blastocyst relationships to the uterus. 1. Relationships between 82 hours and 4½ days. J Embryol Exp Morphol 55:257–277

Spemann H, Falkenberg H 1919 Uber asymmetrische Entwicklung und Situs Inversus viscerum bei Zwillingen und Doppelbildungen. Arch Entwicklungsmech Org (Wilhelm Roux) 45:371–423

Theiler K 1972 The house mouse. Springer-Verlag, Berlin

Walter RJ, Danielson JR, Reyes HM 1990 Characterization of a chemotactic defect in patients with Kartagener Syndrome. Arch Otolaryngol Head & Neck Surg 116:465–469

Wilhelmi H 1921 Experimentelle Untersuchungen uber situs inversus viscerum. Arch Entwicklungsmech Org (Wilhelm Roux) 48:517–532

DISCUSSION

Stern: In animals or people with isomerisms, on the whole the pattern is consistent on the left and on the right. In situs inversus, many organs are inverted but independently. Is there any advantage, in terms of mechanism, in distinguishing organs that become asymmetrical because of rotation, like the gut and perhaps the heart tube at early stages, from organs that become asymmetrical because of interpretation, for example the fourth aortic arch or the lungs?

Brown: There may be practical advantages in making that distinction. For example, we have chosen the lungs as a model system because we find the heart difficult to study owing to the rotation aspects. In contrast, the asymmetry of the lungs is relatively simple. I am not sure that there are any fundamental distinctions between rotational and other asymmetries.

Stern: The pancreas, for example, arises as two midline organs that fuse, then rotation of the whole organ brings it to the left. So the asymmetry of the pancreas is a sort of secondary event. Lung lobes that arise on either the right or the left might interpret information about right or left directly, rather than as a consequence of something else.

Burn: The left and right ventricles of the heart are not lateralized structures, they are sequential structures in the heart tube. The heart loop throws them to left and right. They are not like the left and right atria, which are morphologically lateralized structures.

Brown: Professor Robert Anderson always talks about the ventricles developing in series and the atria in parallel. The point is that the direction in which the heart loops is independent of the specification of the left and right atria. In a single organ those two events can be independent.

Galaburda: I am curious about your wish to do subtractive hybridization. Your model does not require that a particular message is expressed on one side but not on the other. Secondly, timing could be important. At any given time during development, a delay in the expression of a gene on one side with respect to the other would give results that may not address the fundamental question.

Brown: We have thought about this quite a bit. Take the lung buds, for example. The cellular differentiation of the left and the right lung buds is identical on each side, yet their morphogenesis produces quite different structures. I don't believe subtractive hybridization of mRNAs from left and right lung buds will necessarily tell us anything about our model, but there must be something interesting to learn about the different morphogenesis on either side.

Weber: You use a variety of agents to induce situs inversus; it looks like you just went into a pharmacist's shop! Any agent perturbs handedness in these embryos, there is no specificity.

Brown: I agree. The only caveat is that Masahiko Fujinaga's α_1-adrenergic agonists are completely clean; they don't do anything to these embryos apart from cause what is probably randomization of handedness.

Crow: Do you know that it is α_1-adrenergic agonism that is important pharmacologically?

Brown: Masahiko Fujinaga and I have both demonstrated that the effect is blocked by prazosin, which is an α_1-antagonist.

Fujinaga: Even control embryos show situs inversus if they are cultured from earlier than the pre-somite stage. If we culture from the early primitive streak stage, almost 50% of rat embryos develop with situs inversus. During normal *in vivo* development, the shape of the egg cylinder is kept oval until the early somite stage; *in vitro* it rapidly becomes spherical, which leads to premature flattening of the embryonic disc. I think this abnormal flattening is the cause of the loss of control of body asymmetry. This is also an example of 'clean' situs inversus.

Kondepudi: When you sampled the right and left halves of the embryo for chromatographic analysis, at what stage were those embryos?

Brown: A series of stages from primitive streak to about 25 somites. I don't think these studies prove anything other than that there are no obvious differences visible under our conditions.

Galloway: In the mice with situs inversus, the gut becomes inverted. The smooth muscle of the gut follows a helical path around it: does that helical path get reversed when the trajectory of the gut gets reversed or is it invariant?

Brown: I don't know, but it's a great question.

Burn: In humans with isomerism, the mesentery runs straight up and down instead of being set at an angle. When the mesentery does this, there is frequently malrotation of the gut; if the mesentery is inserted at a similar angle to normal but the opposite way, there is complete reversal of the rotation of the gut. I suspect rotation of the gut has nothing to do with the smooth muscle; it gets its cue from insertion of the mesentery.

Layton: Frank Manusek and I have been unsuccessful in our attempts to demonstrate a helical pattern in the fibres of the muscularis externa of the mouse small intestine. However, that is not to say that such a pattern does not exist.

Corballis: You postulate a molecule, which you arbitrarily represent by the letter F to indicate that it is asymmetrical. You then suggest that if this molecule is to signal the difference between left and right sides, it must be 'tethered' to the anteroposterior axis. But in fact the asymmetry is likely to be of helical form. From what we have been told about helices (Galloway, this volume), is tethering that critical?

Wolpert: You have to tether the spiral to the anteroposterior axis or you simply end up with a spiral that doesn't have a front and a back.

Corballis: It only has to be tethered to one axis, not both. Any way you look at a spiral, its left and right are different.

Wolpert: No. You can tell a left spiral from a right spiral, but you can't tell a left-hand side from a right-hand side.

Corballis: A spiral will always have different left and right sides.

Wolpert: Only if you know which is the front end and which is the back, so it has to be tethered.

Galloway: The choice of the F shape for the molecule is interesting because strictly speaking an F is not handed.

Wolpert: Since we chose F we have learned that there has to be a 'sticky' undersurface; the upper surface differs from the lower, and so the molecule is handed.

McManus: In your model, does the state of conversion allow for a gradient that would be in the opposite direction? If you look at flatfish, there are three separate asymmetries: the adult flips over to one side, their optic chiasms cross and they have asymmetrical viscera. In wild-type flatfish there are occasional reports of one of these asymmetries being inverted, but almost no reports of two being reversed. However, every so often all three asymmetries are inverted. The implication is that in the completely inverted flatfish the whole directional system has been reversed, rather than independent processes being read wrongly later in development. Can your model account for that?

Brown: Yes, there are two possible ways in which a mutation in the F molecule could result in reversal of handedness. F might be held back-to-front (flipped over) or, alternatively, it could be held upside down. In either case, the arms of the molecule would point left rather than right and handedness would be reversed. I don't think either option is very plausible, and I know of no mutations that result in consistent reversal of handedness. There are several loss-of-function mutations known, but they all lead to randomization of asymmetry.

Wolpert: One of the virtues and weaknesses of a model like this is that it has many parameters. It could be the interpretation that was being affected.

McManus: But the model describes a situation in which there is one principle asymmetry, whereas in flatfish there are three separate asymmetries, which makes it more complex.

Lawrence: Did you add the gradient that goes from extreme left to extreme right to your model because of the mutant *iv/iv* phenotype?

Brown: Partly, yes. If there were no circumstances under which asymmetry was random, we wouldn't have needed to postulate that gradient. In reality, there are many examples, including the *iv* mutation.

Lawrence: In general, it is very dangerous to build models around a mutant phenotype. The ways of mutations are so weird and wonderful.

Wolpert: There are many cases where asymmetry is randomized, by all sorts of different techniques.

Lawrence: Your model requires a way of producing asymmetry but I am not sure that you need a gradient.

Wolpert: The problem with asymmetry is that if you do something on one side, you have to stop it happening on the other side. A gradient is one way of doing this; there are other ways.

Stern: When I read the original proposal of the model I felt a slight prejudice against the gradient idea. Now I feel that it fits rather nicely with Joe Yost's wounding results (this volume). One possible interpretation is that there is a left and right difference in the behaviour of the gradient. If a left–right gradient is wounded on one side close to the source, there will automatically be a left–right difference, provided the gradient always runs in the same direction. If the gradient runs from left to right and the wound is on the left-hand side of the embryo, the wound lies close to the source of the gradient. It therefore has a relatively small amount of space in which to even out before the wound closes. When the wound is on the right-hand side, there is a considerable distance from the source, which goes across the midline and the gradient may collapse.

Yost: That predicts that wounding on one side of the embryo would be much more effective than on the other. There are earlier reports that that is the case.

In my system I haven't seen any difference between wounding on the left side and on the right side.

Stern: Recent results from Chen & Solursh (1991) seem to fit this hypothesis. They find that a bead soaked in retinoic acid has different effects when grafted on the left or on the right side of a chick embryo. On the left side, secondary embryonic axes are induced; on the right side they are not. According to the hypothesis we just discussed, the results could be explained by assuming that retinoic acid prevents or delays wound healing. However, Solursh also obtained a comparable result when he grafted Hensen's nodes on the left or the right. It is more difficult to explain this, but it is worth noting that Hensen's node is itself asymmetrical; it has a dent on its left side (Hara 1978).

Morgan: In Spemann & Falkenberg's (1919) descriptions of conjoined twins in *Triturus*, the right-hand twin was not just the mirror-image of the left, it was less well developed on its left-hand side.

Yost: I don't find any dominance of the right side or the left side in *Xenopus*. I have looked at naturally formed conjoined twins and I saw no situs inversus on either side.

Wolpert: The point of that argument is that if separation of left and right sides occurs early, the resulting twins will be perfectly normal. If, however, separation occurs at the mid-blastula stage, the left side develops into a normal animal and the right side develops with random situs.

Morgan: Does a possible size difference fit your model in any way?

Brown: No, it wouldn't.

Layton: Spontaneous situs inversus appears to be much more common in urodeles than in anurans. Spemann worked with urodeles, Joe Yost is working with anurans. I have not heard of spontaneous situs inversus in anurans; it is reported to have an incidence of 0.9–2.6% in *Triturus alpestris* (Wehrmaker 1969).

Frankel: In my paper, I emphasized the contrast between strictly local and global effects of asymmetry in ciliates. There is one possible link that was discovered by Gary Grimes, which is quite compatible with Nigel Brown's model. In the model, a tethered asymmetrical molecule has a long-range asymmetrical effect. Gary Grimes has evidence for something he calls an intracortical communication system (Grimes 1982).

Take a ciliate that has a row of structural cirri at the margins and during development produces both a large central oral primordium and marginal cirral primordia. If one cuts off one of the margins, including all the marginal cirri, Maria Jerka-Dziadosz (1974) has shown that a new marginal row of cirri can develop nonetheless. Grimes & Adler (1978) later showed that the oral primordium is affected by the absence of the marginal row. The oral primordium sends out a finger of basal bodies toward the margin where the marginal cirri were removed. Grimes & Adler propose that this actually produces the marginal primordium. The behaviour of the basal bodies in this oral primordium is different when the marginal row of cirri is absent compared to when it is present. Grimes deduced that the presence of ciliary structures has long-range effects within the cortex. This I think is rather similar to what you are postulating.

References

Chen Y-P, Solursh M 1991 A comparison of Hensen's node and retinoic acid in secondary axis induction in early chick embryos. Development, in press
Galloway JW 1991 Macromolecular asymmetry. In: Biological asymmetry and handedness. Wiley, Chichester (Ciba Found Symp 162) p 16–35
Grimes GW 1982 Pattern determination in hypotrich ciliates. Am Zool 22:35–46
Grimes GW, Adler JA 1978 Regeneration of ciliary pattern in longitudinal fragments of the hypotrichous ciliate, *Stylonychia*. J Exp Zool 204:57–80
Hara K 1978 Spemann's organizer in birds. In: Nakamura O, Toivonen S (eds) Organizer: a milestone of a half-century from Spemann. Elsevier, Amsterdam p 221–265
Spemann H, Falkenberg H 1919 Uber asymmetrische Entwicklung und Situs Inversus viscerum bei Zwillingen und Doppelbildungen. Arch Entwicklungsmech Org (Wilhelm Roux) 45:371–423

Establishment of left–right asymmetry in vertebrates: genetically distinct steps are involved

*Martina Brueckner, °James McGrath, †Peter D'Eustachio and °‡Arthur L. Horwich

*Departments of Pediatric Cardiology and °Human Genetics and ‡Howard Hughes Medical Institute, Yale University School of Medicine, 333 Cedar Street, New Haven, CT 06510 and †Department of Biochemistry and Kaplan Cancer Center, New York University Medical Center, 550 1st Avenue, New York, NY 10016, USA

Abstract. Vertebrates exhibit a characteristic pattern of asymmetrical positioning of the visceral organs along the left–right axis. A remarkable developmental step establishes this pattern—primitive organs migrate from symmetrical midline positions of origin into lateral positions. The first organ to pursue such movement is the cardiac tube, which forms a rightward 'D' loop; other organs follow concordantly. The signals and mechanisms directing such organ migration can be studied by analysis of heritable defects of humans and mice. In general, these defects behave as loss-of-function mutations that lead to random determination of visceral situs: for an affected embryo there is an equal chance of correct situs or situs inversus. Distinct phenotypes and patterns of inheritance of these defects suggest that at least three genes are involved in left–right determination, apparently members of a developmental pathway. These genes should be amenable to molecular analysis. We are studying a recessive allele of the mouse called *inversus viscerum* (*iv*). Using linkage analysis with cloned restriction fragment length polymorphism markers, we have genetically mapped the *iv* gene to the distal portion of mouse chromosome 12. We are now pursuing isolation of the gene using methods of positional cloning. Analysis of the *iv* gene product and of its site and timing of expression may offer clues to how left–right lateralization occurs.

1991 Biological asymmetry and handedness. Wiley, Chichester (Ciba Foundation Symposium 162) p 202–218

One of the most remarkable events during vertebrate development is the creation of left–right asymmetry by lateralization of previously symmetrical primordial organs. This event produces, both ontogenetically and phylogenetically, a dramatic increase in the complexity of spatial design. The details of lateralization are at present unknown, but if development of asymmetry along the antero-posterior and dorsoventral axes is any indication, it seems likely that a network of genes and products is involved in this process. For anteroposterior and dorsoventral developmental determination, study of *Drosophila* mutants and the corresponding genes and products has elucidated a sequence of early embryonic steps involving both maternal and zygotic genes (for review, see Driever & Nusslein-Volhard 1988, Roth et al 1989, Manseau et al 1989).

While *Drosophila*, by virtue of its anteroposterior and dorsoventral axes obligatorily has a 'right' and 'left', it has only limited asymmetry along the left–right axis. However, in another genetically tractable animal, *Caenorhabditis elegans*, recent work of William Wood (1991, Wood & Kershaw, this volume) identifies features of left–right asymmetry established as early as the six-cell stage of the embryo. Given the powerful tools for studying *C. elegans*, including knowledge of the lineage of every cell of the animal (Sulston & Horvitz 1977, Sulston et al 1983), the cloning of its genome in cosmid contigs and yeast artificial chromosomes (Coulson et al 1986, 1988), plus the ability to transform the animal with cloned sequences (Fire 1986), this nematode may prove to be the ideal organism for a saturation genetic attack on an animal system of left–right asymmetry.

Yet an understanding of the vertebrate system of left–right determination may, despite the more difficult problems of genetics and animal transformation, be well worth pursuing. The mechanisms operative in vertebrates may differ from those in *C. elegans*: features of left–right asymmetry in vertebrates arise much later in development, after organ formation instead of before it (Copenhaver 1926). Cues for such lateralization may be set up as soon as anteroposterior and dorsoventral axes are established; however, the critical signal for production of asymmetry clearly comes much later. When lateralization *does* occur, instead of single cell movements, entire primordial organs migrate away from initial positions of symmetry. Thus, both the timing and nature of the acquisition of left–right asymmetry in vertebrates differ from those of *C. elegans*. A second attraction of vertebrates is that the effects of mutation on the process of lateralization may have clinical significance—human beings with these defects often suffer from congenital heart disease. Study of the process in vertebrates may thus offer some of the first molecular clues to the nature of a condition that affects one in every 200 human births. Additionally, the issue of determination of cerebral hemispheric dominance may be approachable in vertebrates. At present only primates offer ready identification, by anatomical means, of cerebral dominance, but if molecular markers become available for observation of left–right asymmetry in the vertebrate central nervous system, then access to understanding hemispheric dominance might be gained in a manipulable animal like the mouse. Thus overall, while genetic approaches to mammalian development are still difficult relative to those in *Drosophila* and *C. elegans*, the study of left–right determination in vertebrates potentially offers unique understandings.

Production of mammalian mutants in which specific developmental steps are affected is not, as yet, a practical undertaking. However, the mammalian left–right determination pathway can be studied by examining existing human and animal mutations that result in defective left–right lateralization. Several such mutations have been observed and their phenotypes characterized. The variation in the observed phenotypes and in the patterns of inheritance, which we discuss below, allow us to conclude that, like the formation of anteroposterior and dorsoventral patterns, formation of left–right asymmetry in vertebrates depends on the function of multiple genes.

Cardiac looping

In vertebrates the organ whose function is most greatly disturbed by failure of normal left–right lateralization is the heart: it appears to be the organ most dependent on normal development of left–right asymmetry for acquisition of its correct structure and function. In mammals the earliest visible production of left–right asymmetry involves the heart: the cardiac tube, previously symmetrical along the midline, bends toward the right, forming a so-called D cardiac loop (Copenhaver 1926). This places the future left ventricle to the left and posterior and the future right ventricle to the right and anterior. These are the final relationships of these two pumping chambers in the adult. Looping also establishes the normal relationships between the ventricles and the atrial chambers and great vessels.

Abnormalities of the formation of left–right asymmetry result in major cardiac defects of three types: (1) position abnormalities, with the heart in the right side of the chest instead of the left, so-called dextrocardia; (2) differentiation abnormalities, for example, failure to differentiate left and right atria, resulting in atrial isomerism; (3) intracardiac defects, with atrial or ventricular septal defects or abnormal relationships between the cardiac chambers and great vessels, such as malposition of the great vessels. The last two defects lead to mixing of venous and oxygenated blood, which results in either flooding of the lungs (heart failure) or deprivation of oxygen to the body (cyanosis). It is these symptoms that usually bring an affected individual to medical attention. Overall, there appears to be a 20-fold increased risk of such congenital heart defects in humans with dextrocardia.

Most cases of abnormal cardiac looping are sporadic in occurrence. This leaves the aetiology of such disturbances undetermined, i.e. it is unclear whether purely environmental or genetic factors are involved, or whether some combination is operative. Yet, a few cases where the aetiology is identifiable are informative about the normal process. One environmental factor is the conjoining of twins. Where conjoining occurs through the hearts, the heart of the left twin exhibits normal situs, while that of the right twin displays situs inversus (Cunniff et al 1988). The viscera of the respective sides are concordant, with situs solitus in the left twin and situs inversus in the right. It appears that mechanical effects of cardiac joining lead to abnormal, inverted, looping of the heart of the right twin. It was suggested that this early event leads to the subsequent inverted situs of the visceral organs of the right twin. As a control observation, in twins where conjoining does not involve the hearts, there appears to be no increased occurrence of cardiac or visceral situs inversus. Thus the heart might play a lead role in establishing the situs of the visceral organs, consistent with its being the first organ to become asymmetrical. This suggests that the viscera may not express or respond to the molecular signals that lead to cardiac looping but only to the physicomechanical signals of cardiac looping itself.

Genetic factors have also been implicated in lateralization by the identification of pedigrees in which lateralization defects are observed to be inherited. The inheritance patterns suggest defects in single genes. Here, as with the conjoined twins, the concordance of cardiac and visceral situs is often observed. However, the pedigree data cannot distinguish between the models of cardiac-driven visceral situs determination and autonomous independent determination of situs of the heart and viscera.

Human mutations affecting lateralization

A pedigree with an inherited lateralization defect is illustrated in Fig. 1 (Arnold et al 1983). The first identified affected individual, VII-21, presented to medical attention at five weeks of age with both respiratory distress and cyanosis, and died the following day of severe congestive heart failure. Post mortem examination revealed dextrocardia, bilateral left atria, abnormal pulmonary and systemic venous drainage, abnormal aortic arch, bilateral left-sided lungs and multiple spleens. These abnormalities all indicated a failure of correct lateralization, in particular bilateral left-sidedness. Examination of other members of this pedigree revealed defects in four additional family members that were consistent with the same type of lateralization defect. Their abnormalities were less severe and included: dextrocardia, anomalous great vessel anatomy and septal defects. The presence of consanguinity in this Amish family suggested that autosomal recessive inheritance might be operative. In individuals predicted to be homozygous, cardiac defects can apparently occur regardless of whether the individual has situs inversus.

Another pedigree affected with a lateralization defect is shown in Fig. 2 (Mathias et al 1987). Once again, a new-born infant was the first affected individual to be identified in the family. III-1 presented with cyanosis early after birth and was found to have transposition of the great vessels, a single ventricle, abdominal situs inversus and asplenia. He died at seven weeks of age. Examination of this pedigree revealed eight additional individuals with situs inversus, seven of whom exhibited congenital heart defects and four of whom exhibited either polysplenia or asplenia. All eight were males, related to each other via the maternal lineage, strongly indicating an X-linked mode of inheritance. The presence of affected females in the pedigree of Fig. 1 argues

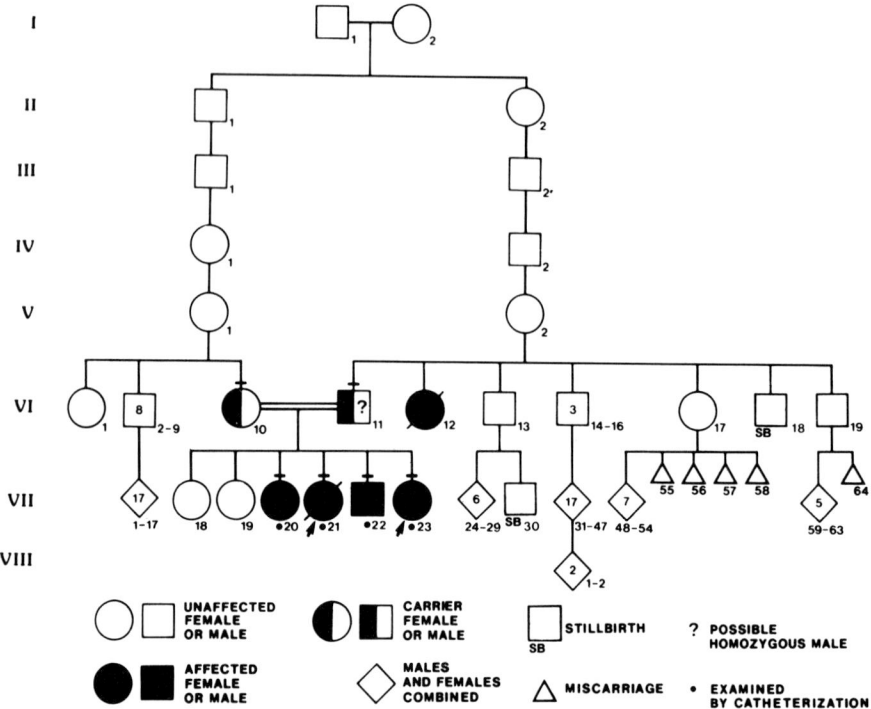

FIG. 1.   Pedigree of an Amish family suggestive of autosomal recessive lateralization defect. From Arnold et al (1983) with permission.

that the X-linked defect is distinct from the one proposed to be autosomal recessive.

A third pedigree illustrates another gene potentially involved with lateralization (Niikawa et al 1983) (Fig. 3). In this Japanese kindred, the proband III-2 presented as a new-born with cyanotic heart disease and was found to have a single ventricle. He died of heart failure at three weeks of age. At post mortem he was found to have trilobed lungs on both right and left sides (a manifestation of bilateral right-sidedness), asplenia, a bilobed liver, a midline stomach and abnormal great veins. Three years later a male sib was born with almost identical disease. At post mortem this infant was found to have dextrocardia, a single ventricle, anomalous venous return, malpositioned great vessels, two right-sided spleens, a symmetrically positioned liver and situs inversus of other visceral organs. The father was subsequently found to have dextrocardia and complete situs inversus above and below the diaphragm. Given a lack of reported consanguinity in this pedigree, the most likely pattern of inheritance is autosomal dominant, although one cannot completely exclude recessive inheritance if the father were a homozygote and the mother a heterozygote. If this were

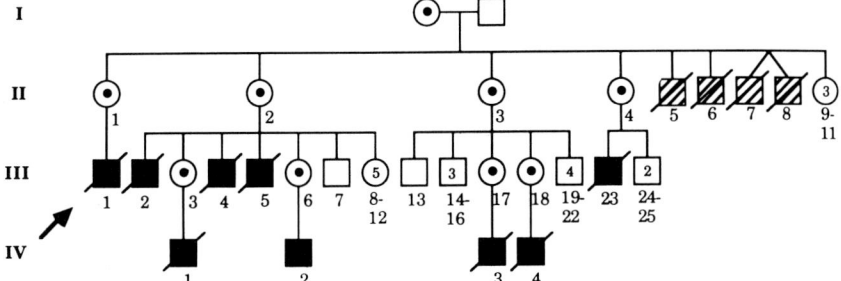

FIG. 2. Pedigree consistent with X-linked lateralization defect. From Mathias et al (1987) with permission. ■, affected male; ▨, unconfirmed affected individual; ☉, obligate carrier.

a dominant disorder that was usually as severe as in the affected new-borns, such rarely seen vertical transmission would not be surprising. The early lethal dominant disorder would arise as the result of new mutation but then fail to propagate; thus only isolated cases would be observed unless mildly affected individuals survived to procreate.

An additional recessively inherited human condition affecting lateralization is Kartagener's syndrome (McKusick 1988). Patients presenting with this condition exhibit chronic sinusitis and pulmonary disease, and males are infertile. These clinical features were found to result from defective motility of cilia, owing to deficiency of the structures known as dynein arms that connect outer

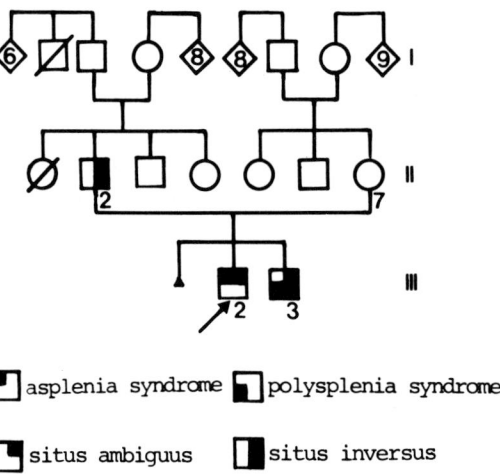

FIG. 3. Pedigree consistent with autosomal dominant lateralization defect. From Niikawa et al (1983) with permission.

doublet microtubules and catalyse the ATP-dependent sliding of microtubules past each other which constitutes ciliary bending (Afzelius 1976). In the respiratory tree, ciliary motility is critical to clearance of secretions. Cilia are also responsible for sperm motility, crucial for reproductive function. An additional phenotype in approximately 50% of patients with Kartagener's syndrome is situs inversus. This frequency suggests that in these patients situs determination is random. Indeed, ciliary function could be critical to the proper directional organ movements that produce situs solitus lateralization; loss of such function in Kartagener's syndrome could lead to randomization of situs.

A mouse mutation affecting lateralization

In 1959 Katharine Hummel and Dorothy Chapman at the Jackson Laboratory were examining an outbred stock of mice when they noticed that one litter contained several animals whose milk-filled stomachs were clearly localized to the right instead of the left side of the abdomen (Hummel & Chapman 1959). Post mortem examination showed that as well as the stomach the heart and visceral organs of affected animals were also inverted in position, i.e. there was virtually 100% concordance of situs of the heart with that of the viscera. The original parents of this litter and the affected siblings were mated to generate additional animals that exhibited situs inversus. Crosses, both at the Jackson Laboratory (Hummel & Chapman 1959) and by William Layton (1976), established that a single genetic locus is involved, called the *inversus viscerum*, or *iv*, locus. The mutant *iv* allele behaves as a recessive: heterozygous animals exhibit normal situs, while homozygous animals exhibit randomization of situs with approximately 50% situs solitus and 50% situs inversus. The phenotype of *iv/iv* mice has been investigated in a number of ways. First, the question of whether the mice have a defect homologous to that of Kartagener's syndrome has been addressed. Physiologically, the mice do not suffer pulmonary disease and males are fully fertile, indicating that ciliary function is intact in the respective organ systems. Cilia obtained from the respiratory tree of affected adult homozygotes have been examined by electron microscopy and found to contain normal dynein arms (Handel & Kennedy 1984). Thus, the step of lateralization affected in *iv/iv* animals is distinct from the ciliary function impairment of Kartagener's syndrome.

To assess whether situs is determined in a humoral or cell-autonomous manner, Brown et al produced aggregation chimeras of *iv/iv* embryos with wild-type embryos (1990). If situs determination were humorally determined, then the genotype of only a few critical cells should influence the outcome of cardiac lateralization. If situs determination were cell autonomous, the outcome would be influenced by the overall genotype of the collective of cardiac cells. The number of animals examined was not large, but in those chimeras that exhibited situs inversus over two-thirds of the heart tissue was composed of *iv/iv* cells. This supports cell-autonomous determination of cardiac situs.

The frequency of associated intracardiac defects has also been investigated in *iv/iv* mice. Layton & Manasek (1980) examined *iv/iv* mouse embryos at the time of cardiac looping and found that even when the direction of looping was normal, in approximately 15% of the embryos the shape of the loop appeared abnormal. This incidence correlated strikingly with the incidence of intracardiac defects, suggesting that abnormal looping might lead to such defects. This frequency of intracardiac defects represents a 20-fold increase over the baseline rate of occurrence of congenital heart disease. It agrees with results from human clinical studies which reveal that dextrocardia is associated with a 20-fold increase in the risk of congenital heart disease (Layton & Manasek 1980).

The nature of the *iv* gene and its product are unknown, but recent technological developments have made it possible to embark on a molecular genetic approach to the gene. This is the result of progress in three areas: acquisition of sufficient, usefully spaced, cloned polymorphic markers in the mouse to enable linkage analysis of matings; development of positional cloning technology; and the development of means either to transfer genes into mice or to disrupt endogenous genes.

We have used cloned polymorphic markers to map the chromosomal position of the *iv* gene and are now engaged in positional cloning of the locus. Linkage mapping of *iv* was accomplished by taking advantage of DNA polymorphisms that exist between different strains of mice (Fitch & Atchley 1985). The *iv* allele was placed within a polymorphic background using the Jackson Lab-derived *iv*-bearing strain called SI/Col that is simply the product of further inbreeding of homozygotes of the original outbred stock first examined in 1959 (Brueckner et al 1989). SI/Col mice were crossed individually with two inbred strains that are predicted to be genetically quite different, C57BL/6J and SJL/J. Offspring of these independent matings were predicted to be obligate heterozygotes for *iv* and all SI/Col-derived markers. The animals were phenotypically normal, as predicted. These F1 animals were backcrossed with homozygous SI/Col–*iv/iv* animals and the progeny scored for situs inversus, detected by right-sided position of the milk-filled stomach. One would predict that 25% of the backcross progeny would exhibit situs inversus, but in fact only 17% did so, probably because of intrauterine and perinatal loss from cardiac malformation.

Genomic DNA was prepared from the livers of situs inversus progeny and analysed by blot hybridization with a number of single-copy cloned mouse DNA probes. Wherever restriction fragment length polymorphisms were detected between SI/Col and either of the backcross strains, they were used to score for linkage between the locus being tested and *iv*. We were remarkably fortunate that with the fourth probe examined, p3′α, representing the immunoglobulin heavy chain gene constant (*IgH-C*) region, 79 of 80 backcross mice with situs inversus, obligately homozygous for *iv*, were also homozygous for the

SI/Col-derived allele of *IgH-C*. This placed the *iv* gene within 2 cM (centimorgans) of *IgH-C*.

IgH-C has been localized to the distal tip of mouse chromosome 12 (D'Eustachio 1984, Blank et al 1988, Taylor et al 1975). We next checked for recombinants with another marker of distal chromosome 12, *Aat*, alpha-1-antitrypsin, which had been mapped 5–10 cM proximal to *IgH-C* (see D'Eustachio 1984, Blank et al 1988, Taylor et al 1975), and found one recombinant in 28 backcross progeny. Not surprisingly, when the backcross progeny were scored with a much more proximal chromosome 12 marker, *D12Nyu1*, many more recombinants were observed. Thus we concluded that the *iv* locus maps to the distal portion of mouse chromosome 12.

More recently, we have refined the localization of *iv* by scoring the situs inversus backcross progeny with additional markers previously mapped to distal mouse chromosome 12, including *Ck-3*, creatine phosphokinase isozyme B (Mariman et al 1987); *D12N1*, an Abelson MuLV proviral integration site (Bauer et al 1988); *Akt-1*, a murine oncogene (Staal et al 1988); *IgH-V*, immunoglobulin heavy chain variable region; and HCGA, human chromogranin A (Murray et al 1987). The animal that was recombinant with *IgH-C* proved also to be recombinant with *IgH-V* and with *Akt-1*. No additional recombinants were observed with these markers. Thus, *iv* lies near to the *IgH* locus, most likely between this locus and *Aat*. Further studies are underway to improve this mapping information with additional markers as the relevant region is being isolated in molecular clones.

Once the *iv* gene is identified, a great deal can be learned about this step in mammalian lateralization, e.g. whether the product bears a relationship to any known protein, where and when the product is expressed, and whether mutations in human pedigrees with situs inversus do affect the human homologue. An even greater challenge lies ahead in the analysis of other steps of the presumed genetic pathway of lateralization in vertebrates. At present, the isolation of suppressor genes that allow second-site correction of a primary genetic defect, and which thus enable analysis of a genetic pathway by 'moving' from one locus to another, is not a practical undertaking in mice. Yet with additional model vertebrate systems for study, such as the zebrafish, it may be possible either to move from a homologous *iv* locus to additional steps or to isolate additional mutants *de novo*.

Acknowledgements

We thank Steven Reeders, Norman Talner and David Ward for advice and encouragement. M.B. is an NIH Physician–Scientist Trainee. A.H. is an investigator of the Howard Hughes Medical Institute.

References

Afzelius BA 1976 A human syndrome caused by immotile cilia. Science (Wash DC) 193:317–319

Arnold GL, Bixler D, Girod D 1983 Probable autosomal recessive inheritance of polysplenia, situs inversus and cardiac defects in an Amish family. Am J Med Genet 16:35–42

Blank RD, Campbell GR, Calabro A, D'Eustachio P 1988 A linkage map of mouse chromosome 12: localization of Igh and effects of sex and interference on combination. Genetics 120:1073–1083

Bauer SR, D'Hoostelaere LA, Huppi K 1988 Restriction fragment length polymorphism near the IgH locus on mouse chromosome 12. Nucleic Acids Res 16:8200

Brown NA, McCarthy A, Wolpert L 1990 The development of handed asymmetry in aggregation chimeras of *situs inversus* mutant and wild-type mouse embryos. Development 110:949–954

Brueckner M, D'Eustachio P, Horwich AL 1989 Linkage mapping of a mouse gene, *iv*, that controls left–right asymmetry of the heart and viscera. Proc Natl Acad Sci USA 86:5035–5038

Copenhaver WM 1926 Experiments on the development of the heart of amblystoma punctatum. J Exp Zool 43:321–371

Coulson A, Sulston J, Brenner S, Karn J 1986 Towards a physical map of the genome of the nematode *Caenorhabditis elegans*. Proc Natl Acad Sci USA 83:7821–7825

Coulson A, Waterston R, Kiff J, Sulston J, Kohara Y 1988 Genome linking with yeast artificial chromosomes. Nature (Lond) 335:184–186

Cunniff C, Jones KL, Jones MC, Saunders B, Shepard T, Benirschke K 1988 Laterality defects in twins: implications for normal asymmetry in human embryogenesis. Am J Med Genet 31:669–677

D'Eustachio P 1984 A genetic map of mouse chromosome 12 composed of polymorphic DNA fragments. J Exp Med 160:827–838

Driever W, Nusslein-Volhard C 1988 A gradient of bicoid protein in Drosophila embryos. Cell 54:83–93

Fire A 1986 Integrative transformation of *Caenorhabditis elegans*. EMBO (Eur Mol Biol Organ) J 5:2673–2690

Fitch WM, Atchley WR 1985 Evolution in inbred strains of mice appears rapid. Science (Wash DC) 228:1169–1175

Handel MA, Kennedy JR 1984 Situs inversus in homozygous mice without immotile cilia. J Hered 75:498

Hummel KP, Chapman DB 1959 Visceral inversion and associated anomalies in the mouse. J Hered 50:9–13

Layton WM 1976 Random determination of a developmental process. J Hered 67:336–338

Layton WM, Manasek FJ 1980 Cardiac looping in early *iv/iv* mouse embryos. In: Van Praagh R, Toao A (eds) Etiology and morphogenesis of congenital heart disease. Futura Press, New York, p 109–126

Manseau LJ, Schupbach T 1989 The egg came first, of course! Anterior–posterior pattern formation in Drosophila embryogenesis and oogenesis. Trends Genet 5:400–405

Mariman ECM, Broers CAM, Claesen CAA, Tesser GI, Wieringa B 1987 Structure and expression of the human creatine kinase B gene. Genomics 1:126–137

Mathias RS, Lacro RV, Jones KL 1987 X-linked laterality sequence: situs inversus, complex cardiac defects, splenic defects. Am J Med Genet 28:111–116

McKusick VA 1988 Entry 24440 In: Mendelian inheritance of man. Johns Hopkins University Press, Baltimore, p 1023–1025

Murray SS, Deaven LL, Burton DW, O'Connor DT, Mellon PL, Deftos LJ 1987 The gene for human chromogranin A (CgA) is located on chromosome 14. Biochem Biophys Res Commun 142:141–146

Niikawa N, Kohsaka S, Mizumoto M, Hamada I, Kajii T 1983 Familial clustering of situs inversus totalis and asplenia and polysplenia syndromes. Am J Med Genet 16:43–47

Roth S, Stein D, Nusslein-Volhard C 1989 A gradient of nuclear localization of protein determines dorsoventral pattern in the Drosophila embryo. Cell 59:1189–1202

Staal SP, Huebner K, Croce CM, Parsa NZ, Testa JR 1988 The AKT1 proto-oncogene maps to human chromosome 14, band q32. Genomics 2:96–98

Sulston JE, Horvitz HR 1977 Post-embryonic cell lineages of the nematode, *Caenorhabditis elegans*. Dev Biol 56:110–156

Sulston JE, Schierenberg E, White JG, Thomson JN 1983 The embryonic cell lineage of the nematode *Caenorhabditis elegans*. Dev Biol 100:64–119

Taylor BA, Bailey DW, Cherry M, Riblet R, Weigert M 1975 Genes for immunoglobulin heavy chain and serum prealbumin protein are linked in mouse. Nature (Lond) 256:644–646

Wood WB 1991 Evidence from reversal of handedness in *C. elegans* embryos for early cell interactions determining cell fates. Nature (Lond) 349:536–538

Wood WB, Kershaw D 1991 Handed asymmetry, handedness reversal and mechanisms of cell fate determination in nematode embryos. In: Biological asymmetry and handedness. Wiley, Chichester (Ciba Found Symp 162) p 143–164

DISCUSSION

Layton: I would like to congratulate Martina on this paper because this was a race to map the *iv* locus and we lost it! We didn't know it was a race; this was a tedious job and it is nice to see that somebody finally won.

Wood: Don't the additional markers, such as *Ck-3*, place the *iv* gene distal to the *IgH-C* cluster?

Brueckner: The animal that shows recombination between *IgH-C* and *iv* is derived from the backcross with the SJL/J strain. Because this strain shows no polymorphism with SI/Col (the background strain containing the *iv* mutation) for the markers that are located centromeric to *IgH-C*, it has been impossible to localize definitively *iv* centromeric or telomeric to *IgH-C*. We are currently examining other DNA markers that should clarify this. The creatine phosphokinase locus turns out to be not very helpful, because it is very close to *IgH-C*.

Collins: Leroy Stevens of The Jackson Laboratory discovered that teratomas invariably occurred twice as frequently in the left testis of male mice as in the right. When he backcrossed the *iv* gene into the inbred strain he was studying, he found the normal asymmetry in males with normally positioned viscera and the reverse pattern in mice with reversed viscera (Stevens 1982).

Brueckner: Is that connected with varying descent of the testes? Is there a correlation between temperature and the development of these testicular tumours?

Collins: Stevens suggested that this effect may reflect the level of origin of the spermatic artery or factors associated with it. In normal mice, the origin of the right spermatic artery is anterior to that of the left. I believe this pattern is reversed in *iv/iv* males.

Layton: We have found that the *iv* gene can be expressed in *iv/+* mice on the inbred SWV background, but at low penetrance. We have also found in about 1500 *+/+* SWV mice three with situs inversus. This raises an important but anecdotal issue. I have found that experienced anatomists or pathologists can examine an open mouse with situs inversus and miss it completely. I suspect that situs inversus and heterotaxis occur much more often than is reported. For instance, Fred Biddle et al (1991) have found heterotaxia in WB/ReJ mice involving only the azygos vein.

Peters: In right–left inversion of the heart, what happens to the vagus nerve? Does the right branch go to the sinoatrial node and the left one to the atrioventricular node?

Layton: In unreported work I did some years ago, I followed the path of right and left vagus nerves in reversed and non-reversed mice that did not have heterotaxia. As far as I could determine, the distribution of the vagus in the animals with situs inversus was a mirror-image of that in animals with situs solitus.

Loren Field (1988) has made transgenic mice that carry fusions between the transcriptional regulatory sequences of the atrial natriuretic factor and SV40 T antigen. The right atrium of these animals undergoes a several hundred-fold increase in mass, while the left atrium remains approximately normal in size.

More recently, he has made *iv/iv* mice transgenic for the same fusion gene (personal communication). As might be expected, most mice with situs solitus develop an enlarged right atrium and most of those with situs inversus develop an enlarged left atrium. A few of these mice develop bilateral atrial enlargement and thus probably have atrial isomerism. Loren Field also did electrocardiograms on many of these animals and found that many of the *iv/iv* control mice with situs inversus had cardiograms that would be expected, i.e. with mirror-image reversal of a wild-type heart. However, the transgenic mice with atrial enlargement had markedly abnormal cardiograms and eventually died of cardiac arrhythmias.

Berg: Does *Drosophila* have situs inversus?

Lawrence: I asked Michael Ashburner who told me to ask Michael Morgan, who told me there are no known mutants that give consistent situs inversus. There is not much bilateral asymmetry in flies, but there is the rotation of the genitalia in the male which is 180° counterclockwise and there is some kind of spiral in the gut. I doubt that anyone has systematically looked for mutants that affect this, and it might be worth doing.

Brueckner: How easy is it to see that rotation?

Lawrence: Very easy. You have to do an F2 screen since you will be looking for a recessive sterile mutation. Such a screen in *Drosophila* would probably be easier than your proposed screen in the mouse.

How many transcripts will you have to screen to cover one megabase in a mouse?

Brueckner: It can be anything from five to more than 25. How difficult identifying the transcript of the *iv* gene proves to be also depends on the size of the gene. If the *iv* gene is large, reintroducing it into a mouse would be technically very difficult.

Jefferies: There is one species of mitrate (*Peltocystis cornuta*) which shows situs inversus—there are 12 specimens known and four of them are the wrong way around.

Brown: Martina, could you tell us the *legless* story?

Brueckner: McNeish et al (1988) did random insertional mutagenesis of mice by injecting a DNA construct consisting of a *Drosophila* Hsp70 gene and a herpes virus thymidine kinase gene into 1-cell mouse embryos. If that DNA integrates into an expressed sequence, it will disrupt that sequence. The mice they generated were called *legless*. They had a complex phenotype with severe limb malformations, cranial defects and facial defects, and survived for only about 24 hours.

They later reported that not only do these mice have limb defects but they also have a 50% incidence of situs inversus (McNeish et al 1990). The location of the insertion was mapped and found to be in the same region as where we mapped the *iv* gene. This suggests that one of the insertional events disrupted the *iv* gene. It should be possible to clone the DNA sequences that immediately surround the insertional event. The problem is that the phenotype of the *legless* mice is very different from that of the *iv* mice. It may be that *legless* is the null mutation and *iv* is a point mutation or the *legless* phenotype may result from disruption of several genes.

Galaburda: Once you have identified the location of the *iv* gene, what are the next steps? Eventually, you want to know how the product of that gene produces the situs inversus phenotype.

Brueckner: Once somebody isolates the gene, the first steps are very straight forward. You sequence the gene, then analyse the nucleotide sequence by computer to derive the amino acid sequence and a prediction of the structure of the protein. Sometimes you can get clues to the function of a protein from the amino acid sequence. There are known sequence motifs for trans-membrane domains and various binding sites. One can compare it to all known sequences and see whether it has homology with any known proteins, which may give more clues to function.

The other approach is to express the cloned gene in a vector, raise antibodies to the protein and do localization studies to see when and where it is expressed within the embryo.

Wolpert: If you are lucky, you can make rapid progress. If you are unlucky, you are in real difficulty. Phil Leder, using a virus, picked up a mutation in the limbs of mice. His group isolated the gene and found that it codes for a protein like no other known one. Five or six years later they know a little about where the protein is made in the limb, but in terms of understanding its role they have made virtually no progress.

Brueckner: In contrast, the progress made on the neurofibromatosis gene has been spectacular. This gene was very difficult to isolate, but knowing the sequence has given some very interesting insights into its function.

McManus: When people first started to try and find genes, they worked backwards from proteins towards the gene sequence. That was easy if you had, say, a tumour cell producing 99% of its protein as a single protein. It is rather more difficult with laterality. There is, however, one condition called hemi-hypertrophy, which is produced by a tumour, classically by Wilms' tumour of the kidney. Is it possible that something on chromosome 14 of the mouse is producing a similar protein to whatever the Wilms' tumour is producing?

Burn: That's upside down, Chris. Wilms' tumour often arises on the side where there is hemihypertrophy, but the hemihypertrophy is almost certainly determined long before a kidney even develops in which one could have a nephroblastoma. There are other types of embryonal tumours in children who are prone to Wilms' tumour.

McManus: Do we know anything about the mechanism by which one side of the body can grow much more than the other?

Burn: The fact that the growth is asymmetrical is fascinating. A condition that is even more interesting is Proteus' syndrome, the condition from which the Elephant Man suffered, where there is massive overgrowth of parts of the body, which can be extremely lateralized.

Wood: Do any of the human syndromes related to situs inversus map to chromosome 14?

Brueckner: To date, none of these syndromes has been mapped. The only human example of inherited defects in the development of left–right asymmetry that follows the same sort of inheritance pattern as *iv* in the mouse is the Amish pedigree (Fig. 1), which is an autosomal recessive. No one has mapped that. The difficulty is statistics. If you looked very hard, you might be able to identify 50–60% of the genetically homozygous individuals. The remaining 40% are phenotypically normal. You need a tremendous number of individuals to get a good Lod score. I have discussed this with the people in Indiana who have this pedigree and we may do these studies.

Burn: Secondly, it is extremely rare to see heart malformation in a pedigree of individuals with Kartagener's syndrome. Half of the homozygotes have situs inversus; that's in marked contrast to the *iv/iv* mouse.

Brueckner: I know of one Kartagener's pedigree which refutes that. The patient has severe cardiac malformations; he has a sister with dextrocardia and a brother who was totally normal. A nasal biopsy on this patient, who had respiratory difficulties that were unrelated to Kartagener's syndrome, showed that he does have Kartagener's syndrome, as does his sister.

Morgan: Torgeson's data on the whole adult population of Norway showed that isolated dextrocardia was much less common than situs inversus viscerum et cordis. Is that typical or is it just Norwegians?

Layton: Kartagener's syndrome is a subset of the immature cilia syndrome and not a synonym. It is a triad that includes situs inversus, sinusitis and bronchiectasis. It was described years before the immotile cilia syndrome was discovered. Only about half of patients with immotile cilia syndrome have situs inversus and thus Kartagener's syndrome.

Brown: Afzelius (1976) proposed that the dextrocardia seen in patients with Kartagener's syndrome is due to a defect in ciliary function. I think it is extremely unlikely to be this straightforward. Some Kartagener's patients have ciliary dynein with an abnormal morphology, for example lacking arms, but that's not true of all patients with this syndrome: there are some with no observable ciliary abnormality. The primary defect in Kartagener's syndrome that is responsible for dextrocardia and situs inversus may be in a subcomponent of dynein. Dynein is an extraordinarily complex group of at least 13 proteins. Some of the ciliary dynein components are shared by cytoplasmic dyneins, and the primary defect in Kartagener's syndrome could be in one of the cytoplasmic dyneins, leading to a cellular dysfunction, perhaps in movement.

Burn: Immotile cilia syndrome is sometimes called Polynesian bronchiectasis because there is such a high occurrence of dynein arm-related ciliary abnormalities. Waite et al (1978) looked at the Polynesians and didn't find dextrocardia in that population. There were not many individuals with bronchiectasis, and by serial radiology Waite et al didn't find bronchiectasis in the few people with dextrocardia. This suggests that random determination of situs is not a consequence of the defect in cilial function.

Fujinaga: SI mice have normal cilia (Handel & Kennedy 1984); on the other hand, situs inversus does not occur in the *hpy/hpy* mutant strain of mice in which there is a high incidence of dynein arm defects (Bryan 1983).

Wolpert: Where do the components of dynein map?

Brueckner: None of its components are known to map to distal mouse chromosome 12 or distal human chromosome 14. We've looked at every gene that has been mapped to that area of the mouse chromosome. Not only were those probes not very helpful, none of the genes represents anything that could lead to this phenotype. We asked Dr Frank Ruddle whether any of the mouse homeobox genes have been mapped to this region or whether the human homeobox genes have been mapped to human chromosome 14 and they have not.

Brown: McNeish's group have obtained a certain amount of information from the flanking sequences of their insert. They have found a homeobox motif and a sequence that is conserved from *Drosophila* to humans, as identified by low stringency Southern blotting.

Wood: Nigel, you mentioned earlier that there might be extraembryonic uterine effects on setting up asymmetries. Could you say more about this?

Brown: Smith (1980, 1985) described the orientation of the axes in mouse embryos relative to the uterus. The mouse embryo implants with its antero-posterior axis in either of two specific directions that are 180° opposed to each other. Smith (1980) suggested that by aligning the axes in a specific orientation relative to the maternal uterus, the uterine environment could be used for positional information.

The *iv* mutation does not seem to be related to the uterine environment. Homozygous *iv* preimplantation mice transferred to a wild-type uterus, or vice versa, develop according to the genotype of the embryo, with no influence of uterine environment (Brown et al 1990).

Brueckner: T.W. Sadler has grown mouse embryos in culture. His explanted embryos don't develop situs inversus. They are grown in a rotating chamber, obviously not in a uterus and they don't implant. He grows them well past the stage of cardiac looping and that appears to occur normally.

Collins: We observed an interesting negative parent–offspring association. An *iv/iv* mother with normal situs tends to have a slight excess of abnormal offspring, whereas an *iv/iv* mother with expressed situs inversus tends to have more normal offspring (R. L. Collins, unpublished observations).

Brown: We don't find that in our lab. We don't find any correlation between maternal phenotype and offspring phenotype.

References

Afzelius BA 1976 A human syndrome caused by immotile cilia. Science (Wash DC) 193:317–319

Biddle FG, Jung JD, Eales BA 1991 Genetically-determined variation in the azygos vein in the mouse. Teratology, in press

Brown ŅA, McCarthy A, Wolpert L 1990 The development of handedness in aggregation chimeras of situs inversus mutant and wild-type embryos. Development 109:949–954

Bryan JHD 1983 The immotile cilia syndrome. Mice versus man. Virchows Arch A Pathol Anat Histopathol 399:265–275

Field LJ 1988 Atrial natriuretic factor–SV40 T antigen transgenes produce tumors and cardiac arrhythmias in mice. Science (Wash DC) 239:1029–1033

Handel M, Kennedy JR 1984 Situs inversus in homozygous mice without immotile cilia. J Hered 75:498

McNeish JD, Scott WJ, Potter SS 1988 *Legless*, a novel mutation found in PHT-1 transgenic mice. Science (Wash DC) 241:837–839

McNeish JD, Thayer J, Walling K, Sulik KK, Potter SS, Scott WJ 1990 Phenotypic characterization of the transgenic mouse insertional mutation, *Legless*. J Exp Zool 253:151–162

Smith LJ 1980 Embryonic axis orientation in the mouse and its correlation with blastocyst relationships to the uterus. Part I. Relationships between 82 hours and 4½ days. J Embryol Exp Morphol 55:257–277

Smith LJ 1985 Embryonic axis orientation in the mouse and its correlation with blastocyst relationships to the uterus. Part II. Relationships from 4.5 days to 9.5 days. J Embryol Exp Morphol 89:15–35

Stevens LC 1982 Genetic influences on teratocarcinogenesis in mice. In: Tsukada Y (ed) Genetic approaches to developmental neurobiology. University of Tokyo Press, Tokyo (Japan Medical Research Foundation Publication No. 17) p 87–94

Waite D, Steele R, Ross I, Wakefield SJ, Mackay J, Wallace J 1978 Cilia and sperm tail abnormalities in Polynesian bronchiectasis. Lancet 2:132–133

Asymmetries of cerebral neuroanatomy

Albert M. Galaburda

Department of Neurology, Harvard Medical School, Beth Israel Hospital, 330 Brookline Avenue, Boston, MA 02215, USA

Abstract. The mammalian cerebral cortex is asymmetrical. One hemisphere does not contain cortical areas or architectonic patterns, histological features, ultrastructural characteristics, or connectivities of the neurons that are not present in the other: homologous areas on the two sides may differ only in size. Asymmetry has directionality: two-thirds of human brains have plana temporale that are larger on the left. Conversely, roughly the same number of non-human brains show asymmetry in one direction as in the other. Asymmetry has magnitude: some brains show a large asymmetry, others show no asymmetry in a given area. Symmetrical areas are larger than their asymmetrical counterparts, which reflects fewer neurons in the latter. Indirect evidence points to variable asymmetry in the germinal zones in the production of symmetrical or asymmetrical cortical areas. These areas differ in their patterns of callosal connections. Fewer connections are seen in the asymmetrical cases, paralleling the smaller number of neurons. The symmetrical cases contain connections that are more widely distributed. These findings of different numbers of neurons and different proportions of callosal connections suggest that symmetrical and asymmetrical cortical areas may have different functional properties.

1991 Biological asymmetry and handedness. Wiley, Chichester (Ciba Foundation Symposium 162) p 219–233

It is now widely accepted that the two hemispheres of the human brain are differentially involved in cognitive, attentional and emotional tasks and that they are anatomically asymmetrical. Functional specialization and lateralization show substantial individual variability. For instance, the exact size and location of a left hemisphere lesion that will produce a given form of aphasia vary among individuals (Mohr et al 1978). Non-right-handers are likely to be functionally more variable and to respond less predictably to lesions. They also have anatomically more symmetrical language areas or otherwise anomalous patterns of cerebral asymmetry. For example, the pattern of asymmetry of the Sylvian fissures that is most often found among right-handers, which reflects the greater development of the parietal and temporal opercula on the left, is less common in non-right-handers (Hochberg & LeMay 1975). A bias of left-handers toward exhibiting more symmetrical brain areas is seen in computed tomographic (CT) scans, whereby the typical left occipital petalia (a protrusion of one lobe of the brain producing a change on the inner surface of the skull) is found less often (LeMay & Kido 1978).

Human cerebral asymmetry

Geschwind & Levitsky (1968) reported consistent anatomical asymmetries in human cerebral hemispheres. They examined the outside border of the planum temporale (a language area on the posterior superior temporal plane) and found that 65 of the 100 brains they examined had a larger left planum, 11 had the reverse asymmetry and 24 had no bias. These results have been confirmed in several studies (Wada et al 1975, Witelson & Pallie 1973, Rubens et al 1976, Teszner et al 1972). Asymmetries in the lengths of the Sylvian fissures assessed directly on brains (Rubens et al 1976, Yeni-Komshian & Benson 1976) and in endocranial cast markings and fossil skulls (Rubens et al 1976, LeMay 1977) showed that the left Sylvian fissure is longer in the majority of human brains.

Cytoarchitectonic asymmetries have also been demonstrated in the superior temporal plane. Cytoarchitectonic area Tpt (an auditory association cortex located partly within the planum temporale) shows striking asymmetries (Galaburda et al 1978) that correlate with the planum asymmetry.

Wada, Clarke and Hamm found that the area of the convexity of the frontal operculum was greater on the right side (Wada et al 1975). However, they observed greater folding on the left and stated that the asymmetry could well be in the opposite direction, if buried cortex was included. Area 44 within the pars opercularis of the frontal operculum shows an asymmetry in favour of the left (Galaburda 1980).

The parietal lobe, too, is asymmetrical. Area PEG, which is architectonically linked to the non-language-related areas of the superior parietal lobule, is larger on the right, while language-related area PG on the angular gyrus is larger on the left (Eidelberg & Galaburda 1984). In addition, there is a significant positive correlation between the asymmetry of the planum temporale and that of area PG—both language related—but no relationship between asymmetry of area PEG and that of a language-related area.

Non-human anatomical asymmetries

Functional cerebral lateralization has been demonstrated in non-human species. Generally, compared to humans lateralization is less biased to the right or left as a population but can be marked in individual animals. Anatomical asymmetries are seen as well. Sylvian fissures in chimpanzees and other great apes are significantly longer on the left. The skull of the mountain gorilla is asymmetrical. The right frontal pole of baboon brains is also longer, which is similar to the finding in humans of right frontal protrusion. Morphological cerebral asymmetries were also found in Old World monkeys (for a review see Sherman et al 1982).

Because our work on mechanisms of asymmetries involves rodents, I will stress their asymmetries here. Diamond, Johnson and Ingham examined the forebrains of male Long–Evans rats at several ages ranging from 6 to 300 days and found that the right neocortex was thicker than the left at all ages (Diamond et al 1975). Female rats had a slightly thicker left neocortex, although not significantly so. Females ovariectomized at birth developed a significantly thicker right neocortex, thus mimicking the male pattern. The same laboratory has reported asymmetries in cortical oestrogen receptors (Sandhu et al 1986). We found that male rats have larger right neocortical volumes, whereas females have an insignificant bias to the left (Sherman & Galaburda 1984). Architectonic cortical parcellation indicated that the primary visual and sensorimotor cortices tended to be larger on the right side, whereas the motor region was symmetrical (G. F. Sherman, A. M. Galaburda, unpublished observations 1984). Additional asymmetries of the posterior cortex have been associated with lateralization of emotionality (Crown et al 1987, Denenberg 1981). Kolb et al (1982) found that the right hemisphere weighed more in adult and 15-day old Long–Evans rats, and that the right hemisphere was longer, wider and taller, and the cortex thicker. Several other asymmetries at the level of neurochemistry have also been reported in the rat (Drew et al 1986, Glick & Ross 1981).

Most studies, in both animal and human brains, have stressed the direction of the asymmetry, that is, biases in populations toward a particular direction. In human brains the biases are usually striking, whereas in animals the direction of asymmetry tends to be more evenly distributed. However, in both human and animal brains an important question is what mechanisms underly the magnitude of asymmetry, regardless of direction. Although animals such as rodents do not show large population biases in the direction of asymmetry of neocortical areas, the presence of asymmetrical areas in individuals is the rule rather than the exception. The magnitude of this asymmetry varies in interesting ways among individual animals—ways that can illuminate the process of asymmetrical development of neural structures and ultimately the functional significance of asymmetry.

Recent studies of brain asymmetry

The planum temporale

A priori analysis of the relationship between asymmetrical and symmetrical brain areas yields three possibilities, which are illustrated in Fig. 1. Symmetrical brains can result from (1) an increase in the size of the normally smaller side, (2) a decrease in the normally larger side, or (3) a combination of these.

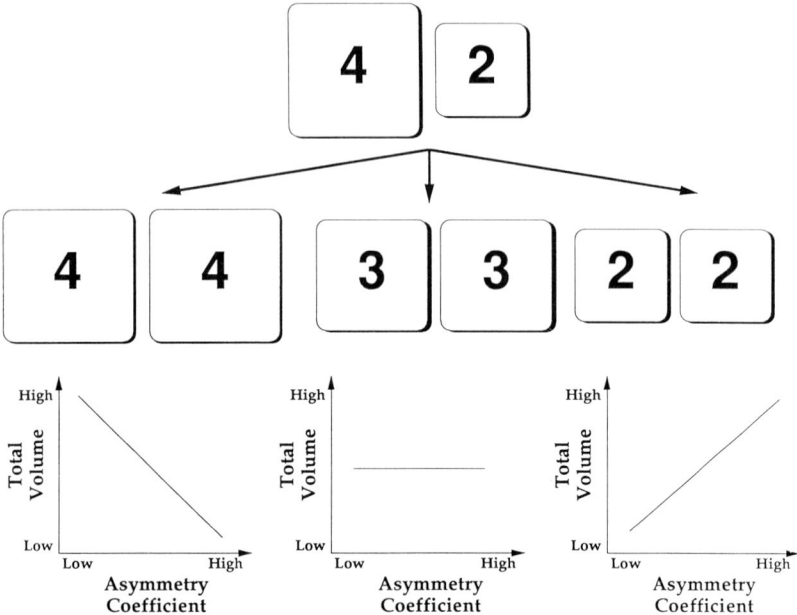

FIG. 1. Schematic illustration of the three hypothetical possibilities when comparing an asymmetrical brain substrate (top) to a symmetrical one (middle). The symmetrical substrate can be made up of two large sides (left), two small sides (right), or two medium-sized sides (middle). Each of these possibilities predicts a different relationship between total amount of substrate $(R+L)$ and degree of asymmetry [asymmetry coefficient $\delta = |R-L| \div \{(R+L)(0.5)\}$]. Thus, when the symmetrical brain region is made up of two large sides, one would expect a negative correlation between these two variables; a positive correlation would be predicted if symmetrical brain regions were composed of two small sides; in the case of the medium-sized symmetrical brain regions, there would be no change in total volume with changes in asymmetry coefficient.

Measurements of the right and left planum areas (rather than external lengths) were made in the same 100 brains studied by Geschwind & Levitsky (1968). The total planum area (right + left) correlated negatively with the asymmetry coefficient (δ), indicating that as asymmetry increased, the total planum area decreased ($r = -0.530$, $t = 6.19$, df (degrees of freedom) $= 98$, $P < 0.001$). The area of the smaller of the two plana significantly predicted the degree of asymmetry ($r = -0.831$, $t = 14.79$, df $= 98$, $P < 0.001$). There was no correlation between the degree of asymmetry and the size of the larger planum ($r = -0.065$, $t = 0.64$, df $= 98$, not significant). These results support the hypothesis that the asymmetrical case may represent a unilaterally reduced version of the symmetrical case (Galaburda et al 1987).

Histological features

There are two means by which a cortical area can be reduced in size, and, therefore, three hypothetical causes of the difference between asymmetrical and symmetrical brain areas: (1) unilateral decrease in cell numbers in the small side of the asymmetrical case, (2) unilateral increase in cell packing density in the small side of the asymmetrical case, or (3) a combination of the two. We counted neurons in visual area 1 of the rat and found no relationship between the asymmetry of cell packing density and that of architectonic volume. Thus, asymmetrical substrates must differ in the number of their neurons (Galaburda et al 1986). Rakic & Williams (1986) showed that reduction in the cortical cytoarchitectonic perimeter resulted from lower cell numbers, not increased cell packing density. Asymmetry of neuronal numbers has also been documented in the lateral geniculate nucleus (Williams & Rakic 1988).

It is not surprising that differences in cell numbers, not cell packing density, account for the bulk of volumetric asymmetry. Some architectonic asymmetries in the human brain, for example, may reach a sixfold difference between the sides. Differences in cell packing density large enough to produce this degree of volumetric asymmetry would be expected to distort the cytoarchitectonic appearance of the areas to such a degree that their identification as hemispheric homologues would not be possible.

Connectivity and asymmetry

The corpus callosum of eight rats was surgically sectioned to look at callosal connections in relation to the asymmetry of architectonic areas. The average density of callosal terminations was determined for visual area 1 and the proportion of the neocortical area receiving callosal terminals was computed. The non-directional asymmetry coefficient was calculated for the visual area. There was an inverse relationship between this coefficient and the average density of terminations in this area ($r = -0.776$, $t = 3.01$, $df = 6$, $P < 0.05$). This indicated that the more symmetrical visual areas had a greater density of callosal terminations than the more asymmetrical ones. Also, the greater the asymmetry, the smaller the percentage of the area that received callosal terminations (Rosen et al 1989).

Ontogenesis of asymmetry

These asymmetries are present early in life, as documented by studies of fetal and infant brains (Chi et al 1977, Wada et al 1975, Witelson & Pallie 1973). Asymmetry of the planum temporale is visible by 31 weeks of fetal age (Chi et al 1977). At that time, generation of neurons destined to form the neocortex and neuronal migration are virtually completed; it is possible that asymmetry arises as a result of unequal early generation of neurons. Time-mated Wistar rats were injected intraperitoneally with [^{3}H]thymidine on embryonic Day 13, 15, 17, 19 or 21. The pups from these mothers were killed on postnatal Day 5, 10, 30 or 60, and the brains were processed for autoradiography. The numbers of labelled and unlabelled neurons were counted and the ratio of these numbers was computed to give an estimate of neuronal division during the period just before migration to the incipient cerebral cortex. The results showed no difference between the rates of neuronal proliferation on either side (Rosen et al 1991). Therefore, any differences in neuronal numbers must reflect different rates of neuronal division early in the germinal zones, a period that cannot be studied using current techniques. Brain cells may also disappear after proliferation and migration, either by neuronal death or by assignment of neurons to neighbouring areas. Preliminary results do not support the presence of significant differences in neuronal disappearance between the two sides.

Conclusion

The only anatomical feature that could explain the functional differences between the hemispheres is the difference in size of specific types of cortex on the two sides. Individual variability in cognitive and emotional functions could be explained, at least in part, by individual variability in the degree of these size differences. Cortical areas that are more asymmetrical have a smaller total number of neurons and more restricted patterns of callosal, and possible other, connections. The origin of these differences is in early brain development, when the number of neurons and number and patterns of long connections are established. Present research is addressing whether the degree of asymmetry is also related to varying proportions of different neuronal types, which would support the idea that this measure could explain individual functional differences.

Acknowledgements

Some of the work reported here was supported by NIH grant NS27119, a grant from The Orton Dyslexia Society and a grant from the Herzog Foundation. The author thanks his collaborators, Gordon Sherman and Glenn Rosen.

References

Cain DP, Wada JA 1979 An anatomical asymmetry in the baboon brain. Brain Behav Evol 16:222–226
Chi JG, Dooling EC, Gilles FH 1977 Gyral development of the human brain. Ann Neurol 1:86–93
Crown DP, Richardson CM, Dawson KA 1987 Lateralization of emotionality in right parietal cortex of the rat. Behav Neurosci 101:134–138

Denenberg VH 1981 Hemispheric laterality in animals and the effects of early experience. Behav Brain Sci 4:1–49

Diamond MC, Johnson RE, Ingham CA 1975 Morphological changes in the young, adult, and aging rat cerebral cortex, hippocampus, and diencephalon. Behav Biol 14:163–174

Drew KL, Lyon RA, Titeler M, Glick SD 1986 Asymmetry in D-2 binding in female rat striata. Brain Res 363:192–195

Eidelberg D, Galaburda AM 1984 Inferior parietal lobule: divergent architectonic asymmetries in the human brain. Arch Neurol 41:843–852

Galaburda AM 1980 La région de Broca: observations anatomiques faites un siècle après le mort de son découvreur. Rev Neurol (Paris) 136:609–616

Galaburda AM, Sanides, F, Geschwind N 1978 Human brain: cyto-architectonic left-right asymmetries in the temporal speech region. Arch Neurol 35:812–817

Galaburda AM, Aboitiz F, Rosen GD, Sherman GF 1986 Histological asymmetry in the primary visual cortex of the rat: implications for mechanisms of cerebral asymmetry. Cortex 22:151–160

Galaburda AM, Corsiglia J, Rosen GD, Sherman GF 1987 Planum temporale asymmetry: reappraisal since Geschwind and Levitsky. Neuropsychologia 25:853–868

Geschwind N, Levitsky W 1968 Human brain: left-right asymmetries in temporal speech region. Science (Wash DC) 161:186–187

Glick SD, Ross AD 1981 Lateralization of function in the rat brain: basic mechanisms may be operative in humans. Trends Neurosci 4:196–199

Hochberg FH, LeMay M 1975 Arteriographic correlates of handedness. Neurology 25:218–222

Kolb B, Sutherland RJ, Nonneman AJ, Whishaw IQ 1982 Asymmetry in the cerebral hemispheres of the rat, mouse, rabbit, and cat: the right hemisphere is larger. Exp Neurol 78:348–359

LeMay M 1977 Morphological cerebral asymmetries of modern man, fossil man, and non-human primate. Ann NY Acad Sci 280:349–366

LeMay M, Kido DK 1978 Asymmetry of cerebral hemispheres on computed tomograms. J Comput Assisted Tomogr 2:471–476

Mohr JP, Pressin MS, Finklestein S, Funkenstein HS, Duncan GW, Davis KR 1978 Broca aphasia: pathologic and clinical. Neurology 28:311–324

Rakic P, Williams RW 1986 Thalamic regulation of cortical parcellation: an experimental perturbation of the striate cortex in rhesus monkeys. Soc Neurosci Abstr 12:1499 Abstr 405.9

Rosen GD, Sherman GF, Galaburda AM 1989 Interhemispheric connections differ between symmetrical and asymmetrical brain regions. Neuroscience 33:525–533

Rosen GD, Sherman GF, Galaburda AM 1991 Ontogenesis neocortical asymmetry: a [3H]thymidine study. Neuroscience 41:779–790

Rubens AB, Mahowald MW, Hutton JT 1976 Asymmetry of lateral (Sylvian) fissures in man. Neurology 26:320–324

Sandhu S, Cook P, Diamond MC 1986 Rat cerebral cortical estrogen receptors: male-female, right-left. Exp Neurol 92:186–196

Sherman GF, Galaburda AM 1984 Neocortical asymmetry and open-field behavior in the rat. Exp Neurol 86:473–482

Sherman GF, Galaburda AM, Geschwind N 1982 Neuroanatomical asymmetries in non-human species. Trends Neurosci 5:429–431

Teszner D, Tzavaras A, Gruner J, Hécaen H 1972 L'asymmétrie droite-gauche du planum temporale: a propos de l'étude anatomique de 100 cerveaux. Rev Neurol (Paris) 126:444–449

Wada JA, Clarke R, Hamm A 1975 Cerebral hemispheric asymmetry in humans. Arch
    Neurol 32:239–246
Williams RW, Rakic P 1988 Elimination of neurons from the rhesus monkey's lateral
    geniculate nucleus during development. J Comp Neurol 272:424–436
Witelson SF, Pallie W 1973 Left hemisphere specialization for language in the newborn:
    neuroanatomical evidence of asymmetry. Brain 96:641–646
Yeni-Komshian GH, Benson DA 1976 Anatomical study of cerebral asymmetry in the
    temporal lobe of humans, chimpanzees and rhesus monkeys. Science (Wash DC)
    192:387–389

DISCUSSION

Corballis: There is much more functional asymmetry in the human brain than
in the rat brain, arguably to the point of a discontinuity. To what extent do
rules that you have from the rat brain apply to the human brain?

Galaburda: Many of the studies we do in the rat we cannot do in humans,
so that is a limitation. We do, however, see the same relationship between size
of areas and the coefficient of asymmetry in all the species we have looked at,
so we feel comfortable about pursuing that particular relationship. I have no
reason to believe that if the number of neurons determines asymmetry in one
species, it is not going to do so in other species.

Corballis: The human brain is also more directionally asymmetrical than the
rat brain. A given area is nearly always larger on one particular side.

Galaburda: Not always, but, as you imply, more often than in the rat.
Directional asymmetry is an important issue. It is not known why in humans
there is a population bias in a particular direction. Our research has been focused
on the degree or magnitude of asymmetry.

Annett: Can you clarify your work on the corpus callosum? You cut the
corpus callosum, then you measured the quantitative degeneration of axons and
found there was more degeneration of the larger side, is that correct?

Galaburda: There is no quantitative or qualitative difference between the two
sides. The differences I described are between symmetrical and asymmetrical
areas in individual animals. In development, the callosal connections finally
settle down to occupy the vertical meridian between area 17 and area 18; most
other callosal connections are eliminated. Before that stage, callosal fibres 'wait'
under the cortical plate over a much broader region. The difference between
asymmetrical and symmetrical areas seems to lie in the size of the alleged vertical
meridian—the symmetrical cases have a larger one anatomically.

Peters: Scheibel (1984) did some work on the tertiary branching of the
dendrites and claimed that there were differences between the left and right sides
of the brain in terms of the richness of dendritic branching.

Galaburda: That was a Golgi study. Arnie Scheibel took blocks from
topographically symmetrical areas of the brain, which doesn't mean that he

was necessarily taking blocks from the same architectonic area, let alone equivalent positions of a given architectonic area. Therefore, I am not convinced that there are histological asymmetries between asymmetrical areas of the cortex. But this is still an open question.

McManus: You say the differences between the two halves of the brain are principally quantitative rather than qualitative in terms of the morphology. Yet in functional terms the differences between the two sides of the brain are qualitative. One side has a function that the other does not.

You were looking at area 17, the primary visual cortex. If I was asked to predict which area of the brain was *least* likely to show any functional asymmetries, it would be area 17. Have you any evidence that the morphological asymmetries you are finding in visual cortex have any relation to the asymmetrical functional abilities in those animals? Have you tested them behaviourally before examining them histologically?

Galaburda: I think the qualitative asymmetries seen in brain function are epiphenomena of the way this problem is studied. I believe there are bilateral areas that are connected to each other and form unified networks that may or may not be asymmetrical. In the asymmetrical case, damage to the larger side of the network is apt to produce a worse deficit. In the more symmetrical cases, the effect is milder; it may also be qualitatively different and recovery is more likely from the intact (contralateral) part of the network. There is a point at which so much of the network has been lost that the function is abolished. It is like clipping a wing—suddenly it is too short for the bird to fly.

If I view lateralization in this way, I have no need to defend our study of area 17. As far as I know, all asymmetrical brain areas behave in a similar way: the planum temporale doesn't have a primary area, it has high order association areas and it has the same relationship between asymmetry and total size as does area 17.

Wolpert: Are you saying that there is no functional significance to brain asymmetries?

Galaburda: No; I am saying that it is unlikely that area A on the left does one thing while homologous area A on the right does another. I believe there are bilateral networks biased to one side or the other such that network A is more prominent on the left while network B is more prominent on the right. This confers asymmetrical functions on the two hemispheres.

McManus: You are implying that there is some critical mass at which a quantitative change flips over into an apparently qualitative one.

Galaburda: Yes; this is partly because patterns of connections change with quantitative changes in cell numbers. Take the wing again: a single quantitative lengthening leads to the emergence of a new property—flight.

Lewis: What about functional asymmetries measured by positron emission tomography that monitors cerebral blood flow? If you give a person a language-intensive task and compare brain activity on the two sides of the brain, does the laterality revealed in that way correlate with your anatomical results?

Galaburda: Most of those blood flow studies show bilateral enhancement, which supports what I am saying. The resolution of those techniques may not be capable of showing the quantitative asymmetries I have shown. Secondly, some of those studies are very difficult to interpret, because they depend on statistical coherence after comparing many subjects. There is a tremendous amount of variability and shifting topography in these brain areas. In one person the language area may be situated a little further forward than in another, which would confound the results. If the variability is mainly in the size of the non-dominant side, as we have shown, then it is more likely that the participation of that side in the function being studied will be missed by these averaging techniques.

Lewis: Are you saying there is no consistent laterality that you can demonstrate and no consistent localization of function with respect to language?

Galaburda: Not in such strong terms, but to some extent that is correct. It has to do with the fact that the minor (non-dominant) side, which has a smaller volume in the asymmetrical cases and a larger volume in the more symmetrical ones, will be less statistically coherent in population studies. Thus, it will tend not to be enhanced in PET studies. In such studies it is only when enough subjects show enhancement in the same area that there is a colour change in the algorithm, causing the area to be identified as participating in a particular function. If there is variability in the topography between individuals, that masks the signal. So there are sometimes qualitative asymmetries, but I think they are artifactual.

Wolpert: I have a terrible feeling you are telling us that all our nice ideas about left brain, right brain function have to be abandoned.

Galaburda: You've got it!

Galloway: The coefficient of asymmetry is derived by subtracting left from right then dividing by right plus left. You say that quantity is proportional to one divided by right plus left. This suggests that the right–left difference is constant.

Galaburda: You have a point. It is difficult to know how to handle this. On the one hand, if one does not correct for size, then 2 versus 1 is less asymmetrical than 4 versus 2 (see Fig. 1). On the other hand, there must be some significance in the fact that some brains are 4/2 and others are 2/1, even when one brain is not twice the size of the other in total. Fortunately, we have found that whether we correct for size or not, the relationships between asymmetry and size, or connectivity, still hold.

Frankel: The inverse relationship between the degree of asymmetry and the total volume of a given brain region on the two sides means that when the brains are symmetrical, there is more brain volume on both sides.

Galaburda: More on one side, the usually smaller side. The usually larger side is always invariant, irrespective of the degree of asymmetry.

Frankel: Then defects such as dyslexia are associated not with having less of something but with having more of something.

Galaburda: We have many examples of exuberance of neurons and connectivity, including preservation of transient connections which may produce more brain tissue (not necessarily more good brain tissue). We have lesioned neonatal rat brains and Innocenti & Berbel (1991) have lesioned neonatal kitten brains; in both cases there was preservation of transient developmental connections.

Chothia: You indicated that loss of asymmetry may have an effect on function. Do people with dyslexia show loss of brain asymmetry to some degree?

Galaburda: Yes, the planum temporale of the nine dyslexics in our autopsy studies has been consistently symmetrical. This pattern occurs in only roughly 25% of the general population. In studies using neuroimaging devices a change in the pattern of asymmetry, including the planum temporale, has also been demonstrated quite frequently in dyslexics.

Peters: What about the idea that there are some ectopias in dyslexia, involving disorderly patterns of development and migration of neurons in the area of the angular gyrus?

Galaburda: We have seen migration anomalies affecting language cortex in eight of nine cases. There are ways to interrelate the different anatomical findings. To summarize, proper elimination of developmental errors appears not to be taking place in the brains of dyslexics. Cells, by various mechanisms, may end up out of place, but there are mechanisms to eliminate them. Likewise, proper asymmetry may reflect asymmetrical elimination of exuberant neurons. Thus, lack of elimination may be the common theme in symmetry and migration anomalies.

Morgan: Dyslexia as a diagnostic category is a terrible mess. What were your criteria for dyslexia and what were your control groups?

Galaburda: The patients we studied were all diagnosed as having dyslexia while they were alive. The same test batteries were not used in all cases, but the diagnosis appears valid. The patients were all intelligent and, in some cases, very successful people. The control group is the normal population that we and others have studied. We know the distribution of asymmetry in the general population; the patients with dyslexia have a very different distribution of asymmetry.

For the ectopias, our control cases are normal brains in the Yakovlev Collection in Washington, which is a series of similarly prepared histological brain sections. In the latter, the instances of migration anomalies were rare.

Annett: You said that it is easy to produce these ectopias in the brain by any sort of damage. How can you be sure they are distinctive of dyslexic brains?

Galaburda: They may not be distinctive of dyslexic brains; it just happens that dyslexics have lots of ectopias and they have them in a peculiar distribution.

In our animal models we have mutants that produce ectopias spontaneously. We have other strains in which we produce ectopias by a variety of injurious methods. Ectopias may not be specific to dyslexia, but the cause of their existence in dyslexia may be unique—we don't know yet.

Corballis: Why do you think dyslexia is singled out, as distinct from language generally?

Galaburda: I don't believe it is singled out. It is more common than developmental dysphasia, so we have had a chance to study dyslexia. Specific language disorders of children may be associated with similar changes, but perhaps more severely. I can't say anything about other types of learning disability because I haven't studied them.

Corballis: There was an old and very influential theory that's now largely discredited, proposed by Samuel Torrey Orton, that dyslexia is due to the lack of cerebral dominance.

Galaburda: I am not going suggest what the lack of anatomical asymmetry in the brains of dyslexic people may have to do with cerebral dominance in the way we usually think of it. Because of the different size and connectivity of symmetrical areas, I think a child with dyslexia who has this kind of brain brings to the task of learning to read a very different machine that would favour the use of certain cognitive strategies over others. Say, for instance, that to learn to read one has to rely on phonological knowledge early on, but as one grows up one can use other strategies, such as semantics and pragmatics. It may be that the dyslexic brain is not well suited to phonological knowledge. There are computational models one could devise that would support the hypothesis that alterations in numbers of processing units and different patterns of connections at the outset are not compatible with certain cognitive strategies.

Crow: I would like to present some results on schizophrenia and represent this as a primary disturbance of the human brain. The clue to the pathogenesis of schizophrenia that we have had for 15 years is that there are morphological changes in the brains (Johnstone et al 1976). The cerebral ventricles of patients with chronic schizophrenia are slightly enlarged. We have recently uncovered a clue to what this means—the changes are in the temporal lobe and they are asymmetrical.

We have done a post mortem analysis of formalin-fixed brains of patients with schizophrenia compared to age-matched controls and of patients with Alzheimer's dementia compared to their age-matched controls (Crow et al 1989). We looked at the percentage increase in ventricular size by examining the various components of the lateral ventricle using X-ray photography after the ventricle had been filled with radioactive material (Fig. 1).

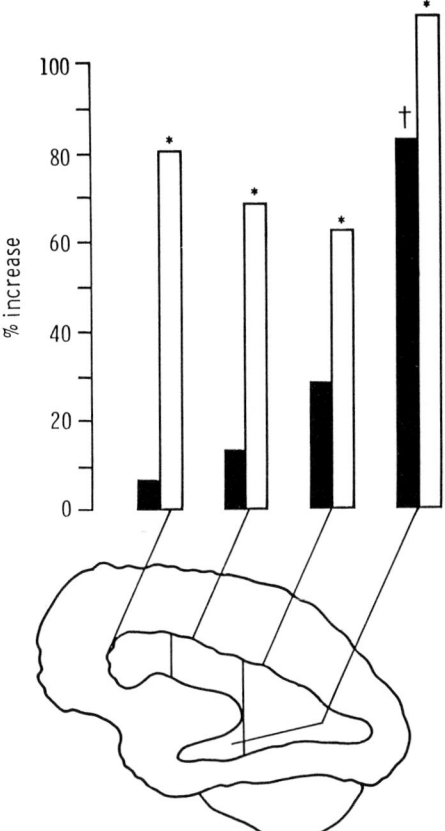

FIG. 1. (*Crow*) Percentage increase in the components of the lateral ventricle in post mortem brains of patients with Alzheimer-type dementia (open columns) and schizophrenia (filled columns). Adapted from Crow et al (1989).

In Alzheimer's dementia the ventricular enlargement is widespread. In schizophrenia the enlargement is in the temporal horn more than anywhere else. That replicates what we had already shown (Brown et al 1986).

We then compared the left and right temporal horns. In Alzheimer dementia there is an increase on the left and an increase on the right of almost 100% (compared to controls). In schizophrenia there is an increase on the left and really no change on the right. The difference is significant, (ANOVA $P<0.001$).

The use of another disease, Alzheimer dementia, that causes ventricular enlargement bilaterally as a control, whereas in schizophrenia the enlargement is on only one side, pushes you towards the conclusion that schizophrenia and human cerebral asymmetry are intimately linked (Crow 1990). A further

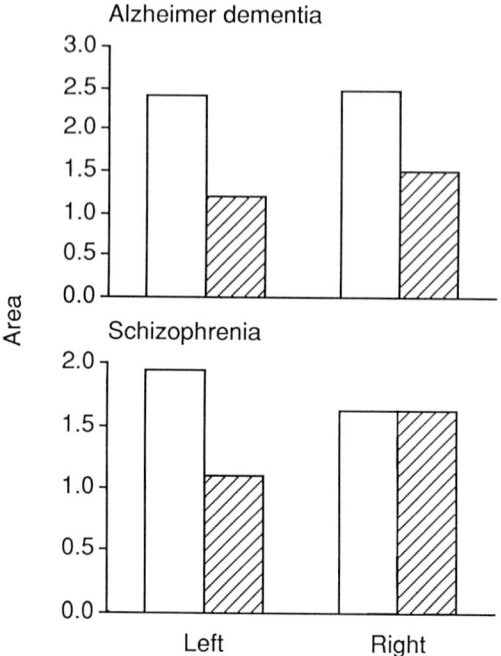

FIG. 2. (*Crow*) Area of the temporal horn (assessed from the lateral aspect) in post mortem brains of patients with Alzheimer-type dementia and schizophrenia (open columns, top and bottom, respectively) compared with age-matched controls (hatched columns). Adapted from Crow et al (1989).

conclusion is that if you assume the cause of schizophrenia is genetic then the gene involved is the cerebral dominance gene.

Galaburda: There are also differences in absolute size. I believe the nervous system cares less about asymmetry than about the detailed architecture and the number of components and types of neurons that constitute neural networks. One could reinterpret your data in that light. For example, one could say that the left temporal lobe of patients with schizophrenia has abnormal circuitry. Why is it the left? Because you select those patients by their behaviour. There may be another group with different behaviours in whom you would find the abnormalities on both sides, or on the right.

References

Brown R, Colter N, Corsellis JAN et al 1986 Post-mortem evidence of structural brain changes in schizophrenia. Arch Gen Psychiatry 43:36–42

Crow TJ 1990 Temporal lobe asymmetries as the key to the etiology of schizophrenia. Schizophr Bull 16:433–443

Crow TJ, Ball J, Bloom SR et al 1989 Schizophrenia as an anomaly of development of cerebral asymmetry. Arch Gen Psychiatry 29:247–253

Innocenti GM, Berbel P 1991 Analysis of an experimental cortical network. II. Connections of areas 17 and 18 after neonatal injections of ibotenic acid. J Neural Transplant, in press

Johnstone EC, Crow TJ, Frith CD, Husband J, Kreel L 1976 Cerebral ventricular size and cognitive impairment in chronic schizophrenia. Lancet 1:848–851

Scheibel AB 1984 A dendritic correlate of human speech. In: Geschwind N, Galaburda AM (eds) Cerebral dominance. Harvard University Press, Cambridge, Massachusetts p 43–52

The asymmetrical genetic determination of laterality: flatfish, frogs and human handedness

Michael J. Morgan

Laboratory for Neuroscience, Department of Pharmacology, University of Edinburgh, 1 George Square, Edinburgh EH8 9JZ, UK

Abstract. The determination of the left–right body axis is unlike that of the two other axes because left–right positional information is not required to specify mirror-image structures on the two sides. When the left and right sides of the body are not mirror symmetrical such positional information is required, as is a mechanism for reading that information. There are several possible gradient schemes for left–right information, including symmetrical gradients from which the information is extracted by spatial differentiation. The genetic mechanisms for the control of handedness are not known. There is no evidence for 'left-handed' and 'right-handed' genes, only for mutations that can interfere with handedness in a non-specific manner. Such mutations never produce situs inversus with a frequency greater than 50%. The situs of individual organs shows a strong correlation, suggesting a global mechanism such as a gradient of left–right positional information. Many asymmetries in vertebrates follow a pattern in which growth on the left is favoured over growth on the right. This may be related to the 'dexiothetism' of chordate ancestors postulated by Jefferies.

1991 Biological asymmetry and handedness. Wiley, Chichester (Ciba Foundation Symposium 162) p 234–250

The fundamental question concerning biological laterality is how the information about left–right asymmetry is transmitted across generations in such a way as to allow structures on the left and right sides of the body to develop differently. A clear example of a directional asymmetry is shown in a transverse section through the diencephalon of the frog *Rana temporaria* (Fig. 1). Just under the stalk of the pineal and to either side of the III ventricle are the habenular nuclei. There are two nuclei on the left (the left lateral and the left medial) but there is only a single nucleus on the right. This asymmetry is seen early in the development of the tadpole (Morgan et al 1973); it is characteristic of all members of the species and is found in many other amphibian species. How do the cells on the left side of a developing embryo get their instructions to develop differently from those on the right side?

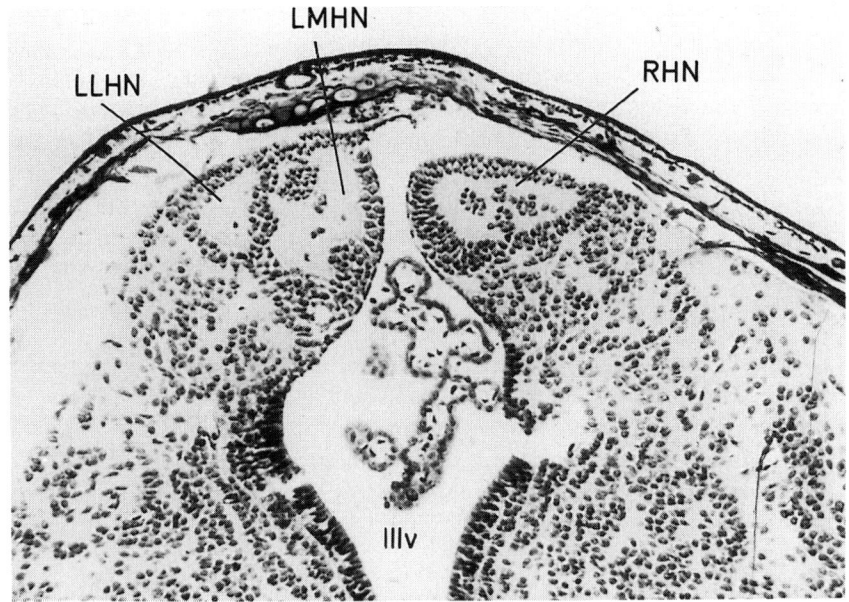

FIG. 1. A frontal section of the habenular region of the adult frog, *Rana temporaria*, taken from a Nissl preparation. RHN, right habenular nucleus; LMHN, left medial habenular nucleus; LLHN, left lateral habenular nucleus. The nuclei border on the III ventricle.

When I reviewed what was known about this question (Morgan 1977), I came to a number of conclusions and advanced some conjectures that were intended to stimulate debate about this 'deep and neglected question' (Brown & Wolpert 1990). In this paper I shall summarize these conclusions and conjectures and say how they stand at present. The main points are the following:

(a) A left–right asymmetry should be defined as a case where a structure on the left is not the mirror image of a corresponding structure on the right. A bilaterally symmetrical organism has no need of left–right positional information.

(b) The development of left–right asymmetry requires left–right positional information, possibly in the form of a left–right gradient.

(c) A reading mechanism is required to translate the gradient into a left–right asymmetry. For example, the gradient may contain growth instructions that lead to more rapid development on the left-hand side (Corballis & Morgan 1978, Morgan & Corballis 1978).

(d) Situs inversus arises not because the mirror-image phenotype is specified genetically but because laterality is no longer determinate. This can happen either because the gradient is lost or because it is no longer translated in a determinate way. Environmental and teratogenic influences are similar to genetic ones in this respect.

(e) 'It all goes together when it goes.' The asymmetries of individual organs like the heart and gut are not independently determined. In situs inversus there is a strong tendency for *all* the asymmetries to be reversed. This would be expected if the direction of an underlying gradient had been reversed. There are no known examples of a genetic mutation causing the reversal of one particular organ.

(f) Left–right positional information is already present in the oocyte; transcription of the embryonic genome is not required for asymmetry to develop.

(g) Individual genes do not themselves determine the direction of a lateral asymmetry. They can affect asymmetries only by working against a background that is already asymmetrical.

(h) There is something systematically peculiar about the way in which single genes affect the direction of asymmetries. There are very few cases, if any, in which alleles are associated with mirror-image phenotypes. I referred to this as the 'asymmetrical determination of asymmetry'.

(i) Left–right differences may be enhanced by reciprocal inhibition between the two sides, the model being the asymmetrical claw development of certain crustaceans.

(j) Many asymmetries in chordates favour a more rapid or more extensive development of structures on the *left-hand side* of the body. There may thus be an 'Ur-asymmetrie' in our ancestors.

Some of these ideas, particularly the more vacuous ones, seem to me still to be correct. Others were mistaken: in particular, I do not think an asymmetrical gradient is required for left–right differentiation.

A bilaterally symmetrical organism does not require left–right positional information

This was not an original idea, but it often gives rise to misunderstanding. I remember a scientist at a previous conference on laterality taking off his shoe and waving it at me in a Kruschevian manner. His point was that a cobbler has to know the difference between left and right to make a left shoe! But it is a fallacy to suppose that an organism has to know this to make a left-handed limb. There are three orthogonal dimensions in Euclidean space and thus two degrees of freedom in labelling their axes. One axis can be called the dorsoventral and labelled 'top' and 'bottom'; the next can be called anteroposterior and labelled 'front' and 'back'. But that's it. The remaining axis has to be left–right and the directions of left and right are already determined. It would be gratuitous to have labels saying which is which because they are already completely defined with respect to the other two axes.

The limbs of a bilaterally symmetrical organism can be constructed using only the instructions 'grow up' (or 'down'), 'grow frontwards' (or 'backwards') and 'grow outwards' (or 'inwards'). No reference to left and right is needed. If the same instructions are followed on the two sides of the body, mirror-image limbs will result. This is equally true in the polar coordinate model of limb development (French et al 1972, Bryant et al 1981). The magnitude of the radius vector represents the proximodistal dimension and its angle represents dorsoventral. Let the labels associated with each angle be $a,b,c....n$ with a at the top and the sequence proceeding *outwards* from the midline. This will automatically construct a clockwise-reading gradient on one side of the body and an anticlockwise gradient on the other. Mirror-image limbs on the two sides will result. To make it otherwise, the gradients on the left and right would *both* have to be clockwise (or both anticlockwise), which would introduce an asymmetry according to the strict definition given in (a) above.

A left–right agnosic cobbler could very easily make a pair of shoes, simply by following the rule that one has to be the mirror image of the other.

The development of left–right asymmetry requires both an asymmetrical gradient to give positional information and a mechanism for reading this gradient

I no longer think this is true. First, I shall define my terminology, because it is easy to confuse gradients, slopes and derivatives. Gradient refers to systematic change over space in some quantity, say the concentration of a morphogen. The slope of such a gradient will be its first spatial derivative, x'. Higher derivatives will be labelled x'', x''' and so on.

I distinguish the mechanism for laying down a gradient from those whereby cells respond to or 'read' a gradient (Morgan 1978, McManus 1984, Brown & Wolpert 1990). A mutation could increase the probability of a left–right reversal without disturbing the left–right gradient by interfering with the ability of cells to read the gradient (Morgan 1978). For left–right asymmetry to emerge, at least one of the two mechanisms has to be asymmetrical. There could be an asymmetrical gradient combined with a symmetrical reading mechanism, or vice versa. Thus gradients do not have to be asymmetrical to permit left–right differentiation, provided that cells can read the sign of the gradient's spatial derivative. I shall argue that the important biological information in gradients may lie in their spatial derivatives rather than in their point-values, thus echoing a familiar notion in sensory physiology. First, I shall consider the types of gradient that might exist, setting aside questions of biological plausibility. The first classification is between gradients that are asymmetrical across the median plane and those that are not.

Asymmetrical gradients

The important property of asymmetrical gradients is that a cell can read its position from the gradient without necessarily being able itself to distinguish left from right. It suffices to read either the absolute value of the gradient or one of its derivatives to obtain a positional signal. Asymmetrical gradients can be subdivided according to whether they are linear or contain higher-order terms. Only the latter can contain positional information in their spatial derivatives.

Simple linear gradients. A notation for this and other kinds of gradient is established in Fig. 2. At the top of the figure is a schematic animal viewed from the dorsal aspect; below is a transverse section. The arrow shows the direction of the putative gradient. Associated with each diagram is a graph showing how the concentration of a hypothetical morphogen changes along the direction indicated by the arrow. Fig. 2a shows the simple case where a linear gradient is the same at different points on the dorsoventral axis. Fig. 2b shows a polar coordinate version where the concentration varies linearly around the circumference of the animal, which necessarily produces a discontinuity in the gradient at one point along the circumference. This can be avoided by the spiral gradient shown in Fig. 2c. The spiral gradient could repeat itself at every turn, as octaves have been claimed to do along the cochlear nucleus of the ear,

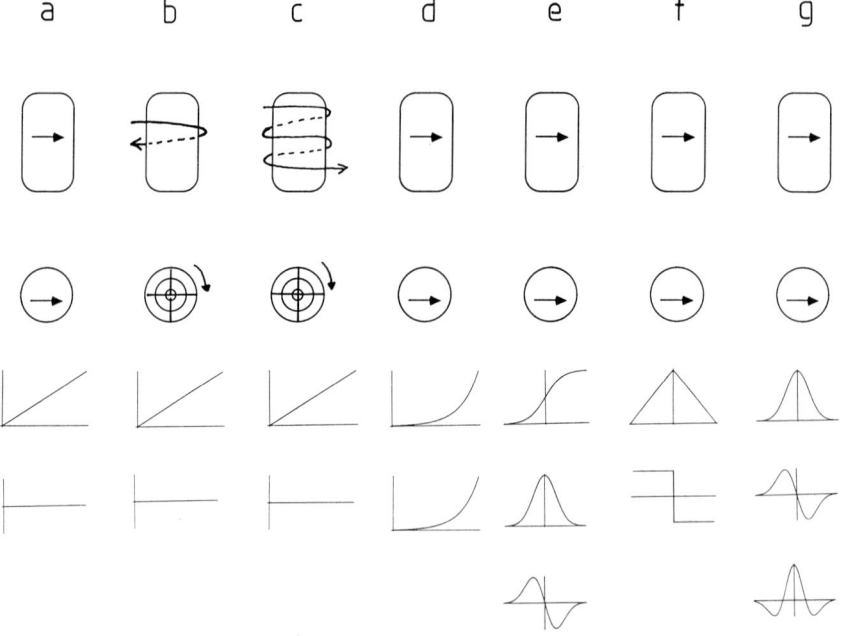

FIG. 2. Schematic representations of the various left–right gradients described in the text. The top row shows a dorsal view of the animal with the direction of the gradient marked by an arrow. The second row represents a transverse section. The third row represents the gradient as a plot of concentration (x) against left–right position; the bottom row shows the first spatial derivative of that function. In (e and g) the second derivative is represented as well.

or a single gradient could run along the spiral from anterior to posterior. The important formal property of all linear gradients is that positional information must be read from the absolute value of the gradient at a particular point. There is no information in the slope of the gradient or in any of its higher derivatives (bottom set of graphs in Fig. 2a–c).

Non-linear gradients. I shall assume that non-linear gradients can all exist in rectangular, polar or helical versions. For present purposes the distinction between these types has no important formal consequence. There are two types of non-linear gradients: monotonic or non-monotonic.

Monotonic gradients. With an exponential gradient (Fig. 2d) a positional signal could be read from either the absolute value or the first derivative, x', which is changing exponentially. Fig. 2e is a cumulative Gaussian, which could be read by its absolute value but its first derivative is symmetrical and therefore useless for establishing left–right asymmetry. To use a derivative of the cumulative Gaussian to distinguish left from right it would be necessary to extract its second spatial derivative (x''), which is negative on one side of the midline and positive on the other.

An important difference between first and second derivatives of gradients is that to extract the former an asymmetrical operation is required. This can be understood by considering the nature of the local operators required to extract a derivative. To extract a first derivative, a cell would have to compare the value of the gradient on one side of its membrane with the value on the other side, for example, by having a hyperpolarizing input from the gradient on the left and a depolarizing input on the right. The net voltage change in the cell would then be proportional to the gradient of the morphogen across it. For this to work the cell has to be appropriately oriented with respect to the anteroposterior and dorsoventral axes, and to have a different distribution of the appropriate ion channel on its left and right (cf Brown & Wolpert 1990).

Second derivative operators are symmetrical and cells have to compare the value of the gradient at three points. For example, a cell carrying out a second derivative operation could have hyperpolarizing inputs on its anterior and posterior surfaces and hypopolarizing inputs on both left and right.

Thus, if the nature of the gradient requires a first derivative operation to extract positional information, the 'reading units' themselves have to be asymmetrical. It will be shown below that if the units themselves are asymmetrical, an asymmetrical gradient is not required. There would thus be little point in having an asymmetrical gradient and extracting a first derivative. If the gradient is asymmetrical, and if the underlying units are symmetrical, a second-derivative positional signal would be indicated.

Non-monotonic gradients. These do not seem very likely. A simple example would be a sine wave of an undamped, unforced oscillator, caused by emitting waves of morphogen from one end of the gradient. Such a gradient carries no unique positional signal either in its point values or in any of its derivatives (which are all the same shape). Only slightly more promising is the damped, unforced linear oscillator, which decays across the gradient. There is no unique positional signal in this or in its derivatives, but some means of registering the peak-to-trough amplitude at any point could be envisaged.

Symmetrical gradients

Fig. 2f shows a symmetrical gradient, a triangle function centred on the midline (cf Brown & Wolpert 1990). Fig. 2g is a Gaussian equivalent. Positional information is carried in these gradients not by their absolute point values but by their spatial derivatives. The first spatial derivative of a triangle or a Gaussian is of opposite sign on either side of the midline, so any unit that could extract the first derivative could appropriately localize itself. This mechanism depends upon differentiation, which raises the question of whether differentiation of a gradient is a plausible biological mechanism.

Presumably, the sensing of a gradient depends upon receptors on the cell surface. There is some exciting recent evidence that α_1-adrenergic receptors are involved in the determination of asymmetry (Fujinaga & Baden 1991). A widespread principle in sensory pathways is that absolute signals are rapidly replaced by differential signals. For example, the degree of hyperpolarization of a photoreceptor is related only loosely to the absolute amount of light falling on it, because of the existence of local gain-regulation mechanisms. The number of photons falling on the receptor is compared to the number falling in the vicinity in the recent past. The comparison is continued in the retinal ganglion cells, which subtract the signal arriving from photoreceptors in the receptive field surround from those falling in the centre. In effect, the retinal ganglion cell computes the local second spatial derivative of the retinal light distribution. Similarly, in hearing, signals arising from the basilar membrane are differentiated by a process of mutual inhibition.

One reason difference signals are so popular in sensory pathways is that the initial receptors are very widely tuned so it is necessary to sharpen the response by taking difference signals. Another is that the precision of nerve cells in representing signals quantitatively is too poor to provide accurate absolute values. It is far better to search for *qualitative* signals that depend upon comparison of activities in different cells. The position of a luminance discontinuity (edge) in the retinal image, for example, is efficiently represented by the position in the zero-crossing of its second derivative, as computed by retinal ganglion cells.

A similar reasoning may be applied to morphogenetic gradients. If the positional signal in a gradient were the absolute concentration of a morphogen, any slight increase or decrease in the amount of morphogen would have disastrous consequences. If, on the other hand, a derivative signal were extracted, the absolute amount of morphogen would not matter, provided the slope of the gradient were unaffected. Adding or subtracting a constant to each point in the gradient would have no effect on a positional signal based on the derivative. Even adding a random amount at each point (noise) would have little effect if the derivative were computed after some local smoothing, as is apparently the case with retinal ganglion cells.

For asymmetrical cells to decode a gradient accurately they should be correctly oriented within the field. This presents no difficulty if dorsoventral and anteroposterior positional signals are assumed to be present. Cells could cooperate by selectively attracting one another by their unlike surfaces, like small magnets. Cells that did not already have a clear chirality could have it imposed upon them by neighbourhood interactions. In this way, an initially weak distribution of oriented subunits could be reinforced by a cooperative process.

It may be objected that having chiral cells *and* having an external gradient is a bit like keeping a dog and doing one's own barking. If the cells can distinguish their left from right what is the point of the gradient? The answer is that knowing the difference between its *own* left and right does not tell the cell which side of the body it is on. For that, the cell needs a positional signal, albeit a symmetrical one. This has the important consequence that individual cells will show regulation. If moved from one side of the body to the other before they begin to differentiate, the cells will differentiate according to their new position, not their old. This might appear to be contradicted by Wood's recent report of reversal of handedness in *Caenorhabditis elegans* caused by interchanging the relative position of the ABl and ABr cell pairs (Wood 1991, Wood & Kershaw this volume). However, in this case it was the relative anteroposterior positions that were altered, not the left–right positions.

Situs inversus is an indeterminate condition of laterality

Situs inversus could result from a gradient-reading mechanism in two ways. One is inversion of the gradient, which would alter the sign of the first derivative on the two sides of the body, making cells on the left 'think' they were on the right and *vice versa*. A single mutation that inverted the gradient could cause 100% situs inversus in the phenotype. I still believe this never happens (Morgan 1977, 1978). Neither mutations nor teratogens produce situs inversus with a frequency higher than 50%. (See, for example, the effect of phenylephrine, Fujinaga & Baden 1991). It is more plausible that situs inversus happens when the gradient is abolished or flattened to the point where it is unreadable, so that the direction of asymmetry is determined by chance. Alternatively, the reading mechanism rather than the gradient may be affected: individual cells might be misoriented so that they read the gradient in the 'wrong' direction.

If each cell takes up its position by chance, the different asymmetries will become discordant. However:

'It all goes together when it goes'

I was originally struck by von Woellwarth's evidence (cf Morgan 1977) that newts with naturally occurring situs inversus of the abdominal organs *also had reversed asymmetry of the habenular nucleus of the brain.* The same was true of animals with artificially induced situs inversus. This finding led von Woellwarth to speak of a 'general asymmetry determining mechanism in the embryo'. Isolated dextrocardia in human populations is much less common than situs inversus (reviewed by Morgan 1977). Brown & Wolpert (1990), however, challenge the view that there is a strong correlation between organs with respect to their situs. They find numerous heterotaxias in Layton's *iv/iv* mice; for example, normal aortic arch asymmetry but reversed stomach and spleen. They conclude that in *iv/iv* mice there is no overall biasing mechanism and in its absence individual asymmetries are fluctuating and independent.

An alternative explanation is that the lateral gradient has been weakened and the individual organs have different thresholds for reading it. To distinguish between these models, we need to know whether there is any systematic tendency for some organs to be normal more often than others. Fujinaga & Baden (1991) looked at situs inversus induced in rat embryos by the α_1-adrenergic agonist phenylephrine. Table 1 reproduces their data on the number of embryos with various combinations of left-sided placenta, left-sided tail and inverted heart. Fujinaga & Baden state that there is 'no strong tendency' for the positions of the organs to correlate, but statistical analysis seems to me to show the contrary. To analyse the data, I have assumed that the probability of having a left-sided placenta is given by the number of animals having this condition, as a proportion

TABLE 1   Numbers of rats found with various combinations of reversed placenta, tail and heart

Placenta	Tail	Heart	Actual	Expected
+	+	+	31	23
−	−	+	31	15.5
+	−	+	10	20
+	+	−	9	6
−	+	−	8	4.6
+	−	−	2	5.2
−	+	+	1	17.6

+ indicates a reversal; − the unreversed condition. Expected numbers are calculated from the assumption that the three asymmetries are independent. Modified from Fujinaga and Baden (1991).

of the total N; similarly for the other two traits. Let the probabilities of the three traits be $P(1)$, $P(2)$ and $P(3)$. Assuming independence (the null hypothesis), the probability of the $+ + +$ condition is $p(1) \times p(2) \times p(3)$; the probability of the $+ - +$ condition is $p(1) \times (1 - p(2)) \times p(3)$, and so on. The probabilities have been scaled to unity before the predicted frequencies are calculated and χ-squared is computed; this has the value 42.333, df (degrees of freedom) = 6. This value is highly significant, so the assumption of independence between the probabilities may be rejected. The proportion of $+ + +$ cases is higher than expected, supporting the assertion that 'it all goes together when it goes'.

Transcription of the embryonic genome is not required for left–right asymmetry

The chirality of asymmetries resulting from spiral cleavage is evident before transcription of the embryonic genome. The best-understood case of a straightforward Mendelian factor determining asymmetry is the direction of coiling in *Lymnaea peregra*, where it is the maternal genome that determines the direction of asymmetry. Freeman & Lundelius (1982) transferred cytoplasm from wild-type, dextral eggs into uncleaved sinistral eggs and found that these eggs cleave in a dextral pattern. Cytoplasm from sinistral eggs had no effect on dextral eggs. Sinistrality is thus the 'default option', which may be reversed by dextral gene products.

Paternal gene products may be involved in asymmetries that appear later in development. The paternal effect upon human handedness is only slightly weaker than the maternal (McManus, this volume). Policansky (1982) argued for a paternal effect in crosses between dextral and sinistral races of the starry flounder, *Platichthys stellatus*. In crosses of three males with the same two females, sinistral males produced more sinistral offspring than did dextral males. The numbers of fish involved were too small to be statistically significant and further data are needed. Unfortunately, it has not been possible to pursue the flounder breeding programme to the F2 generation (Policansky 1984), so this question remains open. The flatfish case will be discussed further in the following section.

Genes don't know the difference between left and right

I originally argued that genes do not themselves encode the difference between left and right but can influence laterality only by working in the context of a pre-existing asymmetry in the cytoplasm. It is a truism that any genetic influence can be mediated only by a complex interaction between the gene and the 'environment' in the widest sense of that word. The question can be focused by enquiring whether there is any case where two alleles both produce active products, one of which codes for dextrality and the other for sinistrality. Following an earlier suggestion by Dahlberg (1943), I proposed that this case is never found. Apparently contrary instances, such as rotation of the abdomen in *Drosophila* (Stern 1955) prove on closer examination to represent mutants that simply interfere, in a non-handed way, with a normal chiral process of development. Similarly, the inheritance of coiling in *L. peregra* seems to show that the so-called 'sinistral' gene may simply be the absence of the wild-type 'dextral' gene product. The function of the latter seems to be to *reverse* the otherwise normal cytoplasmic tendency towards sinistrality. This supports my suggestion that 'genetic influences upon asymmetry are asymmetrical'.

Policansky (1982, 1984) has argued strongly for dextral and sinistral genes in flatfish. The asymmetry of these animals is intriguing: some species flop over onto their left sides during development, some flop to their right, others are inconsistent floppers. The downwards pointing eye migrates to the new dorsal surface and there is some evidence that this migration uncrosses, rather than twists, the asymmetrical optic chiasm (Hubbs & Hubbs 1945, Parker 1903). Quite closely related species can have opposite handedness, and these may even interbreed. Starry flounder populations are 100% sinistral in Japan and approximately racemic in the United States.

Policansky (1982) states that if laterality is species specific, we don't have to look any further for genetic influences on laterality. He raises the interesting possibility that changes in chirality are involved in speciation, possibly as a mechanism for reproductive isolation. (There may be mechanical impediments between incompatible screws.) His table of evolutionary relationships is compatible with an originally dextral percoid ancestor, from which a sinistral branch emerged, to split once more into the sinistral Cynoglossidae and the dextral Solidae, and so on (Fig. 3). The situation in *P. stellatus* may represent speciation in progress.

Coiling has also been suggested as a mechanism for sympatric speciation in snails. However, as coiling is determined by the maternal genotype there remains substantial gene flow between emerging sinistral and dextral groups even if there is no mating between sinistral and dextral phenotypes (Johnson et al 1990).

Returning to the flatfish, it is still an open question whether there are active genes responsible for both dextrality and sinistrality. At the species level, dextrality may be the normal cytoplasmically induced condition, with a single gene that can reverse this direction. Policansky argues against this 'asymmetrical' model on the basis of his *P. stellatus* crosses, but it is difficult to draw statistically valid conclusions from experiments with only four females and nine males (for a contrary view, see Boklage 1984, Policansky 1984). It would be interesting to carry out cytoplasmic injection experiments on these fish.

Qualitative asymmetries involve reciprocal left–right inhibition

The snapping shrimp *Alpheus heterochelis* has a large size difference between the chelae on left and right. If the larger member is removed, it regenerates

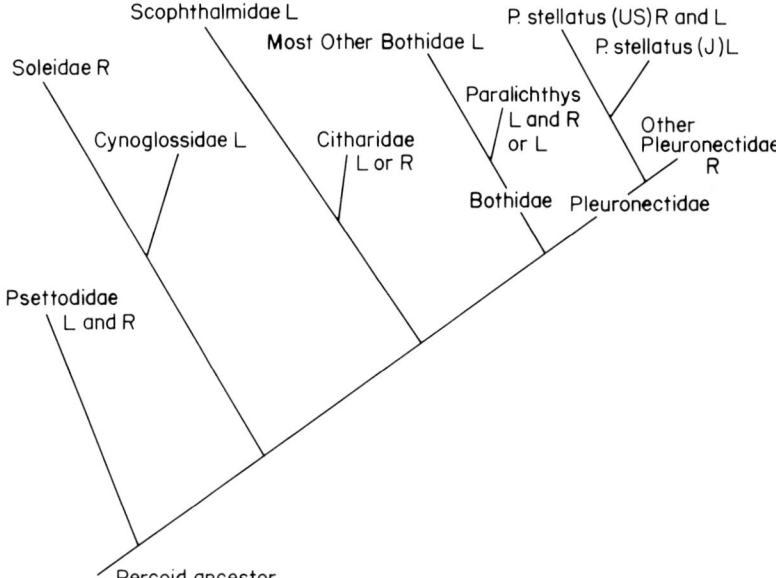

FIG. 3.   The possible evolutionary relationships between the different families of flatfish.
From Policansky (1982).

as a small one and the originally smaller one grows at the next moult into a
large one. It looks as if any member getting a 'claw-start' will inhibit the other.
The mechanism appears to be under neural control (cf Morgan 1977).

   Corballis & Morgan (1978) proposed reciprocal inhibition as a mechanism
for amplifying initially small developmental asymmetries into the qualitative
asymmetries seen, for example, in cerebral laterality. We argued that the
development of a speech mechanism in the right hemisphere is actively prevented
by the left hemisphere. I am not aware of any recent evidence that tests this
speculation.

Is there an 'Ur-asymmetrie'?

The echinoderms, which probably share a common ancestor with hemichordates
and chordates (see Jefferies, this volume), have bilaterally symmetrical larvae
with three pairs of coelomic cavities, but the right-hand cavities atrophy and
the organ systems of the adult develop mainly from those on the left. Many
asymmetries in chordates favour more rapid or complete development of
structures on the left-hand side. In vertebrates, examples of left-sided dominance
include: the amphibian habenulae, the tusk of the narwhal (which develops from
a left tooth); the ovaries of birds (left only functional); the control of bird song
by the left side of the brain; and the left cerebral dominance in humans (for

further discussion see Morgan 1977, Morgan & Corballis 1978). In Spemann & Falkenberg's conjoined amphibian twins (1919), the right-hand twin was generally smaller than the left, as well as being prone to situs inversus.

Not all asymmetries can be fitted into a dominance framework. 'Left-cerebral dominance' has become an unpopular idea because of the discovery that the right hemisphere is 'dominant' for some spatial abilities. This even appears to be the case in monkeys (see the convincing behavioural investigation by Hamilton & Vermeire 1988). However, I believe the idea of an Ur-asymmetrie has been given new life by Jefferies' fascinating work (this volume) on the origin of vertebrates. Jeffries suggests that an ancestor similar to the colonial hemichordate *Cephalodiscus* lay down on its right-hand side, a condition he calls 'dexiothetism'. This organism gave rise to the echinoderms (with their fundamental left-sided asymmetry) as well as to the chordates. Dexiothetism may be responsible for an overall bias favouring the left-hand side of the body.

Acknowledgements

I am grateful for support from the Darwin Trust of the University of Edinburgh. Linda Partridge made many useful comments and convinced me that left–right gradients could be useful without being asymmetrical.

References

Boklage CE 1984 On the inheritance of directional asymmetry (sidedness) in the starry flounder, *Platichthys stellatus*: additional analyses of Policansky's data. Behav Brain Sci 7:725–730

Brown NA, Wolpert L 1990 The development of handedness in left/right asymmetry. Development 109:1–9

Bryant SV, French V, Bryant PJ 1981 Distal regeneration and symmetry. Science (Wash DC) 212:993–1002

Corballis MC, Morgan MJ 1978 On the biological basis of human laterality: I. Evidence for a maturational left–right gradient. Behav Brain Sci 1:261–269

Dahlberg G 1943 Genotypic asymmetries. Proc R Soc Edinburgh Sect B (Biol Sci) 62:20–31

Freeman G, Lundelius JW 1982 The developmental genetics of dextrality and sinistrality in the gastropod *Lymnaea peregra*. Roux's Arch Dev Biol 191:69–83

French V, Bryant PJ, Bryant SV 1972 Pattern recognition in epimorphic fields. Science (Wash DC) 193:969–981

Fujinaga M, Baden JM 1991 Evidence for an adrenergic mechanism in the control of body asymmetry. Dev Biol 143:203–205

Hamilton CR, Vermeire RA 1988 Complimentary hemispheric specialization in monkeys. Science (Wash DC) 242:1691–1694

Hubbs CL, Hubbs LC 1945 Bilateral asymmetry and bilateral variation in fishes. Mich Acad Sci 30:229–310

Jefferies RPS 1991 Two types of bilateral symmetry in the metazoa: chordate and bilaterian. In: Biological asymmetry and handedness. Wiley, Chichester (Ciba Found Symp 162) p 94–127

Johnson MS, Clarke B, Murray J 1990 The coil polymorphism in *Partula suturalis* does not favour sympatric speciation. Evolution 44:459–464

McManus IC 1984 The inheritance of asymmetries in man and flatfish. Behav Brain Sci 7:731–733

McManus IC 1991 The inheritance of left-handedness. In: Biological asymmetry and handedness. Wiley, Chichester (Ciba Found Symp 162) p 251–281

Morgan MJ 1977 Embryology and inheritance of asymmetry. In: Harnad SR, Doty RW, Goldstein L, Jaynes J, Krauthamer G (eds) Lateralization in the nervous system. Academic Press, New York p 173–194

Morgan MJ 1978 Genetic models of asymmetry should be asymmetrical. Behav Brain Sci 1:325–331

Morgan MJ, Corballis MC 1978 On the biological basis of human laterality: II. The mechanisms of inheritance. Behav Brain Sci 1:270–277

Morgan MJ, O'Donnell JM, Oliver RF 1973 Development of left–right asymmetry in the habenular nuclei of *Rana temporaria*. J Comp Neurol 149:203–214

Parker GH 1903 The optic chiasma in teleosts and its bearing on the asymmetry of the Heterosomata (flatfishes). Bull Mus Comp Zool Harv Univ 40:221–242

Policansky D 1982 Flatfishes and the inheritance of asymmetry. Behav Brain Sci 5:262–265

Policansky D 1984 Do genes know left from right? Behav Brain Sci 7:733–735

Spemann H, Falkenberg H 1919 Über asymmetrische Entwicklung und Situs Inversus Viscerum bei Zwillingen und Doppelbildungen. Arch Entwicklungsmech Org (Wilhelm Roux) 45:371–422

Stern C 1955 Gene action. In: Willer BH, Weiss PA, Hamburger V (eds) Analysis of Development. Hafner, New York p 151–169

Wood WB 1991 Evidence from reversal of handedness in *C. elegans* embryos for early cell interactions determining cell fates. Nature (Lond) 348:536–537

Wood WB, Kershaw D 1991 Handed asymmetry, handedness reversal and mechanisms of cell fate determination in nematode embryos. In: Biological asymmetry and handedness. Wiley, Chichester (Ciba Found Symp 162) p 143–164

DISCUSSION

Wolpert: The model of Nigel Brown and myself is identical to the last example in Fig. 2g.

Morgan: I thought you had an asymmetrical gradient.

Wolpert: No, not to begin with. We are making the left side of the cell different from the right side of the cell because the F molecule transports other molecules across the cell. This biases the gradient or whatever other mechanism makes the two sides asymmetrical.

Morgan: Both models need a cell or molecule that is tethered. I think of the tethering in terms of a sheet of adhering cells that have an anterior and posterior and so on. There is no point in having a cell reading a symmetrical gradient unless it knows the difference between left and right. I would emphasize the informational richness of derivatives of gradients for giving invariant positional information.

Galloway: Bacteria are moving around and can sample gradients at different places and thereby 'calculate' the gradient. How can cells in a multicellular organism, which are presumably fixed in position, sample gradients?

Morgan: They have a finite size so the gradient runs across them.

Galloway: Is that a speculation or do we know that actually happens?

Morgan: Pure speculation.

Wolpert: Can you explain why one has to be asymmetrical to discriminate a symmetrical pattern.

Kondepudi: Just look at the process in a mirror.

Morgan: Think of two enantiomorphs. Describe one to me, such that I could recognize it again, pretending to be a symmetrical animal. I will press the buzzer every time you do something that you are not allowed to do because you are symmetrical.

Corballis: Suppose there is an organism that's symmetrical that can identify a *b* and a *d*. Every time it sees a *b* it says *b*; every time it sees a *d* it says *d*. Now look in a mirror. You see the same organism because it's symmetrical and is not affected by mirror-image reversal. But the organism is now calling a *b* a *d* and a *d* a *b*, which denies its original wiring.

Lewis: But, there is nothing to prohibit a motor response that is, say, directed to the right when the organism is presented with *b* and directed to the left when it is presented with *d*.

Corballis: That is not a true test of mirror-image discrimination.

Lewis: Another objection is that the process of learning to discriminate involves memory.

Corballis: Or wiring.

Lewis: OK, but take the case of memory. There, past asymmetrical experience can be sufficient to break symmetry and allow discrimination.

Corballis: But what creates the asymmetry in the first place?

Lewis: The bodily structure of the organism doesn't need to be asymmetrical.

Corballis: It does if memory is encoded in the bodily structure of the organism, which it surely is.

Lewis: I meant the bodily structure of the naive organism.

Corballis: Having given it asymmetrical memory, it becomes asymmetrical in bodily structure.

Wolpert: I think you had better tell us how you teach pigeons to read!

Corballis: It is difficult to train a pigeon to discriminate, for example, *b* from *d*. The pigeon has to peck at the *b* when it appears but not the *d*; it is important that the letters are not presented together. A pigeon has great difficulty learning this discrimination. But if you make the pigeon asymmetrical by covering one eye, it can do it. If you then cover the other eye, the pigeon reverses the discrimination—if it was trained to peck at *b* it now pecks at *d*—because you have reversed the asymmetry.

If we were all symmetrical, we couldn't have this particular symposium because we wouldn't know what we were talking about! There is also evidence, incidentally, that left-handers are more symmetrical than right-handers, which explains why there are so few left-handers at this symposium. In order to understand the left–right problem, you have to be asymmetrical.

Peters: Michael, I was not surprised to hear your description of the asymmetry of the habenular nuclei. If there is one region in that area of the brain that is asymmetrically duplicated, I would expect it to be the habenular nucleus. If you had said the ventral nucleus of the thalamus or the inferior colliculus, I would have been surprised because these structures are strictly related to the bilateral asymmetry of the body.

I will now speculate wildly as to what this could mean. The habenular nucleus sends a large fibre bundle to the Raphe region. The Raphe nuclei, which are the most important source of serotonergic fibres for the entire cerebrum, play no part in functions that involve particular spatial elements that relate to the left or right sides of the body. The pineal gland, also important in terms of serotonin, lies close to the habenular nuclei: there is no logical left–right body connection here either. That is why Descartes thought the pineal gland was the point where the body communicated with the soul, because it seemed to be the only place that was unitary in structure.

Crow: The gene for the enzyme that converts serotonin to melanin in the pineal gland is in the pseudoautosomal region of the X chromosome. This region is implicated in the inheritance of schizophrenia; it has also been proposed to include the locus for left-handedness.

Peters: My argument can be developed to include structures in the brain that are involved in the regulation of autonomic function. Some of the internal organs, such as the stomach and the heart, have no natural bilateral symmetry. Another example is the innervation of the larynx by the vagus nerve. First, there is clear anatomical asymmetry in that the right and left recurrens nerves take different paths to the larynx. Second, there is a remarkable asymmetry in activation, whereby both recurrens nerves can be controlled by one half of the brain only. This occurs during speech; the left brain half by itself can operate both the right and left outputs of the recurrent laryngeal nerve. We usually think of vocalizations in terms of human speech but, evolutionarily, vocalizations are closely associated with motivation/emotional states, and are therefore closely linked to the autonomic/limbic axis.

I am speculating that the underlying source of cerebral asymmetries which culminate in the asymmetries of areas involved in higher level cognitive and motor functions may lie in the systems responsible for the regulation of activity cycles, motivation and emotion.

Morgan: You are thinking of the Meynert's bundle and you point out the connection to the pineal. I think the pineal is the remaining one of an originally paired set of glands, the pineal and the parapineal. So there's a big asymmetry there which probably explains what we see in the roof of the diencephalon.

Jefferies: But when both the pineal and the parapineal are present, they are not symmetrical. The pineal overlies the parapineal.

Peters: There is no real reason why these glands should be bilaterally symmetrical. Structures that are tied to the bilateral symmetry of the body would be expected to be symmetrical. For instance, as Chris McManus said, one would not expect to find marked asymmetries in primary visual cortex. One would be less surprised to find asymmetries in the higher-order visual areas. Similarly, the planum temporale deals with specialized higher-order functions that are far removed from concerns of body symmetry, and asymmetries are not unexpected in this area.

McManus: And that could easily be implicated in schizophrenia. Modern schizophrenia research is to a large extent dopaminology research. There is a close link in some way between dopamine and schizophrenia.

The inheritance of left-handedness

I C McManus

Department of Psychology, University College London, Gower Street, London WC1E 6BT and Department of Psychiatry, St Mary's Hospital Medical School, Imperial College of Science, Technology and Medicine, Norfolk Place, London W2 1PG, UK

Abstract. Left-handedness occurs in about 8% of the human population. It runs in families and an adoption study suggests a genetic rather than an environmental origin; however, monozygotic twins show substantial discordance. The only genetic models that successfully explain the family and twin data are those of McManus and Annett, which share the feature of incorporating a random component reflecting the biological phenomenon of 'fluctuating asymmetry'. The models have each been modified to explain the greater incidence of left-handedness in males. The McManus model is more successful at explaining the maternal effect— left-handed mothers have more left-handed offspring than do left-handed fathers. Both models explain the association of handedness with cerebral language dominance. The models differ principally in their conception of the phenotypes of handedness: Annett proposes a unimodal continuum, McManus proposes two discrete categories of handedness. Finding the gene for handedness and hence for language dominance would unlock the neurobiology of language. Two ways of finding the gene for handedness are proposed: searching the pseudoautosomal region of the X chromosome or invoking a specific evolutionary model of lateralization in which the handedness gene has evolved from the *situs* gene then searching the human genome for homologues to the mouse *situs* gene.

1991 Biological asymmetry and handedness. Wiley, Chichester (Ciba Foundation Symposium 162) p 251–281

'The existence of a dominant hemisphere raises a number of problems, the most fundamental of which is why it should exist. . . . A question of major interest is the relationship between cerebral dominance for speech and handedness. . . . It may be possible for the geneticists to settle this question . . .'.

Lord Brain, (at the Ciba Foundation Symposium,
Disorders of Language, 21st May, 1963)

About 8% of the human population is left handed, using the left hand preferentially for complex skilled activities such as writing. Art historical evidence suggests the incidence has been constant for five millenia (Coren & Porac 1977); patchier palaeontological evidence suggests that human brain asymmetry has existed for a million years or so. Handedness is therefore a stable, behavioural polymorphism.

In 1865 Paul Broca showed that most individuals have language functions localized in the left cerebral hemisphere. Although left-handers were once thought to show the mirror image of this pattern, the relationship is actually more complex, 5% of right-handers and 30% of left-handers have language functions localized in the right hemisphere. Handedness and language localization are therefore interconnected, albeit not simply.

If any cognitive ability is typical of mankind, it is language; understanding language's biological bases would illuminate much of cognitive neuroscience. Handedness is another peculiarly human characteristic. Rats, mice, cats and dogs all show individual handedness (or strictly pawedness) in that each animal has an individual preference for using the right or the left paw. However, in those species there is no population preference, 50% prefer the left paw and 50% the right paw (e.g. Collins 1977). Recent claims that primates show population handedness preferences (MacNeilage et al 1987) have been much criticized and are not supported by thorough naturalistic studies such as those in the mountain gorilla (R. W. Byrne, J. M. Byrne, personal communication 1991).

The unique occurrence of handedness and language in humans, the lateralization of language and its association with handedness suggest that understanding the biology of handedness should illuminate the biology of language. Handedness is intrinsically simpler to study than language dominance, because of its ease of measurement. It is therefore an excellent surrogate for language dominance in the study of one of neurobiology's big problems.

The incidence of left-handedness

A meta-analysis of 284 665 individuals described in 88 studies in the literature (Seddon & McManus 1991) found an overall incidence of left-handedness of 7.78%. Variation between studies (Fig. 1) was not related to measurement methods (such as the numbers of items on inventories). The incidence was related to two background measures: the number of subjects, smaller studies reporting somewhat higher incidences, and the age of subjects, older subjects showing less left-handedness. Age and year of birth were highly correlated with each other across studies; the data did not allow disentanglement of an age effect from a cohort effect, although date of publication did not relate to the incidence. Of particular interest is the absence of geographical differences, a finding compatible with handedness being a balanced polymorphism present in all cultures. No cultural group has been found in which the incidence of left-handedness deviates substantially from 8%. As a single example of cultural uniformity, Professor Kevin Connolly (personal communication, 1991) has measured handedness in a pre-literate culture in a remote area of the Western Highlands of Papua New Guinea: of 188 individuals 21 (11.2%; 95% confidence interval = 6.7% to 15.7%) showed preferential use of the left hand.

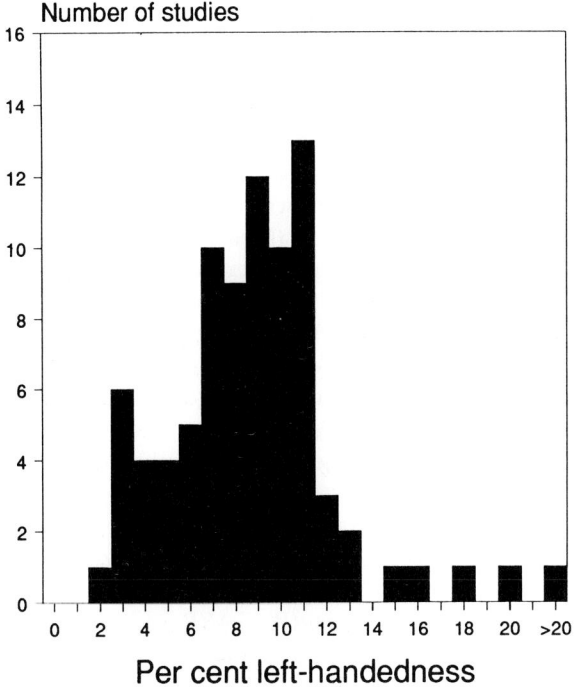

FIG. 1.   Overall incidence of left-handedness from a meta-analysis of 100 populations examined in 82 studies. From Seddon & McManus (1991).

The measurement of handedness

To the person in the street, handedness is straightforwardly described; individuals are right handed if they write with the right hand, with the proviso that a natural left-hander who is forced by social pressure to write with the right hand is really left handed. However, the description of handedness has been complicated by more sophisticated measures (see Bishop 1990 for a good review).

Handedness is assessed as either *preference* or *skill*. These are highly correlated in normal individuals (although recently we have found that autistic children show no correlation of handedness/preference and handedness/skill; I. C. McManus, B. Murray, K. Doyle, S. Baron-Cohen, unpublished). Handedness/preference is often measured using a questionnaire that asks which hand is preferred for a range of tasks. In children handedness/preference is more often measured using performance tasks; the observer notes which hand a subject uses for a range of unimanual tasks. Handedness/skill is measured using manual tasks in which subjects use first one hand and then the other, for example to make marks with a pencil or to move pegs across a board:

handedness is expressed as either a difference score $(R - L)$ or a standardized difference score $((R - L)/(R + L))$.

Measures of handedness/preference usually show a bimodal distribution, with few subjects appearing truly ambidextrous (e.g. McManus 1979, McManus et al 1988). Handedness/skill measures are more controversial: some tasks, such as the Annett peg-board (Annett 1970) give a unimodal distribution, suggesting a continuum of handedness; whereas other tasks, such as putting dots in circles (Tapley & Bryden 1985) or squares (McManus 1985a), give a bimodal distribution, implying two discrete categories of individual, analogous to the two modes found in handedness/preference measures. Elsewhere (McManus 1985a), I have argued that measures of handedness/skill differ principally in the extent to which they share skills used in the most highly developed skill asymmetry–writing. Peg-moving appears to be a unimodal distribution because it is less dependent upon asymmetrical processes, so that the modes have become closer, one being subsumed within the other; nevertheless, statistical decomposition of the distributions shows it is actually a mixture of normal distributions, as predicted (McManus 1985a).

Handedness measures can be analysed in terms of *direction* and *degree*. Direction of handedness considers whether a subject is on the right or left side of the distribution; it is defined as $\text{sign}(R - L)/(R + L)$. Degree of handedness assesses whether subjects show weak laterality (i.e. are nearly ambidextrous) or strong laterality; it is defined as $\text{Abs}(R - L)/(R + L)$. Most genetic studies of human handedness have assessed direction of handedness; in this paper I will also do so. Familial studies of degree of human handedness show no parent–offspring correlation (McManus 1979, 1985b) or very weak correlations (Bryden 1982, Coren & Porac 1980). Collins (1985) has selected mice for high and low degrees of pawedness, resulting in two strains (HI and LO). Elsewhere (McManus, submitted), I have suggested that these strains are not evidence that degree of handedness is under direct genetic control. Selection may have been for overall heterozygosis, the LO mice being less heterozygous; this would also explain their lower birth weight, poorer reproductive fitness and the lower proportion of males.

Environmental influences on handedness

Although this paper argues principally for a genetic influence on handedness, there have been suggestions that environmental influences are also important. Perhaps most influential is the birth stress hypothesis, which states that pathological left-handedness results from intrapartum anoxia. The meta-analysis of Searleman et al (1989) has effectively disposed of this theory.

Since 1982, Geschwind's hypothesis (see Geschwind & Galaburda 1987) that fetal testosterone levels relate to handedness, cerebral dominance and a range of other conditions has received much attention. The hypothesis is complex and difficult to present as a formal model (see McManus & Bryden 1991a). A final verdict is probably premature, although it is safe to say that many of the theory's postulated correlations have not been replicated (see McManus et al 1990, Marchant-Haycox et al 1991).

Evidence for a genetic basis for handedness

Family studies

Left-handedness undoubtedly runs in families. McManus & Bryden (1991b) reviewed 25 studies, based on 72 600 offspring, relating handedness of children to parental handedness (Fig. 2). Clearly, left-handedness is more common if one parent is left handed, and still more common if both parents are left handed.

Left-handedness is often reported to be more common in males than females. A meta-analysis by Seddon & McManus (submitted) of sex differences in 63 studies found a 27.4% higher incidence in males than females (Fig. 3); the difference was unrelated to any other measures. McManus & Bryden (1991b) also looked at parental and offspring handedness in relation to the sex of parents and offspring; eighteen studies were available (Table 1). Log-linear modelling (of individual studies not grouped data as in Table 1, which is purely for illustrative purposes) showed a higher incidence of left-handedness in male parents and male offspring; additionally, left-handed mothers had more left-handed offspring than did left-handed fathers, although there was no interaction with sex of offspring. This 'maternal effect' has been reported before in the literature (e.g. McGee & Cozad 1980, Annett 1985a); there seems little doubt of its reality and it must be explained by any adequate genetic model of handedness.

Adoption studies

Familial associations need not imply genetic transmission because families also transmit culture and environment. Adoption studies can separate genetic effects from environmental ones. Surprisingly, only a single, adequate adoption study has looked at children adopted early enough and assessed at a late enough age for handedness to be defined. Carter-Saltzman (1980) showed that, in contrast to that of normal, biological children, the handedness of adopted children showed no relationship to that of their adoptive parents. We may therefore interpret familial trends as reflecting genetic effects.

Twin studies

Twin studies have long been seen as the Achilles heel of the genetics of handedness. Table 2, from McManus & Bryden (1991b), summarizes 14 studies, collected since 1930. Data obtained earlier are biased because discordant

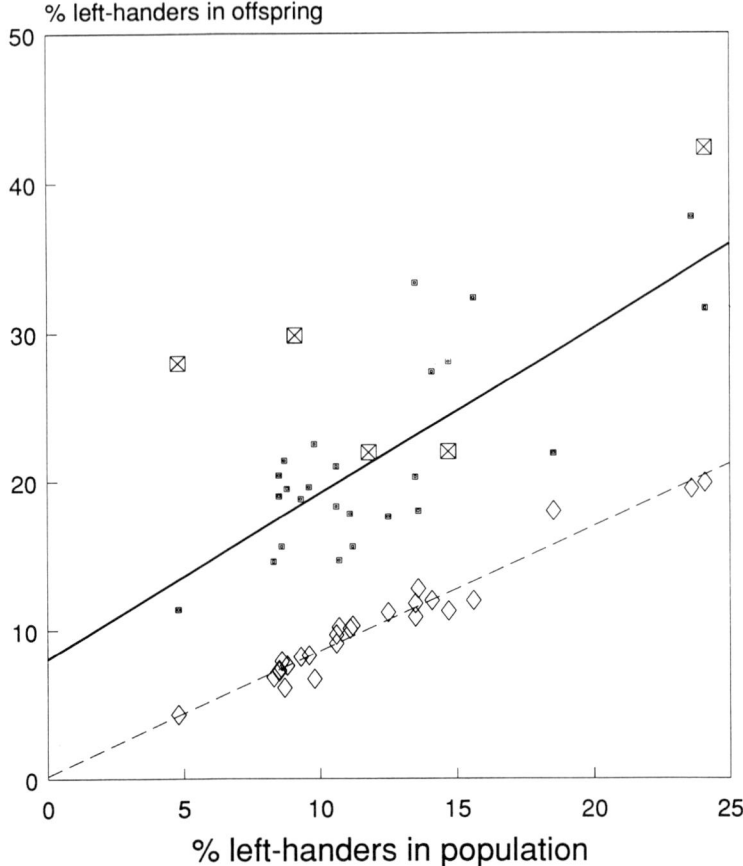

FIG. 2.   Handedness in offspring of R × R, R × L and L × L matings in relation to the overall incidence of left-handedness in the cohort. Offspring, parental and grandparental generations are plotted separately within studies, since the incidence of handedness is typically different in each. Dashed and solid lines represent best-fitting linear regressions through the offspring of R × R (◇) and R × L (■) matings, respectively. Offspring of L × L (⊠) matings are rare and therefore several studies with similar overall incidences of left-handedness have been amalgamated. From McManus & Bryden (1991b).

handedness was used as a criterion of monozygosity, in the belief that it represented 'mirror imaging' (McManus 1980). Discordance of handedness is frequent in monozygotic (MZ) and dizygotic (DZ) twins. Comparison of the expected proportions under a binomial distribution with the observed frequencies shows fewer discordant MZ pairs (90.1% of expected) than DZ pairs (99.3% of expected). MZ twins are therefore more similar than DZ twins. Nevertheless, the substantial amount of discordance requires explanation by any adequate genetic model.

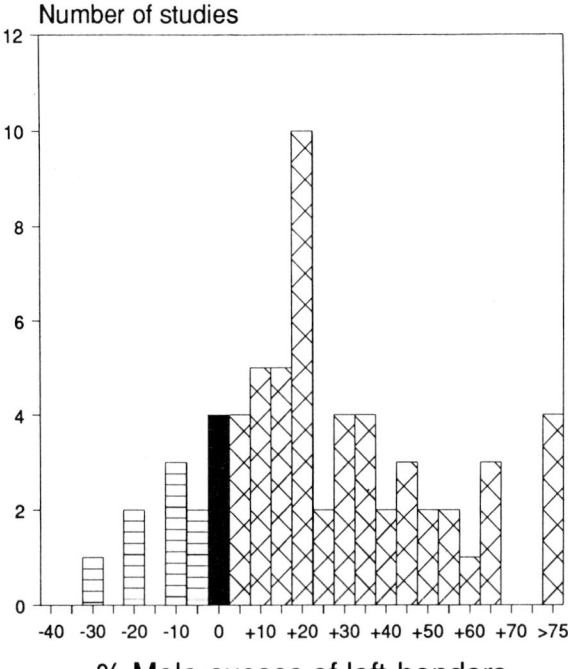

FIG. 3. The excess incidence of left-handedness in males expressed as a proportion of the incidence in females (100 × (Male − Female)/Female) on the basis of a meta-analysis of 63 populations. Hatched, M > F; striped, M < F; black, M = F. From Seddon & McManus (1991).

TABLE 1 Handedness in families, according to maternal/paternal handedness and the sex of offspring

Parental handedness		Left-handed offspring		
Father	Mother	Sons	Daughters	Total
Right	Right	10.4% (30 268)	8.5% (26 020)	9.5% (63 250)
Right	Left	22.1% (1815)	21.7% (1688)	19.5% (8933)
Left	Right	18.2% (2308)	15.3% (2100)	
Left	Left	27.0% (215)	21.4% (168)	26.1% (417)

Overall totals ignoring sex are based on 25 sets of data; entries by sex of parents and offspring are based on 18 studies. From McManus & Bryden (1991b).

TABLE 2 **Frequency of monozygotic and dizygotic twin pairs concordant and discordant for handedness, based on 14 studies**

Handedness of twins	Monozygotic	Dizygotic
Right/Right	2184	1951
Right/Left	629	585
Left/Left	87	53
Observed/expected discordant pairs	0.901	0.993

The ratio of observed to expected disordant pairs is estimated by calculating for each study the ratio of the actual number of discordant pairs and the proportion expected under a binomial distribution, then averaging the ratios across studies. From McManus & Bryden (1991b).

Genetic models of handedness

Many genetic models of handedness have been published (see McManus & Bryden 1991b for a review). Most have failed because they predict that right-handers or left-handers 'breed true', or because they do not predict MZ twin discordance. At present there are only two adequate models: the 'right-shift' model of Annett (1978, 1985a) and my own genetic model (McManus 1979, 1984a, 1985b). These models were developed independently in the late 1970s. The success of each indubitably results from the same key feature—that one genotype does not control handedness in the strict sense, but results in randomness and hence right-handers and left-handers in equal numbers. This situation is known in biology as 'fluctuating asymmetry'; it is seen in the inheritance of *situs inversus* encoded by the *iv* gene of the mouse (Layton 1976) and in Kartagener's syndrome in humans. The 'inheritance' of non-human handedness depends entirely on this randomness that produces a 50% incidence of left-handedness and non-transmissibility to offspring.

McManus' genetic model

This model originally proposed two alleles, *D* (*Dextral*) and *C* (*Chance*), at a single, autosomal locus (Table 3). The *DD* genotype produces right-handedness in all offspring. The *CC* genotype results in fluctuating asymmetry, 50% of offspring being right handed and 50% left handed. The offspring are *not* ambidextrous—each is fully right handed or left handed, lateralization being absent at the level of the population not the individual. Model-fitting (McManus 1985b) showed that the effects of the alleles in the heterozygote, *DC*, had to be additive, producing 25% left-handers. The population gene frequency, p(*C*), is estimated from the family and twin data to be 0.155, and is assumed to be identical in each study population. Differences between populations in the incidence of manifest left-handedness are assumed to represent response biases or criterion problems (McManus 1985b).

TABLE 3   The inheritance of handedness and cerebral dominance predicted by the genetic models of McManus and McManus & Bryden

(a) McManus' (1985a) genetic model

Genotype	% left-handedness	% right cerebral language dominance
DD	0	0
DC	25	25
CC	50	50
	p(C) = 0.155	

(b) McManus & Bryden's (1991b) genetic model

	Modifier gene in males		Modifier gene in females		
	M	m	MM	mM	mm
DD	0	50	0	0	50
DC	25	50	25	25	50
CC	50	50	50	50	50
		p(C) = 0.135	p(m) = 0.045		

D(*Dextral*) and C (*Chance*) are alleles at an autosomal locus. M and m are alleles at a modifier locus on the X chromosome. The entries in (b) show the percentage left-handedness or right cerebral language dominance. p(C) is an estimate of the frequency of allele C in the population; p(m) of the m allele.

The model fits family data well. It predicts that R × R, R × L and L × L matings (R, right-handed; L, left-handed) produce 5.97%, 17.42% and 28.87% of left-handers, values compatible with Fig. 2. Discordance in MZ twins is explained by assuming that chance processes in the *CC* and *DC* genotypes occur *independently* in each monozygotic twin, so that in 25% of *CC* MZ twin pairs both are right handed, in 25% both are left handed and 50% of pairs are discordant. Formal calculations using this model predict 11.02% of MZ pairs and 12.66% of DZ pairs to be discordant, each value being less than predictions under a binomial distribution of 14.29% discordance.

The model is extended to explain cerebral dominance by assuming that the genes also control language dominance; *DD* individuals are all dominant, *CC* individuals are right or left dominant with equal probabilities. The model predicts that 5.98% of right-handers and 28.88% of left-handers will be right-hemisphere dominant.

Originally, the McManus model did not attempt to explain sex differences or the maternal effect. In a recent modification it accounts for these problems (see below).

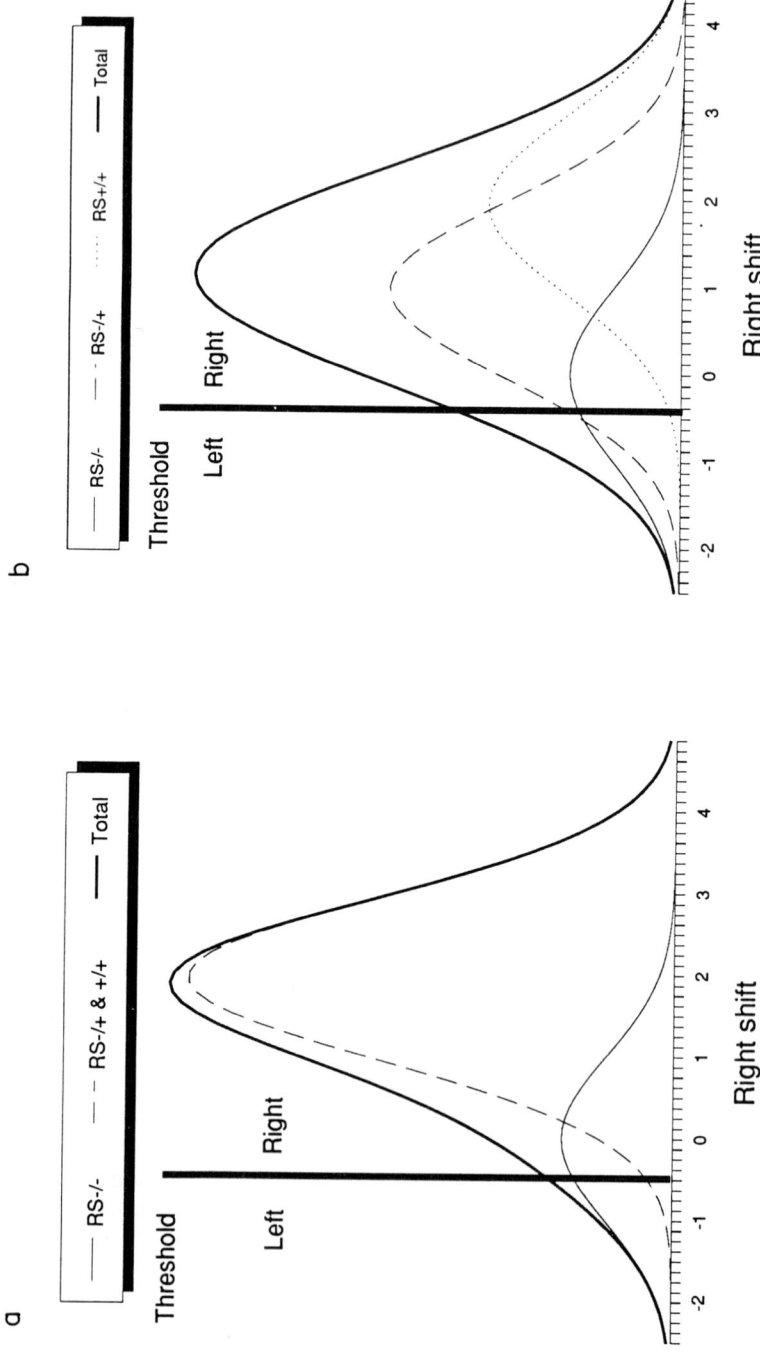

FIG. 4.   The phenotypic distributions of the RS − / −, RS − / + and RS + / + genotypes, as predicted by (a) the Annett (1978) model, and (b) the Annett (1983) model (described in Annett & Kilshaw 1983). Distributions are shown only for singletons: those for twins are shifted somewhat less to the right.

Annett's right-shift model

Annett's right-shift model originates in the observed unimodal distribution of between-hand skill differences. In the 1978 model these were attributed to two populations: one, phenotype RS −, lacked a right shift (RS) and had a mean skill difference of zero; the other, RS +, had its mean shifted 1.96 standard deviations (S.D.s) to the right (see Fig. 4a). Populations differ in manifest incidence of handedness because survey methods use different criteria for 'left-handedness'. Phenotypes were determined by two alleles at an autosomal locus, with + dominant to −; − / + and + / + produced RS + and − / − produced RS − phenotypes. The model fitted various sets of family data, but did not fit twin data well. The problem was circumvented by the *ad hoc* assumption of a right shift of only 1 S.D. in RS + MZ twins. The model accounted for language dominance by assuming a 50% chance of right hemisphere language dominance in RS − phenotypes. The frequency of the − allele was calculated from a knowledge of the incidence of left-handers with right-hemisphere language dominance.

The right-shift model was modified in 1983 (Annett & Kilshaw 1983) to make the − / + genotype additive, so that the R − / −, RS − / + and RS + / + phenotypes had right shifts of 0, 0.98 and 1.96 S.D.s (Fig. 4b). The model was further modified (Annett 1985a) with the proposal that males and females differed in their right shift; the RS − / −, RS − / + and RS + / + genotypes having shifts of 0, 1 and 2 S.D.s in males and 0, 1.2 and 2.4 S.D.s in females. The model then accounted for the excess of left-handed males, albeit by adding an additional parameter. However, as Annett (1985a; p. 327) recognized, the model still cannot adequately explain the maternal effect, the calculated effect being far smaller than the observed one (Table 4a).

The McManus and Bryden (1991) model

McManus & Bryden (1991b) have modified the McManus genetic model to account for sex differences and the maternal effect. The revised model proposes the same autosomal locus with D and C as previously, plus an additional X chromosomal modifier locus, with two alleles, M and m. (Modifier genes have also been invoked in the genetics of Tourette's syndrome (Comings & Comings 1986) and fragile X syndrome (Israel 1987).) The dominant gene, M, does not affect the function of D and C alleles. However, in the m genotype in males or the mm genotype in females, m, the rarer recessive gene, modifies D alleles to act as C alleles, substituting fluctuating asymmetry for directional asymmetry (Table 3b). The C allele occupies about 13.5% of its gene pool and the m allele about 4.5% of its gene pool. Not surprisingly, the model's extra parameter allows it to explain the excess of male left-handedness. More gratifying is that the model also predicts a substantial maternal effect (Table 4b), albeit not quite as large

TABLE 4   The predictions of the Annett (1985a) and McManus & Bryden (1991b) models of the inheritance of handedness

Handedness of parents		% left-handedness in offspring	
Father	Mother	Sons	Daughters
(a) Annett (1985a) model			
Right	Right	10.69	9.07
Right	Left	18.27	16.57
Left	Right	17.41	15.72
Left	Left	28.86	28.09
(b) McManus & Bryden (1991b) model			
Right	Right	10.49	8.63
Right	Left	21.50	19.55
Left	Right	18.49	17.44
Left	Left	29.40	28.39

Calculations are given for a population incidence of left-handedness of 6.7% in parents and 10.83% in offspring, the particular incidences found in Table 1, so that direct comparison is possible. For the model of Annett (1985a) the right shift is 2.4 and 2.0 for $RS+/+$, 1.2 and 1.0 for $RS+/-$ and 0 and 0 for $RS-/-$ in females and males, respectively; the frequency of the $RS-$ allele is 0.4306. The model of McManus & Bryden (1991b) is as given in Table 3b.

as that observed. However, the model has not yet been formally fitted and better parameter estimates may produce a better fit. McManus & Bryden have also noted that there is a cluster of conditions, such as stuttering, dyslexia and autism, characterized by a male preponderance and increased sinistrality; the modifier gene proposed here might explain such associations if it modified not only the cerebral dominance gene but also other autosomal loci involved in language or communication.

Comparing the McManus and Annett models

Both models succeed because they reflect the biology of asymmetry and especially of fluctuating asymmetry. Differences result from their conceptions of the phenotype of handedness: Annett sees a continuum of skill differences as the primary phenotype, preference occurring secondarily, whereas McManus sees a dichotomous, categorical measure of preference as primary, with skill differences of only secondary importance. The question is complex in its details (McManus 1985a, see also Annett 1985b), although I have argued that data on skill differences are more compatible with my own 'symmetric bimodal' model than with the right-shift model.

Formal comparisons of the McManus (1985b) model with Annett's 1983 model have been carried out using maximum likelihood fitting (McManus 1985b); my own model was generally a better fit.

Finding the gene for handedness and cerebral dominance

In the 1990s genetic models are properly tested by finding DNA and assessing its properties. That is difficult if one has no specific idea of where to look in the genome and there are no obvious routes through known chromosomal errors or obvious gene products. More seriously, the fundamentally random component in the models of handedness results in a substantial loss of power when searching for linkage with polymorphic markers. Two routes do seem possible and, I would suggest, worth pursuing: looking for the gene in the pseudoautosomal region and investigating whether it is a homologue of the *iv* gene.

Is the handedness gene in the pseudoautosomal region?

Crow (1990) has provocatively and imaginatively hypothesized that the atypical cerebral temporal lobe lateralization in schizophrenia could be explained if the schizophrenia gene were a mutation of the cerebral dominance gene. Additionally, Crow points out that the excess of schizophrenia in the sex chromosome aneuploidies would be explained if the schizophrenia/cerebral dominance gene were in the X chromosome's pseudoautosomal region. If this is so, schizophrenia should be more likely to occur in siblings of the same sex when transmitted by the father, but there should be no such association when the mother transmits the condition. Crow et al (1989) reported data that supported that prediction. I. C. McManus & T. S. Crow (unpublished) have analysed a similar prediction for handedness. A large sample of families showed an effect in the predicted direction, although it was not statistically significant; however, power calculations suggested the need for extremely large studies. Nevertheless, the pseudoautosomal region may be an interesting area to explore for the handedness gene.

An evolutionary model of the origins of handedness and cerebral dominance: the handedness gene as a homologue of the iv gene

Humans show several independent, uncorrelated asymmetries, all seemingly under genetic control: *situs*, handedness and cerebral dominance, hand-clasping, arm-folding (see McManus & Mascie-Taylor 1979) and eye dominance. Of these, only handedness, *situs* and perhaps eye dominance are of functional importance, the others showing wide geographical variation in incidence. *Situs* is the only asymmetry shared with other vertebrates and hence is probably evolutionarily the oldest of the human asymmetries. Palaeontological evidence suggests that vertebrate asymmetry originated perhaps 500 million years ago in a sub-group of the deuterostomes, the dexiothetica, which eventually formed modern vertebrates, chordates and echinoderms (see Jefferies, this volume). It is probable that the wild-type allele of the mouse *iv* gene (see Brueckner et al, this volume) is closely related to, if not the same as, the gene for dexiothetism (see Fig. 5).

Genes controlling anatomical and functional asymmetry are rare, and hence unlikely to have evolved separately. The handedness/cerebral dominance gene is therefore probably a mutation of the wild-type *iv* gene, which no longer causes cardiac asymmetry (see Brown et al, this volume) but causes cerebral asymmetry, perhaps by subtly altering a protein's tertiary structure. If this is true, then the handedness gene should be very similar to the *iv* gene. Other human asymmetries, of hand-clasping, arm-folding and eye dominance, being of less biological consequence than handedness but still acting on cerebral tissue, are most likely mutations of the handedness gene.

Such an evolutionary scenario suggests a straightforward way of finding the handedness gene. Firstly, find the *iv* gene in the mouse (see Brueckner, this

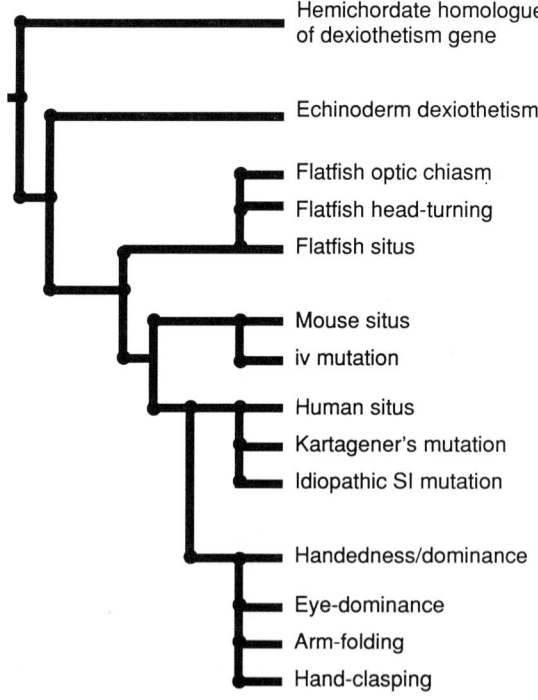

FIG. 5. A hypothetical evolutionary tree for the origins of the genes controlling handedness, hand-clasping, arm-folding, eye dominance, *situs* in humans, mice and flatfish, and dexiothetism.

volume), then hybridize the mouse *iv* gene to human DNA and search for partial homologues; one, probably very similar to the mouse gene, should be the human *situs* gene. Others of lesser homology should be the genes for handedness and the other asymmetries. Once the handedness gene has been found, it will be straightforward to determine its genetics, and to evaluate the McManus and Annett models. Other analyses are also possible. By studying *iv* gene homologues in flatfish, the evolution and inheritance of flatfish asymmetry (McManus 1984b) can be assessed. The evolutionary model of Fig. 5 can also be tested by looking for homologues in modern echinoderms. Last, but not least, identifying the handedness gene would allow *in situ* hybridization in embryos to determine when the gene is expressed and the sites of its action in embryogenesis.

It is probably not hyperbole to suggest that finding the gene for handedness and cerebral dominance would unlock the molecular biology and neurobiology of the characteristic which is most peculiar to humans, that of language.

References

Annett M 1970 The growth of manual preference and speed. Br J Psychol 61: 545–558

Annett M 1978 A single gene explanation of right and left handedness and brainedness. Lanchester Polytechnic, Coventry, Warwickshire, UK

Annett M 1985a Left, right, hand and brain: the right shift theory. Lawrence Erlbaum, London

Annett M 1985b Which theory fails? A reply to McManus. Br J Psychol 76: 17–29

Annett M, Kilshaw D 1983 Right- and left-hand skill. II: Estimating the parameters of the distribution of L-R differences in males and females. Br J Psychol 74: 269–283

Bishop DVM 1990 Handedness and developmental disorder. Blackwell, Oxford

Brain Lord 1964 Statement of the problem. In: Disorders of language. Churchill, London (Ciba Found Symp) p 5–20

Brown NA, McCarthy A, Wolpert L 1991 Development of handed body asymmetry in mammals. In: Biological asymmetry and handedness. Wiley, Chichester (Ciba Found Symp 162) p 182–201

Brueckner M, McGrath J, D'Eustachio P, Horwich AL 1991 Establishment of left-right asymmetry in vertebrates: distinct genetic steps are involved. In: Biological asymmetry and handedness. Wiley, Chichester (Ciba Found Symp 162) p 202–218

Bryden MP 1982 Laterality: functional asymmetry in the intact brain. Academic Press, New York

Carter-Saltzman L 1980 Biological and socio-cultural effects on handedness: comparison between biological and adoptive families. Science (Wash DC) 209: 1263–1265

Collins RL 1977 Toward an admissible genetic model for the inheritance of the degree and direction of asymmetry. In: Harnad S, Doty RW, Jaynes J, Goldstein L, Krauthamer G (eds) Lateralization in the nervous system. Academic Press, New York p 137–150

Comings DE, Comings BG 1986 Evidence for an X-linked modifier gene affecting the expression of Tourette syndrome and its relevance to the increased frequency of speech, cognitive, and behavioral disorders in males. Proc Natl Acad Sci USA 83: 2551–2555

Coren S, Porac C 1977 Fifty centuries of right-handedness: the historical record. Science (Wash DC) 198:631–632

Coren S, Porac C 1980 Lateral preferences and human behavior. Springer-Verlag, New York

Crow TJ 1990 The continuum of psychosis and its genetic origins. Br J Psychiatry 156:788–797

Crow TJ, DeLisi LE, Johnstone EC 1989 Concordance by sex pairs in sibling pairs with schizophrenia is paternally inherited: evidence for a pseudoautosomal locus. Br J Psychiatry 155:92–97

Geschwind N, Galaburda AM 1987 Cerebral lateralization. MIT Press, Cambridge, MA

Israel MH 1987 Autosomal suppressor gene for fragile-X: an hypothesis. Am J Med Genet 26:19–31

Jefferies RPS 1991 Two types of bilaterality in the metazoa: chordate and bilaterion. In: Biological asymmetry and handedness. Wiley, Chichester (Ciba Found Symp 162) p 94–127

Layton WM Jr 1976 Random determination of a developmental process. J Hered 67:336–338

McGee MG, Cozad T 1980 Population genetic analysis of human hand preference: evidence for generation differences, familial resemblance and maternal effects. Behav Genet 10:263–275

McManus IC 1979 Determinants of laterality in man. PhD thesis, Cambridge University, Cambridge, UK

McManus IC 1980 Handedness in twins: a critical review. Neuropsychologia 18: 347–355

McManus IC 1984a The genetics of handedness in relation to language disorder. In: Rose FC (ed) Advances in neurology, vol 42: Progress in aphasiology. Raven Press, New York, p 125–138

McManus IC 1984b The inheritance of asymmetries in man and flatfish. Behav Brain Sci 7:731–733

McManus IC 1985a Right- and left-hand skill: failure of the right shift model. Br J Psychol 76: 1–16

McManus IC 1985b Handedness, language dominance and aphasia: a genetic model. Psychol Med, Monograph Suppl No 8

McManus IC 1991 Is degree of paw preference in HI and LO mice under direct genetic control? Submitted

McManus IC, Bryden MP 1991a Geschwind's theory of cerebral lateralization: developing a formal genetic model. Psychol Bull, in press

McManus IC, Bryden MP 1991b The genetics of handedness, cerebral dominance, and lateralization. In: Rapin I, Segalowitz SJ (eds) Handbook of neuropsychology, Section 10: Developmental neuropsychology. Elsevier, Amsterdam, in press

McManus IC, Mascie-Taylor CGN 1979 Hand-clasping and arm-folding: a review and a genetic model. Ann Hum Biol 6:527–558

McManus IC, Sik G, Cole DR, Mellon AF, Wong J, Kloss J 1988 The development of handedness in children. Br J Dev Psychol 6:257–273

McManus IC, Naylor J, Booker BL 1990 Left-handedness and myasthenia gravis. Neuropsychologia 28:947–955

MacNeilage PF, Studdert-Kennedy MG, Lindblom B 1987 Primate handedness reconsidered. Behav Brain Sci 10:247–303

Marchant-Haycox SE, McManus IC, Wilson GD 1991 Left-handedness, homosexuality, HIV infection and AIDS. Cortex 27:49–56

Searleman A, Porac C, Coren S 1989 Relationship between birth order, birth stress and lateral preferences: a critical review. Psychol Bull 105:397–408

Seddon BM, McManus IC 1991 The inheritance of left-handedness: a meta-analysis. Submitted

Tapley SM, Bryden MP 1985 A group test for the assessment of performance between the hands. Neuropsychologia 23:215–221

DISCUSSION

Fujinaga: I am asking this question because English is not my native language. Is there any difference in nuance between sidedness, handedness and laterality?

Wolpert: Sidedness is where there is a difference between one side and the other. Handedness would be where this sidedness is consistently specified with respect to right and left.

Collins: There is also lateralization, which we use to describe the degree of asymmetry without respect to the direction.

Wolpert: Lateralization is like sidedness—a difference between left and right but no preference for left or right.

Fujinaga: Both models described by Dr McManus ignore ambidextrous people.

Annett: No. There are many people who show no difference between hands in skill. A basic assumption of the RS theory is that there is a continuous distribution of differences between strong right and strong left for hand preference and hand skill.

Frankel: Could you please explain how these models deal with the nearly binomial distribution of handedness in identical twins?

McManus: My model is formally identical to Marian Annett's; I will explain in terms of my own model. Consider a pair of monozygotic twins who have the *CC* genotype. Chance determines whether an individual is right or left-handed: therefore there is an independent chance process going on in each twin, so 25% of twin pairs will both be right handed, 25% both left handed and 50% discordant. Complete calculations give almost exactly the numbers actually observed.

Frankel: What about *DD* twin pairs?

McManus: They are all right handed; therefore they are all concordant for right-handedness and for left-brainedness for language.

Annett: There is no gene for right-handedness in the RS theory; only a gene (*RS*+) which displaces the chance distribution to the right. Some monozygotic twins of *RS*++ genotype could differ in their handedness, but the proportion would be small.

Galaburda: I am not sure one needs a genetic theory for directionality of handedness. We know that Nature has to have these biases. You can disturb the system in many ways—by the products of many genes or by changing the uterine environment—to explain the observed maternal effect. That produces randomness; it produces symmetrical brains which will include those of left-handed people. The anatomical observations could be reconciled with the directionality of functional handedness by proposing a system where one is either right handed or random. Then you could postulate almost anything non-genetic or genetic in the sense of products of other genes to determine that. Why does one need genetics?

McManus: My model doesn't tell us where the directional effect comes from; the system is simply reading a pre-existing asymmetry. The model is therefore strictly a theory of the genetics of the variation in lateralization. It says there *is* variation in lateralization and that variation seems to run in families and therefore to be genetically transmissible. The model does not say why the brain is asymmetrical in the first place.

The brain asymmetry can also be disrupted. That may be due to a failure to read the system, which I am arguing can occur because of a fairly specific major gene. The system can also be overridden by other things; for instance, anything that produces developmental noise seems to increase the incidence of left-handedness. Almost every cause of mental retardation is associated with a higher incidence of left-handedness. But I would argue that those are phenocopies of true left-handers.

Burn: I don't think your genetic models need to explain twin studies. Francis Galton (1876) accepted that twins could be strongly contrasting. I would agree with Bronson Price (1950) who pointed out that twins are different from singletons. There are a whole series of potential primary biases—shared circulation, potential disturbance of laterality and so on—which mean that twins are morphologically different from singletons. You don't have to explain with the single model such as yours or Marian's the fact that monozygotic twins don't quite conform to your predictions. Monozygotic twins have their own reasons for being different. As Lotze said (1937), they are a malformation to whom Nature was kind.

McManus: That's not entirely reasonable. Galton was one of the first to propose the use of statistical tests in evaluating hypotheses. Handedness in monozygotic twins does not occur in binomial proportions; by any significance test the association is highly significant. This association of handedness between monozygotic twins has to be explained.

There may be an additional mechanism operating in twins: Marian Annett's model says that quite specifically. I would prefer to have a single process that doesn't need to invoke these additional factors which we don't know about. We can model the data quite adequately using a single gene.

Burn: Biologically, one would expect there to be other factors that could influence handedness in twins, as they do in malformations in other areas.

McManus: But you have to predict the similar incidence of left-handedness in twins and singletons. If you argue that there is an additional mechanism acting in twins, you have to explain why the incidence of left-handedness is similar to that found in singletons. If there is something different happening in twins, the models for singletons become rather different, because the permissible models for singletons are constrained by the permissible models for twins.

Burn: I doubt whether they need to be.

Annett: Surely the geneticists could be rather worried about this meta-analysis putting all these very different studies together. When you do such a meta-analysis, I am not surprised that you come out with a mean value that looks reasonably appropriate.

McManus: When I am fitting the models, I am not fitting them to average data. I have fitted each separate study individually. I agree that all formal model fitting has to be done on the basis of individual sets of data, not on aggregated data sets.

Peters: I agree with Marian Annett that if you do a meta-analysis, real problems can arise. I looked at the sex ratios in a number of studies on handedness. Using the single criterion of which hand is used for writing, some studies found twice as many male left-handers as female left-handers, other studies found no difference.

Lewis: There was a time when left-handers were taught to write with their right hand, which would account for some of the discrepancy.

Peters: There is still a bias in this direction: just because teachers and parents do not overtly force a left-handed child to write with its right hand does not mean that pressures towards right-handedness are no longer present. For instance, a child may observe that it does things differently from other children and may quietly, without being told to, try to imitate the other children. Because left-handed parents are more accepting of left-handed preferences in their children, it is important that findings of a raised prevalence of left-handed offspring of left-handed mothers are not immediately interpreted solely in terms of genetics.

Burn: Looking at the age-related factor and also the temporal shift over this century, presumably the skill level of the task brings out the left-handedness. If you do something that requires more skill, it will expose your left-handedness; whereas something with a low skill level may not. Many girls in the 19th century were taught to write their name but little more. Writing one's name with the 'wrong' hand is relatively easy. Ask those girls to write fast in an exam and their laterality might become more exposed. The observed sex difference could be related to the degree of literacy.

McManus: Kevin Connolly's study in Papua New Guinea shows that the same prevalence of left-handedness occurs in a non-literate society.

Meta-analysis is a much criticized technique. People argue that if you put garbage in, you get garbage out. Michael Peters has effectively taken a number of studies and made individual *ad hoc* hypotheses that explain the different studies. With *n* studies, if you can find $n - 1$ characteristics, you can explain away all the variance and it is just a matter of deciding which remaining study is the 'correct' one.

The other way of comparing studies is effectively to say that these measurable characteristics do relate to the incidence of left-handedness, then to measure the characteristics of the way the study has been done. Statistical analysis of the results then estimates the association of the characteristic with left-handedness. I was well aware that in early studies there was a lower incidence of left-handedness, so I put the date of the study into the meta-analysis. I also put the estimated response rate to questionnaires into the meta-analysis. If we believe that left-handers are responding differentially, those studies with 100% response rates should have a different incidence of left-handers from those studies with a 20% response rate. But response rate doesn't come out of the analysis as being a significant predictor. Meta-analysis is no more than a formalization of what we do whenever we review any complex set of data.

Annett: The more important test of the model is to go back to the threshold idea which is inherent in my model and ask whether you can predict the outcome equally well if there is an incidence of 3% left-handers in the parents as if there is an incidence of 20% in the parents. Because it is a threshold model, the RS theory successfully predicts all the findings in the literature, wherever the thresholds fall, for parents and children. Because you want to stick with a single incidence in your model, you have to average the data to get it.

McManus: You are right and I have to argue that studies differ in their observed incidences of left-handedness for all sorts of extraneous social reasons; people deciding at different times, for instance, to say they are more likely to be left handed rather than right handed.

Your model says there are certain types of people who are more likely to swing across the threshold. We need to argue about why that threshold is moving. What it is about the threshold that allows such a change? Your model says it is those individuals with small skill asymmetries who can be described as right handed in one study and left handed in another; those are the people around the threshold.

Such a model assumes that the degree of handedness is inherited. I can't convince myself on the basis of the studies that two strongly right-handed parents tend to have strongly right-handed children. The threshold model does make that prediction, as far as I understand it.

Annett: I don't think that is the case at all. I do straightforward Mendelian genetics, having cut the distribution at whichever point, and find that the model predicts the incidence of left-handedness in the offspring.

Professor Wolpert discovered at the beginning of this symposium that only two of the participants are left handed. When assessing handedness, if we ask about more than one action, then we find many more people who prefer to use their left hand. I would like to ask people in this room to imagine they are dealing playing cards.

(Everyone deals imaginary cards)

Wolpert: Three from 28.

Annett: I would have expected about one in six from previous studies on students. I ask people 12 standard questions and find the incidence of left-handers to be about 35–36%.

McManus: But does that mean they are left handers?

Annett: That is semantics.

McManus: No, it is not. If I ask 28 questions on a handedness questionnaire, almost everybody finds some task at which they are left handed. This does not mean they are all left handed or that the phenomenon of right-handedness has disappeared; I think we are measuring two separate classes of individual, right- and left-handers.

Peters: When determining whether or not someone is left handed on the basis of which hand they use for certain activities, it is crucial to consider the nature of the activity. Some actions require no particular skill, such as picking up a book from a table. In those cases, it means little when people say they prefer their left or right hand. Other activities require a degree of skill and the preference expressed is important.

Wolpert: Which skills are the relevant ones?

Peters: Ludwig said the preferred hand moves more quickly, more accurately and more forcefully. Tasks that demand any of these qualities are more likely to reveal the truly preferred hand.

Wolpert: Chris, it would be enormously helpful to those of us not in the field if you could summarize very briefly the difference between your model and Marian's.

McManus: There is a series of differences. Marian proposes that the phenotype is a continuum of skill differences and an individual can be anywhere along it; the arbitrary division of the continuum into two groups divides the population into left- and right-handers. I propose that the phenotype is categorical: there are two types of individual, left-handed and right-handed, and skill differences are secondary to preference differences, owing to practice. The principle difference therefore concerns the definition of the behavioural phenotype. Marian argues that it is principally a skill difference with preference being a secondary phenomenon. I argue that the behavioural phenotype is principally preference and skill differences are secondary.

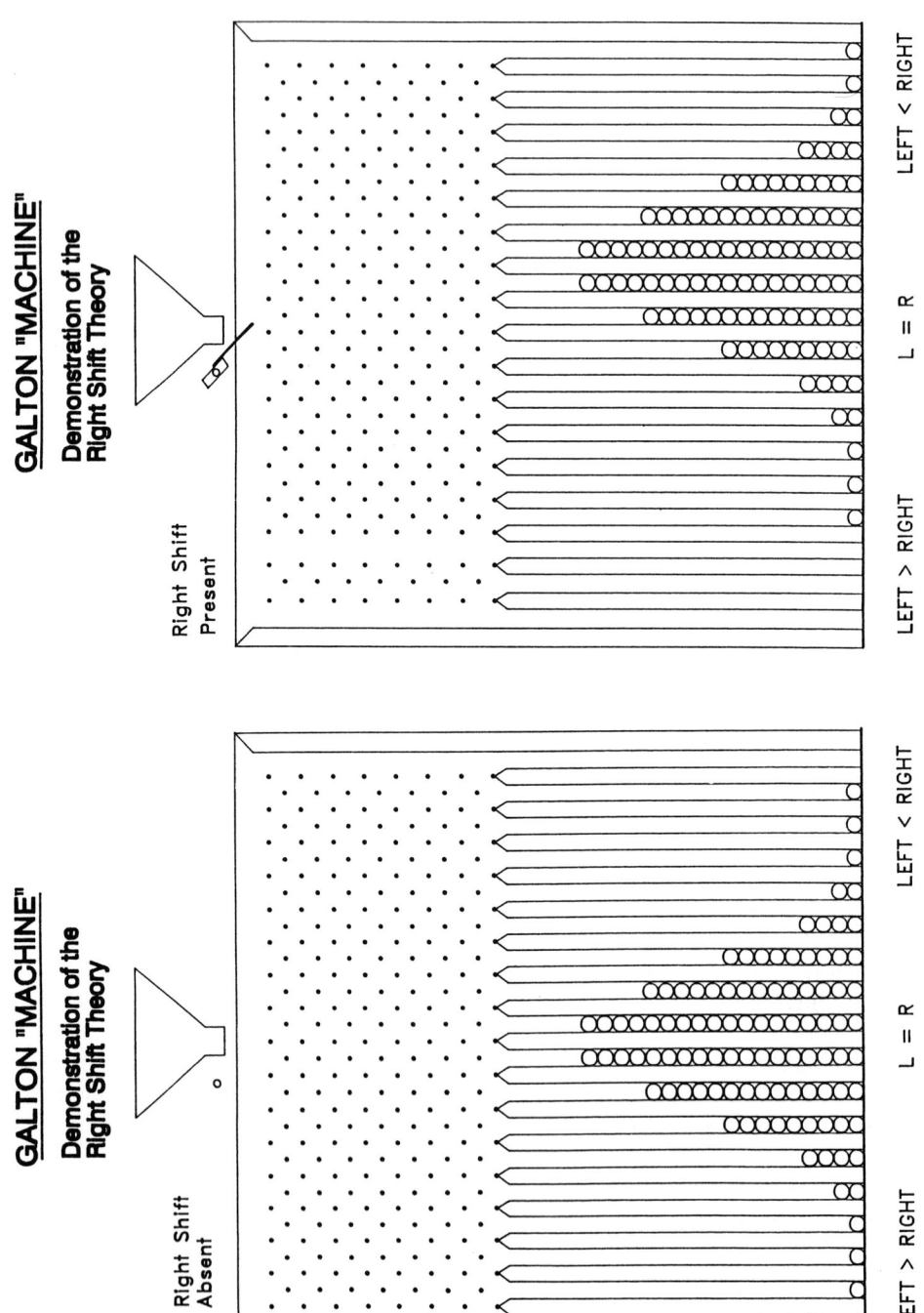

Galaburda: Then you are talking about different things! Marian is talking about magnitude, you are talking about directionality. According to Marian's model, as an individual becomes less right-biased, he or she becomes more random.

McManus: We are both talking principally about direction of handedness. Neither of us says that degree (or magnitude) need not be explained. However, I argue that degree of handedness does not need to be explained by genetic models.

Annett: The twins are a very important issue. Chris says that my model needs an extra postulate to account for the twin data. His model also does not account for twins unless a second randomness postulate is introduced.

McManus: No, I do exactly as you do: I assume that the phenotype of one twin is determined independently of the phenotype of the other twin, according to its genotype. Since in some genotypes there is a chance component in determining the phenotype, those random processes are independent of each other, according to both my model and yours.

Annett: I have a Galton machine which I use to explain the theory (Fig. 1). There is a funnel at the top, you drop ball bearings in at the top and they fall down and make a nice normal distribution.

My right-shift model basically grew from evidence which suggests that our primate cousins have a normal distribution of those who are better on the left and those who are better on the right, with the majority in the centre. For humans, I insert a bias which displaces the distribution to the right. If you have the same threshold for humans and for other primates—this is what convinced me that this is not just a model but a revelation—the percentages of pure left-handers, mixed handers and right-handers are exactly the same as found in monkeys and in humans, provided humans have this shift to the right.

If we assume that the same biasing factor puts speech on the left side of the brain, that is all we need to account for the main phenomena. My parameters for genetic calculations were derived entirely from the literature on patients with dysphasia after unilateral left- and right-sided lesions, not from the genetic data.

Wolpert: So you have a biasing factor where Chris has a phenotype which is random.

McManus: I see the bias as already there. I am not proposing a mechanism for the fundamental asymmetry but for the disruption of that asymmetry which sometimes results in left-handedness and right-brainedness.

FIG. 1. (*Annett*) The 'Galton' machine demonstration of the right shift theory. When the RS factor is absent, differences between the hands in skill are thought to be due to chance alone. Left and right hand preference are equally likely, with the majority having mixed hand preference. When the RS factor is present, the distribution of differences between the hands in skill is due to chance, as before, but is displaced to the right.

Morgan: It seems quite inappropriate to talk about a gene for right-handedness. We have seen no evidence that any such gene exists. I am not even sure it is a meaningful concept. At best, there is evidence for some factor related to genetic variance that can interfere with the system and reduce the amount of right-handedness. The variants could be doing anything—raising the level of a neurotransmitter, decreasing receptor sensitivity, many things that have nothing to do with handedness but which fortuitously interfere with coding.

You accept that there are major environmental sources of variation which could add to the genetic variation. That means there are some people who are *DD* genotypically in your terminology who are phenotypically left handed. In that case your model shouldn't work and the fact that it does work is a bit suspicious because you haven't taken account of it in predicting the relative proportions of left- and right-handers.

McManus: You have to work out the incidence of pathological left-handedness. Dorothy Bishop (1990) has calculated that the proportion of left-handers who are left handed due to pathological factors is probably about one in 20. If one in 20 left-handers is left handed for non-genetic reasons, it hardly alters the calculations and therefore the models continue to show a good fit with the data.

I am not saying that 100% of the variation in handedness is genetic, but that the large proportion is. There is a major gene there. Other factors might be important, but birth stress is not one of those.

Morgan: When you compared the observed frequencies with those predicted by the model, you were talking about differences of 1.9% as evidence against a model. So why doesn't this additional 1/20 randomness coming from the environment make it impossible to discriminate between these different models?

McManus: If I took all the left-handers in my study and randomly made 1/20 of them right-handed, it would have an equivalent effect across all groups, irrespective of the handedness of the parents; thus it would average out in the calculations. The differences between the groups are not affected by an environmental insult which is affecting individuals at random. Unless the environmental insult is itself genetically transmitted, the insult shouldn't be related to the familial patterns.

Crow: I would like to present an argument that the human directional factor, which I assume is genetic, is pseudoautosomal (Crow 1989). The argument is based on the neuropsychology of sex chromosome aneuploidies (Netley & Rovet 1982, 1987).

Patients with Turner's syndrome lack an X chromosome (XO) and those with Klinefelter's have an extra X chromosome (XXY). Patients who are XXX resemble those with Klinefelter's and not Turner's. XXX are females and Klinefelter's are males, which diminishes any interest you might have in hormones and suggests that genes are more important in these differences.

The measures were of verbal and performance IQ. 100 is the expected value: patients with Turner's syndrome have a performance deficit (88, $n = 35$) and no verbal deficit; conversely, those with Klinefelter's have a verbal deficit (83, $n = 24$) and a normal performance IQ. These are substantial deficits. Other studies show deficits that are not as great; this difference between studies may depend on the age of the patients being tested.

Performance tests are mostly right sided and verbal tests are mostly left sided. Therefore, it is difficult to resist the conclusion that on the X chromosome there is some factor that is affecting the relative development of the two hemispheres. One of the things that Marian Annett and Chris McManus agree on is that the right-shift factor is autosomally transmitted, which doesn't fit with this conclusion.

The pseudoautosomal region is the small region at the tip of the short arms of the X and Y chromosomes within which there is recombinational exchange of genetic material. This was postulated by C. D. Darlington 40 years ago and its existence has been clearly established by the work of a number of geneticists. In male meiosis, these regions of the X chromosome and the Y chromosome are apposed and there is exchange of material, so that within this region genes on the X and the Y are homologous. Genes within this region appear to be autosomally transmitted, even though they are on the sex chromosomes.

The pseudoautosomal region is quite small; it is 2.6 Mb, less than 0.1% of the genome. So if there is anything in this hypothesis, the problem of identifying the handedness gene is simplified. There are a limited number of genes in this region; as judged by HTF islands it is probably seven or eight, of which three have been identified. Two of them are quite interesting: the MIC2 gene, which is a primate-specific membrane antigen, and the pineal enzyme, hydroxy indole-*o*-methyl transferase.

Burn: Recent evidence suggests that the pseudoautosomal region is not confined to the upper region of the short arm; other areas on the X chromosome behave in a pseudoautosomal fashion. So the mapping exercise might not be quite as simple as you assumed.

Crow: That's not quite true. There are homologies, but those regions do not exchange between X and Y. Nevertheless, the possibility that the handedness gene is not pseudoautosomal but in a region in which the X and Y chromosomes carry sequence homology deserves consideration.

Wolpert: Tim, you imply there is a significant difference between left and right sides of the brain. Earlier, I had a feeling from several speakers that the idea of left brain, right brain is rapidly disappearing.

Galaburda: We used the language very differently. That patients with Turner's syndrome have difficulty with visual things which ultimately show up in the performance IQ tests doesn't say anything about right hemisphere asymmetry as far as I am concerned. There is no reason to suspect that the only change between the normal karyotype and that in Turner's syndrome is a change in asymmetry. A lot of other things are likely to be happening to the brain in Turner's syndrome.

All you may have to do is make smaller brains and bigger brains. For visual systems and visual analyses you may need big networks; for phonological and syntactic aspects of speech and language you may need small networks because learning of rules is the crucial step. All you may need to do is make the brain bigger in individuals who are XXY, which appears to be the case, and make the brains smaller in Turner's syndrome, which they are. You don't have to change the asymmetry. It is not that all the left and right differences can be dismissed, but they are part of a more interesting biological process that also includes brain size, connectivity and strategies for completing a task.

Crow: Surely it's not true that people with small brains have defective spatial abilities or that people with large brains have defective verbal abilities?

Galaburda: I don't know, but it is an empirical question. I would imagine that people with large visual networks will tend to be more symmetrical and better at visual spatial abilities, and people with severe asymmetry of language areas, which are then smaller, may be good at phonological and syntactic tasks.

McManus: I know the asymmetries in brain sometimes look small and they can sometimes be explained away. We must never forget what Broca found in 1865. If the left middle cerebral artery is occluded, 95% of right handers become aphasic, they stop speaking. If the right middle cerebral artery is occluded, those people probably carry on talking. That to me is the fundamental thing that has to be explained.

Galaburda: I believe my model begins to explain exactly that. It is the asymmetry of the assembly that is important. If you damage most of the assembly, wherever the assembly is, you cause the deficit. In most subjects, this is on the left.

McManus: Nevertheless, the basic phenomenon is indisputable. We cannot argue that there is a random combination of possible brain types and we just happen to be seeing some of them. There is clearly directional asymmetry that has to be explained.

Corballis: We should probably dismiss the pop psychology notion that there are two fundamentally different and opposite kinds of processing on the two sides of the brain. The critical thing is the left dominance in most people (not everybody) for language and perhaps for some kind of praxis that is reflected in handedness. The so-called right hemispheric specialization for spatial representation is not nearly as absolute or pronounced.

Peters: Chris's observation may be rephrased to say that the most salient and clearly observable aspects of language and speech are affected after left hemisphere damage. Because the characteristic changes in linguistic behaviour after right hemisphere damage are much more subtle and less easily detected, it is often asserted that right hemisphere damage has no effect on language.

Wolpert: The story that we left-handers are more liable to autoimmune diseases, more intelligent, etc, is it all nonsense?

Galaburda: There is no evidence to support the claim in the original observations that left-handers are more vulnerable to autoimmune diseases in general, although they may be more susceptible to some types of allergies. Among left-handers there is a greater proportion who are very bad at mathematics and a greater proportion who are very good than in the general population.

Annett: There is something in the theory that left-handers have certain advantages. Some people have to have two copies of the right shift gene, so the bias for them is a bit stronger. My model suggests that these people ($RS+ +$) are at risk. There are about 32% in the population who carry a double dose of the right shift gene. If you take any talented population, remove the 32% who are $RS+ +$, then the incidence of 8% left-handers in the total population becomes 8/68. If the best people are selected from the $RS- -$ and $RS+ -$ genotypes and the $RS+ +$ are missing, you come up with 11–12% left-handed. This explains the increased proportion of left-handed tennis players, other top sports people and surgeons. It is not that being left handed is particularly good for you, but that it is bad to be too right handed.

Collins: I have tried to gain a better understanding of the inheritance of asymmetry through experimental studies on genetically defined mice. I have focused on studies of handedness in the hope that they may lead to a more general perspective of genetic influences affecting other asymmetries of function and structure.

If mice are placed in individual testing cubicles in which rolled wheat is available in a feeding tube attached to the front wall, they will reach for the food. These reaches can be observed easily and scored with patience. I usually assess handedness in mice by measuring 50 reaches for each mouse. The 'right-paw entry score' (RPE) is a useful measure and ranges from 0 to 50. High scores indicate dextrality, whereas those near zero indicate sinistrality. Mice scoring in the region of 25 RPEs may be judged ambilateral. Mice may alternatively be classified as dextral or sinistral if their RPE scores are greater or less than 25 RPEs, respectively. The handedness of mice is extremely reliable when assessed within-session, across several days, or over several months (Collins 1968).

What happens when you observe the hand preferences of many members of a highly inbred strain? I tested 709 C57BL/6J mice that had, at that time, been inbred for more than 100 generations. The distribution of their RPE scores was markedly U shaped (Collins 1975). Approximately half the mice were strongly right handed, and half were strongly left handed; very few were ambilateral. This finding presents us with a puzzle. The maximum phenotypic variance is observed in a population that possesses minimum genetic variance.

It might be suggested that C57BL/6J mice still harboured a residue of genetic heterogeneity despite more than 100 continuous generations of brother–sister inbreeding. If so, this residue should be responsive to bidirectional selection.

I tested this by selectively breeding for right-handedness and left-handedness in C57BL/6J mice (Collins 1969). Through three generations of selection there was no change in the proportions of dextral or sinistral mice or in the RPE averages. Half the offspring from dextral–dextral matings remained left handed. Clearly, the maximum variance in directionality of handedness observed in C57BL/6J mice does not appear to be maintained by a residue of heritable genetic variation that escaped prolonged inbreeding. Nor does it appear that the handedness of mouse pups is learned from their mother, because if there were a cultural inheritance of directionality, the selection programme would have detected it. However, a cultural diffusion of behavioural asymmetry is possible. One example of observational learning for a novel behavioural asymmetry in mice has been presented (Collins 1988).

I observed a sex difference in the degree of asymmetry in the large-sample study of C57BL/6J mice (Collins 1975). Female mice, on average, showed a stronger hand preference than did male mice.

I believe the interest of researchers has been so focused on questions of directionality, we have tended to forget that asymmetry is a mathematical vector with a degree of magnitude as well as a direction. Perhaps the inheritance of the degree of asymmetry will be better behaved. I addressed this by initiating a long-term bidirectional selective breeding study for the degree of handedness in mice.

First, a foundation population possessing large potential genetic variability was established. It was derived from an 8-way crossing plan using the following inbred strains and partially inbred stocks of wild mice: BALB/cJ, C57BL/6J, DBA/2J, LP/J, RF/J, SM/J, *M. molossinus* and *M. castaneus*.

The distribution of RPE scores for the foundation population was markedly U shaped (Collins 1985). To provide a measure of the degree of asymmetry I used the 'preferred-paw entry' (PPE) score. This is a non-monotonic transformation of the RPE score in which the RPE distribution is cut at 25 and the left portion (0–24) is folded and superimposed as a mirror image on the right portion (26–50). The PPE score ranges from 25 to 50 and measures the degree of handedness without regard to its direction.

A HI-selected line was formed by mating mice with PPE scores of 48–50; a LO-line was produced from mice with PPE scores of 25–40. By the third generation of selection, there was a statistically significant separation in mean PPE performance in the selected lines. This difference increased as selection continued. Maximum separation was observed at the tenth generation (G10). At this time, approximately 45% of HI-line mice exhibited strong lateralization (PPEs of 48–50) compared to 8% of LO-line mice. The degree of lateralization of both lines was distinct from that of the unselected HET reference population (Collins 1985).

The selection pressure was relaxed for 17 generations, then selective breeding was reinstated at G28 for three consecutive generations. The HI- and LO-lines retained their distinctive patterns of lateralization throughout this long period of random mating. There was no evidence of counter-selection for average effect or of regression toward the mean (R.L. Collins, unpublished).

Thus far I've considered only mice tested for handedness in unbiased or 'U-worlds', testing situations in which the food tube is located equidistant from the right and left walls. What happens when mice are tested in biased worlds in which the food tube is placed flush against either the right wall as faced by the mouse (R-world) or the left wall (L-world)? The distribution of RPE scores now changes from being bimodal and U shaped, to being unimodal and J shaped. Approximately 90% of a large sample of naive C57BL/6J mice tested in an R-world could be classified as dextral, only 10% as sinistral. The mirror-image pattern was obtained for mice tested in an L-world (Collins 1975). Such a bias in the environment may exert a profound effect on manifest laterality. However, this effect would be impressed on an already laterally dichotomized population. The external bias does not create laterality, it shifts an already lateralized equiprobable population into a non-equiprobable distribution of directions.

Let us now consider genetic models of asymmetry. In developing any genetic model of laterality it is necessary to map phenotype functionally to presumptive genotype by a series of probability statements for each of the genotypes in the model. I propose that as we do this, we map the probability of right as equal to the probability of left for each genotype considered. This is another way of saying that genes maintain an asymmetry lottery in which right and left are stochastic outcomes, and that genes do not specify the outcomes of a given gamble. Any observed uniformity in direction or a non-equiprobable distribution of directions, arises not because of differing gene frequencies for 'right' and 'left' alleles, but rather through an interaction of the stochastic outcomes with an external gradient of asymmetry. Consider the case of situs inversus in mice. In iv/iv mice the side of the midline on which the stomach lies appears to be determined randomly. In $iv/\pm$ and $\pm/\pm$ mice there is apparent uniformity of sidedness. Does this mean that the wild-type allele codes for 'stomach on the left'? I submit that it codes for directionality in the same way as does the iv allele. The wild-type allele could lead to uniformity of directionality in one of two ways. It could code for a receptor that detects the external asymmetry gradient. Or it could code for an effector that translates information about an environmental bias into movement towards polarity.

The advantage of considering the distribution of directional forms arising through interaction with external world biases lies in our ability to manipulate the asymmetry gradients. If a gradient points one way, perhaps it can be made to point in the opposite direction. Or it could be removed. There are two experimental examples of the use of manipulated asymmetry gradients. One is the biased world studies of handedness in mice already discussed (Collins 1975, 1977b); the other is the series of elegant experiments by Leslie Rogers and her collaborators on lateralization of function in the chick forebrain. In these, a null gradient was applied which abolished an existing directional distribution of neural and behavioural lateralization (Rogers & Anson 1975, Rogers 1982, Rogers & Bolden 1991).

Peters: In the initial task where the food was in the middle of the chamber, it might be inappropriate to talk about 'pawedness'. When the mouse reaches into the tube, it has to twist its body to be able to extend its paw and leg into the tube. Could the selection have been for the direction of twisting of the torso? If so, does this have anything to do with 'pawedness'?

Collins: I believe axial torsion does not play a major role in mouse handedness or in the response to selection. For example, if we put the food tube on the side opposite to the mouse's native hand preference, most mice will try to use their preferred paw even though it is very difficult for them posturally. Secondly, when mice were tested in a circular apparatus with the food hole through the floor, the correlation of RPE scores between the two tests was 0.96 (Collins 1970). The motor response topographies and the axial postures of mice in the two tests are each quite different.

Peters: When you said 90%, were those both HI-lines?

Collins: In the original biased world studies, C57BL/6J mice were directly placed into biased worlds and tested twice and then once in biased antiworlds (test sequences R-R-L or L-L-R). In the first two tests approximately 90% of mice exhibited a preference consistent with the world bias, whereas 10% resisted it. But we must remember that each group was an equal mix of dextrals and sinistrals, so that the proportion of mice resisting the bias was much higher. If we first test mice in an unbiased world and then twice in worlds biased opposite to their hand preference (sequences U-R-R or U-L-L), the proportions of mice resisting the bias can be quite high. For example, when we used HI-line mice in this challenge paradigm, the median PPE score remained flat across the initial and two biased world challenges.

Peters: If humans were presented with this sort of task, they would simply reach for the food with the arm that was most appropriate, i.e. for an opening that was to the left of the body midline, even a right-handed human would not twist the body to use the right arm. In this sense, these results reflect an unusually powerful lateral bias in the mice.

References

Bishop DVM 1990 Handedness and developmental disorder. Blackwell Scientific Publications, Oxford p 98

Collins RL 1951 When left-handed mice live in right-handed worlds. Science (Wash DC) 187:181–184

Collins RL 1968 On the inheritance of handedness. I. Laterality in inbred mice. J Hered 59:9–12

Collins RL 1969 On the inheritance of handedness. II. Selection for sinistrality in mice. J Hered 60:117–119

Collins RL 1975 The sound of one paw clapping: an inquiry into the origin of left-handedness. In: Lindzey G, Thiessen D (eds) Contributions to behavior-genetic analysis. Appleton-Century-Crofts, New York p 115–136

Collins RL 1977a Origins of the sense of asymmetry: Mendelian and non-Mendelian models of inheritance. Ann NY Acad Sci 299:283–305

Collins 1977b Toward an admissible genetic model for the inheritance of degree and direction of asymmetry. In: Harnad S, Doty RW, Goldstein J, Janes Y, Krauthamer G (eds) Lateralization in the nervous system. Academic Press, New York p 137–150

Collins RL 1985 On the inheritance of direction and degree of asymmetry. In Glick SD (ed) Cerebral lateralization in nonhuman species. Academic Press, New York p 41–71

Collins RL 1988 Observational learning of a left-right behavioral asymmetry in mice (*Mus musculus*). J Comp Psychol 102:222–224

Crow TJ 1989 Pseudoautosomal locus for the cerebral dominance gene. Lancet 2:339-340

Galton F 1876 The history of twins as a criterion of the relative powers of Nature and Nurture. J Anthropol Inst 5:391–406

Lotze R 1937 Zwillinge einfuhrung in die zwillingsforschung. Ferd-Ran quoted by Gedda L (1961) Twins in history and science. Charles Thomas, Springfield Illinois

Netley CT, Rovet J 1982 Atypical hemispheric lateralisation in Turner syndrome subjects. Cortex 18:377-384

Netley CT, Rovet J 1987 Relations between a dermatoglyphic measure, hemispheric specialisation and intellectual abilities in 47XXY males. Brain and Cognition 6:153–160

Price B 1950 Primary biases in twin studies. A review of prenatal and natal difference-producing factors in monozygotic pairs. Am J Hum Genet 2:293–352

Rogers LJ 1982 Light experience and asymmetry of brain function in chickens. Nature (Lond) 297:223–225

Rogers LH, Anson JM 1979 Lateralization of function in the chicken forebrain. Pharmacol Biochem Behav 10:679–686

Rogers LJ, Bolden SW 1991 Light-dependent development and asymmetry of visual projections. Neurosci Lett 121:63–68

Disturbance of morphological laterality in humans

John Burn

Division of Human Genetics, University of Newcastle Upon Tyne, 19/20 Claremont Place, Newcastle upon Tyne, NE2 4AA, UK

Abstract. Complete situs inversus is well described in humans and may be associated with defective cilia in Kartagener's syndrome, an autosomal recessive trait. Only half of homozygotes display situs inversus, presumably due to 'chance'. Isomerism sequence or Ivemark syndrome involves major disturbances of organ formation, the patterns of which suggest such individuals have either two right sides or two left sides to the body. Defects in an autosomal and more rarely in an X-linked gene are involved, though again chance factors probably influence expression. Isomerism may occur in one of monozygotic twins, particularly in the right half of conjoined twins. It is likely that the twinning process disturbs laterality in one of the pair. This provides one mechanism to explain the marked excess of heart defects in monozygotic twins. Most of these defects are not, however, associated with overt disturbances of laterality. Since the heart is the organ most sensitive to disturbed situs, one explanation is that the excess reflects a lesser degree of disturbed laterality. This would require a model where there is a gradation from normal situs through isomerism to situs inversus.

1991 Biological asymmetry and handedness. Wiley, Chichester (Ciba Foundation Symposium 162) p 282–299

In most humans the heart is on the left side of the chest, the liver is on the right side of the abdomen and all other organs are in a characteristic place. In just over one in 10 000 people the heart is on the right, the liver is on the left and every other organ is on the opposite side to that which is usual. This condition of situs inversus causes no impairment of function (Torgersen 1950). Any biological mechanism which addresses the determination of laterality must account for the usual reliability of this process and the facility for it to be, on occasion, completely reversed. Torgersen found very few familial cases in his population study. However, the group of individuals with situs inversus includes a group where the reversal of body symmetry is associated with bronciectasis, sinusitis, anosmia and sterility in the male. This is Kartagener's syndrome (Kartagener & Stucki 1962), originally described by Siewert in 1904. Afzelius (1976) demonstrated a defect in the dynein arms of cilia that led to impaired ciliary function in patients with Kartagener's syndrome. A variety of defects in the dynein protein were described. Subsequently, a series of other abnormalities in the function of cilia have been reported in patients with immotile cilia syndrome (Sturgess et al 1986). In all cases, the ciliary defect behaves as an autosomal recessive trait but penetrance of the situs inversus is incomplete. Moreno & Murphy (1981) examined the possibility of a chance factor in the determination of situs inversus. They found that the data were compatible with the hypothesis that half the homozygotes had reversed situs, suggesting that there was a chance event and that the homozygotes were, in effect, agnostic for laterality.

It has been proposed that the situs inversus may arise as a result of ciliary dysfunction. This is made less likely by the studies of Waite et al (1978) on Maoris and Samoans; their high incidence of bronciectasis caused by ciliary dyskinesia has led to it being known as Polynesian bronciectasis. Situs inversus did not appear to be associated with ciliary dysfunction in that population; this is supported by the observation that bronchiectasis was not apparent in the small number of individuals with dextrocardia detected by serial radiography. Disturbance of laterality determination seems to be a pleiotropic effect of the single gene defect which is not apparent in all forms of the disorder, such that it is possible to have ciliary dysfunction without situs inversus.

Analogies have often been drawn between Kartagener's syndrome and the chance determination of laterality in the *iv/iv* mouse (Layton 1976, 1981). The disorders are not entirely equivalent since in humans with Kartagener's syndrome heart malformation is almost never seen, whereas it is present in about one fifth of homozygotes in the mouse model. There have been reports of Kartagener's syndrome and polysplenia in the same sibship (Teichberg et al 1982, Schidlow et al 1982). Polysplenia is a feature of ambiguous situs, which suggests that homozygotes for Kartagener's syndrome gene on occasion fail to reverse situs completely.

Isomerism sequence

In the discussion of laterality and our understanding of its determination, the condition of isomerism sequence presents an interesting challenge. This disorder has been known by many names, including Ivemark syndrome, situs ambiguus, visceral heterotaxy, laterality sequence and the asplenia/polysplenia syndrome. Sequence refers to a pattern of anomalies which results from an underlying developmental disturbance. The term isomerism enjoys favour in cardiological circles. It refers to the concept of the two sides of the body representing mirror images of one another, such that in the classic case right isomerism is associated with trilobed lungs bilaterally, together with short eparterial bronchi and a heart with two right atrial appendages. The spleen, which arises from the left side of the mesogastrium, fails to develop; the pulmonary veins, which normally drain into the left atrium, are frequently displaced and drain aberrantly, as there is no morphological left atrium with which to connect. The liver remains in a central position; the mesentery is aligned vertically rather than diagonally, with the result that malrotation of the bowel is common and both testes hang at the same height instead of the usual lower positioning of the left testis. Malformation of the heart is usually severe and may involve a variety of defects in septation, chamber formation, and in connections between the chambers and the outflow vessels (van Mierop et al 1972).

In left isomerism, there are two bilobed lungs, two long hyparterial bronchi with distal branching and two atria with left atrial appendages. The absence of a morphological right atrium leads to failure of the development of the suprarenal inferior vena cava and subsequent drainage via the azygos system; typically, multiple spleens form. Again the liver forms centrally. Heart malformations are common, though in general less severe than those seen in right isomerism. In both types of isomerism, heart malformation may be attributed, by analogy to the situation in the *iv/iv* mouse, to disturbance of heart loop formation, since a primary step in heart formation is the development of right-sided curvature. This results from differential growth of the left and right sides of the early heart tube; when the tube is not clear which side is left, the differential growth is liable to be severely compromised. Since formation of a curve to the right involves an arrest of growth on the left side of the tube, it is not surprising that disturbance of heart formation is more severe in right isomerism.

Table 1 lists cases of isomerism associated with recurrence in sibs or with consanguineous parents or both. Not all authors recognize the pathological diagnosis of isomerism sequence and therefore a variety of terminologies are used but it is evident from the case reports that this was the diagnosis in the probands and in the majority of affected relatives. The high incidence of sib recurrence and parental consanguinity points to an autosomal recessive disorder. There are two reports which go against this interpretation: Niikawa et al (1983) described a child with right isomerism whose sib had partial situs inversus and complex heart defects and whose father had complete situs inversus, which is surprising but not impossible in an autosomal recessive disorder. Mathias et al (1987) described a family in which situs inversus was unequivocally X-linked with nine affected males in two generations. I have seen a similar family in which a child with right isomerism and sacral agenesis, whose mother was normal, had a maternal uncle with a history of disturbed laterality and a heart malformation that resulted in early death. In our own case and in the report by Mathias et al, craniospinal defects were a variable feature, suggesting that the X-linked form of the disease is associated with impairment of neural tube formation. Asplenia and polysplenia have been described in the same sibship (Polhemus & Schafer 1952, Zlotgora & Elian 1981, Mathias et al 1987, Burn et al 1986). This suggests that right isomerism sequence and left isomerism sequence are manifestations of the same genetic defect. Another important feature of the families shown in Table 1 was that on several occasions affected sibs had isolated heart malformations, such as tetralogy of Fallot, without overt disturbance of situs.

Since the heart is the organ most sensitive to disturbance of situs, it may be argued that in such sibships there has been a disturbance of laterality sufficient to impair heart formation but not to cause overt signs of isomerism or situs inversus. The case for autosomal recessive inheritance is supported by an excess of complex dextrocardia with heart malformation in Muslim and other inbred communities in England compared to a population where inbreeding was uncommon (Gatrad et al 1984). Rose et al (1975), however, examined the sibs of 60 consecutive patients with asplenia and polysplenia and found only three affected sibs; this falls far short of the prediction for an autosomal recessive even with chance factors influencing expression.

Our own study (Burn et al 1986, Burn 1991) involved a consecutive series of 47 cases of right isomerism and 51 of left isomerism. There were 32 cases of male right isomerism compared to only 15 female, but this discrepancy was counterbalanced by 21 cases of male left isomerism and 30 female. Overall, this does not support a major contribution by an X-linked gene defect to the population prevalence. A total of 109 sibs and 92 parents were traced. Of these 201 first degree relatives, information was available on pedigree from 66, eight were subjected to clinical examination and in the remaining 127 investigation was possible using chest X-ray, electrocardiography and echocardiography. Our purpose was to discover the recurrence risk in sibs and to establish whether occult disturbance of laterality was apparent without heart malformation. No parent showed any evidence of disturbed situs: only one had a minor heart defect, which is in keeping with expectations from the general population. Of 109 sibs there was one definite case of isomerism and two early deaths from what was probably heart malformation associated with disturbed laterality. One girl whose sib had died of right isomerism had two left bronchi (Fig. 1) on the basis of the high kilovoltage-filtered chest X-Ray which is designed to display the bronchial shadows. Examination of bronchial length in the study population as a whole (Fig. 2) showed a clear segregation between left bronchi and right bronchi. The best fit line of segregation agreed closely with that reported by Deanfield et al (1980). The girl with two left bronchi was reinvestigated using real-time screening to confirm the finding. This suggests that it is possible rarely to have isomerism without major heart malformation. More important, complete situs inversus was not found in any relatives. It may be deduced, therefore, that again the *iv/iv* mouse is unlikely to be a precise model for this human disorder. This is unfortunate, because it means the human equivalent of the chromosome 12 location of the *iv/iv* mouse, the long arm of chromosome 14, cannot be assumed to be the location of the gene for isomerism.

Models of mechanisms of laterality determination must be able to account for isomerism sequence of left or right type within the same family. The model

TABLE 1   Reports of isomerism sequence in sib pairs or children with consanguineous parents

Polhemus & Schafer 1952	Two sisters	1. Asplenia; right isomerism 2. Polysplenia; left isomerism
Schonfeld & Frischman 1958	Single case Consanguineous parents	Asplenia and heart malformation, right isomerism sequence
Badr-El-Din 1962	Single case Consanguineous parents	Right isomerism sequence
Ruttenberg et al 1964	Three affected sibs	Asplenia and heart defects; right isomerism sequence
Neimann et al 1966	Single case Consanguineous parents	Polysplenia and heart defects; left isomerism sequence
Fuhrmann 1968	Case 68 affected sibs	1. Asplenia; right isomerism sequence 2. Ventricular septal defect
Ehlers & Engle 1966	Case eight affected sibs	1. Bilateral superior vena cava, partial anomalous venous drainage, partial AVSD, pulmonory atresia 2. Asplenia; right isomerism sequence
Silver et al 1972	Two affected sibs	1. Asplenia; right isomerism sequence 2. Tetralogy of Fallot
Simpson & Zellweger 1973	Brother and sister	Asplenia and heart defects, right isomerism sequence
Chen & Montelione 1977	Two brothers Male and female first cousins	Asplenia; right isomerism in both Female: polysplenia; left isomerism Male: right-sided hypoplastic spleen, dextrocardia, AVSD, TGA, PA
Jojarte & Fekete 1978	Two sibs	Asplenia; right isomerism
Hallet et al 1979	Two sibs	Polysplenia; left isomerism sequence
Tsuda et al 1979	Two sibs	Asplenia; right isomerism sequence
Kawagoe et al 1980	Two sibs	Polysplenia; left isomerism sequence
Katcher 1980	Brother and sister	Girl: TGA, subpulmonary stenosis Boy: asplenia, common atrium, single ventricle; right isomerism sequence
Zlotgora & Elian 1981	Brother and sister	1. Polysplenia; left isomerism sequence 2. Asplenia; right isomerism sequence
Hurwitz & Caskey 1982	Consanguineous parents in one of six cases from consecutive records search	Asplenia and heart defects; right isomerism sequence
Arnold et al 1983	Three sisters and one brother, paternal aunt	1. Boy: left isomerism sequence 2. Left isomerism sequence 3. Anomalous pulmonary venous return, ASD, hypoplastic left heart, interrupted arch, dextrocardia 4. VSD Aunt: dextrocardia, VSD

(*continued*)

TABLE 1 (*continued*)

de la Monte and Hutchins 1985	Two sisters	Left isomerism sequence
Zlotgora et al 1987	Four brothers	1. Situs inversus, VSD 2. Situs inversus, TGA, VDS, PS 3. Situs inversus 4. Normal situs, AVSD, TGA, PS
DiStefano et al 1987	Brother and sister Consanguineous parents	Boy: dextrocardia, TGA, PDA Girl: partial situs inversus, right pulmonary isomerism
Czeizel 1987	Three sisters and one brother Consanguineous parents	1. Situs inversus, TGA 2. Situs inversus, truncus arteriosus, ASD, VSD 3. Boy: situs inversus, VSD 4. Situs inversus, ASD, VSD, PDA

Families not supportive of autosomal recessive inheritance

Niikawa et al 1983	Father and two sons	Father: complete situs inversus 1. Right isomerism sequence 2. Partial situs inversus, dextrocardia, single atrium and ventricle, right aortic arch, TAPVD, systemic veins on left, two spleens on right
Mathias et al 1987	Nine males in two generations linked through nine normal 'carrier' females	1. Asplenia; right isomerism sequence, AVSD, TGA, PA, anterior aortic arch, arrhinencephaly posterior anus 2. Partial situs inversus, early death 3. Partial situs inversus, early death 4. Partial situs inversus, early death 5. Partial situs inversus, early death, meningomyelocele 6. Situs inversus, TGA, PS, mitral atresia, sacral agenesis, anal stenosis 7. Asplenia; right isomerism sequence 8. Polysplenia; left isomerism sequence, cerebellar hypoplasia 9. Polysplenia; left isomerism sequence

ASD, atrial septal defect; AVSD, atrioventricular septal defect; PA, pulmonary atresia; PDA, persistent ductus arteriosus; PS, pulmonary stenosis; TAPVD, total anomalous pulmonary venous drainage; TGA, transposition of the great arteries; VSD, ventricular septal defect.

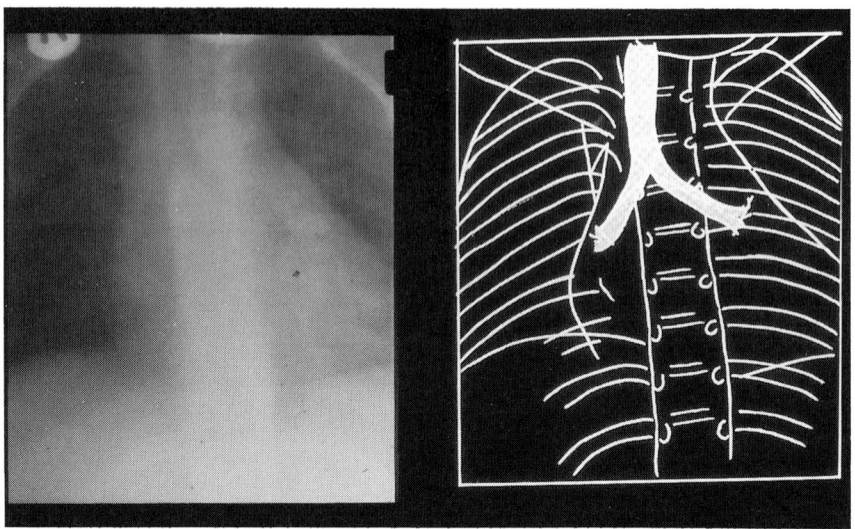

FIG. 1. A high kilovoltage-filtered chest radiograph, designed to show the airways, of the healthy sister of a child with right isomerism. The tracing shows the bilateral long hyparterial 'left' bronchi of left bronchial isomerism.

of Corballis & Morgan (1978) proposes a laterality gradient across the early zygote from the left side. Brown and Wolpert (1990) have further developed this concept. Complete failure of the formation of this gradient or its impairment could result in isomerism or the other varying degrees of disturbed situs. Zlotgora et al (1987) and Czeizel (1987) both described large sibships with situs inversus or situs solitus with or without heart malformations (see Table 1). These seem to be the best analogy to the *iv/iv* mouse model; unfortunately, testing of this is difficult because early death is extremely common in such cases.

In a prospective study of patients who had given birth to children with heart defects and who were undergoing fetal echocardiography, Allan et al (1986) found a familial recurrence risk for isomerism approaching 10%. It remains possible that a chance factor reduces penetrance of this disorder and that it is an autosomal recessive trait. Our own study of consanguinity supports this, in that among 46 pedigrees obtained from white European stock we found one case of isomerism with first cousin parents and one with second cousin parents. Against a background expected level of approximately 1 in 300 white European couples being consanguineous, discovery of two cases of cousin marriage in a series of 46 represents a weakly significant excess made more impressive when compared to the frequent reports of consanguinity in the literature.

In our series of 98 families for the offspring who showed the isomerism there was a significant excess of births in the second quarter of the year and a significant deficit in the first quarter when compared to births of sibs, parents

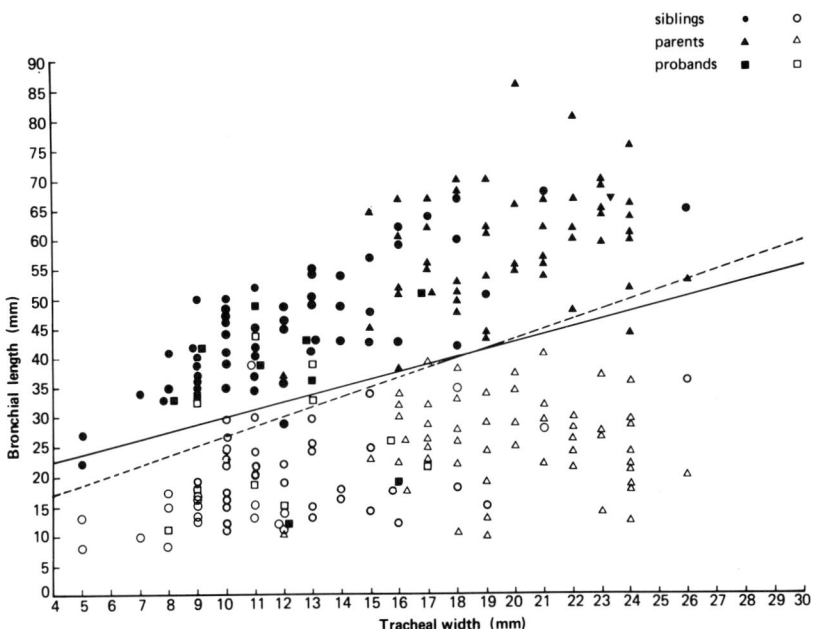

FIG. 2.   Bronchial measurements plotted against tracheal width from available probands and first degree relatives in a series of 98 cases of isomerism sequence. The line which best segregates the two populations of measurements agrees closely with that published by Deanfield et al (1980). This illustrates the clear lateralized definition of bronchial morphology.

and the general population. This raises the possibility of an environmental factor underlying this disorder in some cases (Burn 1991).

Disturbance of laterality and twinning

A principle plank of the left–right gradient hypothesis put forward by Corballis & Morgan (1978) was the occurrence of reversed and impaired symmetry in conjoined twins. Newman (1923) claimed that the original Siamese twins, Chang and Eng, showed such contrast with the right partner having dextrocardia. Noonan (1978) drew attention to the similarity between the types of heart defects in conjoined twins and in individuals with isomerism sequence. There have been several reports of disturbed laterality in conjoined twins, the majority of which involved right isomerism in the right-sided member of the pair (Seo et al 1985, Rossi et al 1987). Fig. 3 shows the case from our own report where the right side of the conjoined heart had bilateral right atria indicative of right isomerism. Following the model of Brown & Wolpert (1990), it may be suggested that the

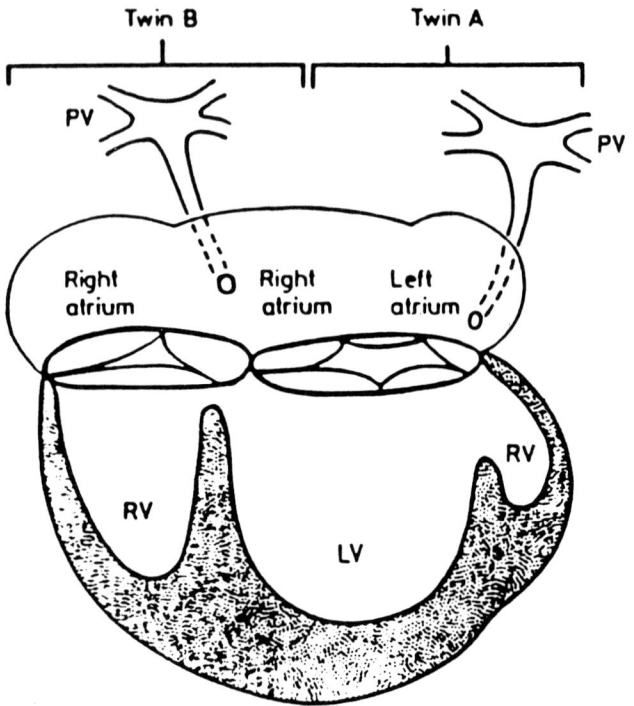

FIG. 3. Diagrammatic representation of the heart in a pair of conjoined twins attached at the lateral thorax. The right half of the pair had a heart with two right atrial appendages, indicating right isomerism sequence in that twin. From Rossi et al (1987) with permission.

right side of the zygote had already become 'fixed' in its right-sidedness. The late twinning event, assumed to have occurred in the second week after fertilization (Brocklage 1981), postdated this laterality determination with the result that the right-sided twin could not form a left side. The report of left isomerism in the left-sided twin by Seo et al (1985) is more difficult to explain on this basis.

Heart malformation in twins

My personal interest in disturbance of laterality arose from a study of heart malformations in twins. Individuals from monozygotic pregnancies were two to three times more likely than dizygotic twins or singletons to suffer a heart malformation (Burn & Corney 1984, Burn 1988, 1991). A variety of explanations could account for this excess: one is the so-called two-hit hypothesis whereby one twin, weakened by the twinning event, was then damaged by an external agent; the second is the effects of twin-to-twin transfusion. I found anecdotal cases to support these hypotheses in my study which involved some 400 twin pairs where one or both had heart malformation from two hospital series and six population studies. Another attractive mechanism to account for the excess was disturbance of laterality.

The idea of laterality disturbance emerged from studying the monoamniotic twins shown in Fig. 4a. They were presumed to result from late twinning on the basis that their placenta had conjoined cords. The mother noted that the twins showed mirror-image patterns of thumb sucking and tooth eruption. The hair whorl was on opposite sides of the head in the two twins. The

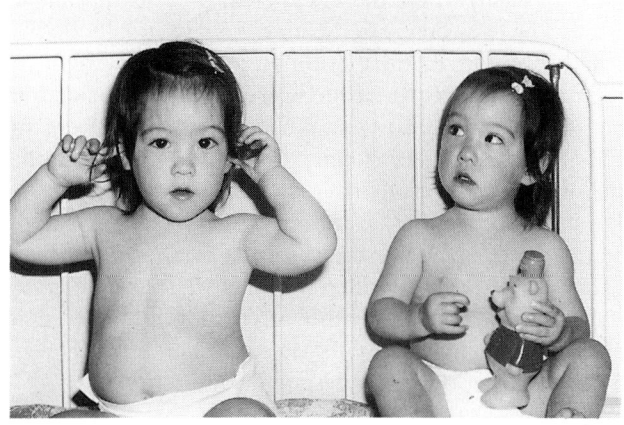

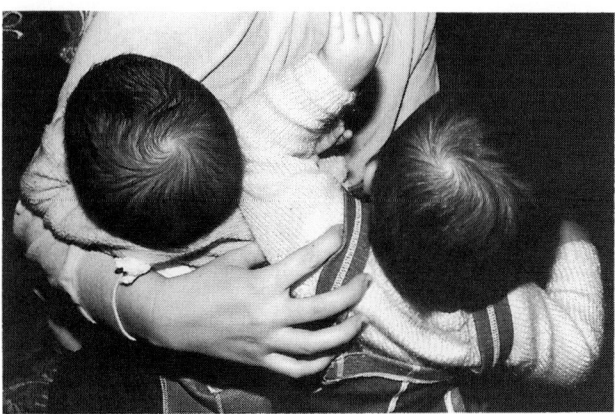

FIG. 4. (a) Monoamniotic twin sisters who showed physical and behavioural 'mirror imaging' making them 'book-end' or enantiomorphic twins. The right half of the pair based on tooth eruption and hair whorl position had tetralogy of Fallot while the co-twin was healthy. From Burn & Corney (1984) with permission. (b) Contrasting hair whorl position in twin boys where the 'right half' had a complex heart defect.

position of the hair whorl (Fig. 4b) is determined in the second trimester by the rapid growth of the occiput; displacement to left or right reflects preferential growth on the left or right side of the head (Smith & Gong 1974).

Among monozygotic twins examined subsequently in the project, there was an excess of twin pairs with contrasting handedness compared to healthy control twins investigated by Griffiths & Phillips (1976) (12/28 versus 9/77, $P < 0.001$). This may reflect a secondary phenomenon where handedness is influenced by the pathological disturbance of circulation. There was no clear association between left handedness and the heart defect in affected pairs. Torgersen (1950) also found that handedness did not appear to be related to situs inversus.

Twelve pairs of twins where there was clear contrast in hair whorl position were identified. The four dizygotic twin pairs were equally divided: two of the 'left twins' had the heart malformation and two of the 'right twins'. Among the eight monozygotic twin pairs, the twin with the right-sided hair whorl had the heart malformation in all cases. While the numbers were small, taken in the context of embryological development of the heart and with the observations in conjoined twins, these results suggest that one mechanism underlying the excess of heart malformations in monozygotic twins may be disturbance of laterality in one of the pair. The twins shown in Fig. 5 were, perhaps, an extreme example of this, where the right-sided twin had right isomerism and his brother was entirely normal. Tragically, the affected boy suffered a dense right hemiplegia owing to a cerebral vascular accident in late childhood. Interestingly, he did not lose the power of speech despite the infarction of the left hemisphere.

FIG. 5. Monozygotic twins discordant for isomerism. The 'right half' had right isomerism sequence while the co-twin was healthy.

This might suggest that the boy had bilateral speech representation in the cortex, which would be compatible with the general symmetry of development.

While it cannot be proven, it is reasonable to suggest that the twinning event disturbed laterality to a sufficient degree that heart loop formation was impaired in one of the pair and this led to a variety of abnormalities of heart septation and development.

Summary

The usual pattern of laterality or situs solitus is occasionally perfectly reversed in situs inversus. This may be associated with autosomal recessive defects of cilia, although the disturbance of laterality is probably not a consequence of cilial dysfunction but another reflection of that gene defect, given that the two disorders are not invariably seen together. The fact that only half of homozygotes have situs inversus reflects the influence of chance in the human, analogous to that seen in the *iv/iv* mouse. The mouse model is not representative of Kartagener's syndrome because the latter patients lack heart defects. Similarly, it is not representative of the human genetic defect best known as isomerism sequence, where situs inversus does not occur in half the homozygotes. Isomerism sequence may be of right or left type and both may occur in one sibship. Lesser degrees of heart malformation may occur in one sibship also. There is strong evidence for isomerism being a frequent autosomal recessive defect with a rarer X-linked form and possible environmental factors in some cases. The major excess of heart malformations in monozygotic twins may be attributed to a variety of causes, including disturbance of laterality in view of the demonstrable discordant impairment of laterality in conjoined twins. In this group, typically the right twin will have isomerism if they are connected side-to-side, though left isomerism in the left twin has been reported once. Complete reversal of symmetry is not common in monozygotic twins, but if it is accepted that the heart is the organ most susceptible to disturbed situs, many of the complex heart defects seen in one of monzygotic twin pairs could result from lesser degrees of disturbance of situs. These observations in humans lead to the conclusion that the gradient of laterality must be amenable to complete reversal and, presumably, to complete absence or a horizontal mode. There is some attraction in the concept of a graded character which varies from a complete left–right, through horizontal wherein isomerism occurs, to the complete right–left of situs inversus.

Acknowledgements

The majority of the work reported here was performed with support from a Medical Research Council project grant. A great many colleagues have contributed to the studies upon which this paper is based; these are listed in my doctoral thesis (Burn 1991). The greatest assistance was given by Gerald Corney, Lindsay Allan, Becky Coffey, Stewart Hunter, Marcus Pembrey and Linda Burn.

References

Afzelius BA 1976 A human syndrome caused by immotile cilia. Science (Wash DC) 1983:317–319

Allan LD, Crawford DC, Chita SK, Anderson RH, Tynan M 1986 Familial recurrence of congenital heart disease in the prospective series of mothers referred for fetal echocardiography. Am J Cardiol 58:334–337

Arnold GL, Bixler D, Girod D 1983 Probable autosomal recessive inheritance of polysplenia, Situs inversus and cardiac defects in an Amish family. Am J Med Genet 16:35–42

Badr-El-Din MK 1962 Syndrome of levocardia, multiple cardiac defects, situs inversus and absent spleen in a case report. Am Heart J 63:115–118

Bocklage CE 1981 On the distribution of non-righthandedness among twins and their families. Acta Genet Med Gemellol 30:167–187

Brown NA, Wolpert L 1990 The development of handedness in left/right asymmetry. Development 109:1–9

Burn J 1988 Monozygotic twins. In: Chamberlain G (ed) Contemporary obstetrics and gynaecology. Butterworths, London p 161–176

Burn J 1991 Cardiovascular malformation: an analysis of genetic contribution. MD thesis, University of Newcastle upon Tyne, UK

Burn J, Coffey R, Allan LD, Robinson P, Pembrey ME, Macartney FJ 1986 Isomerism: a genetic analysis. In: Doyle EF, Engle MA, Gersony WM, Rashkind WJ, Talner NS (eds) Pediatric cardiology. Springer-Verlag, New York p 1126–1128

Burn J, Corney G 1984 Congenital heart defects and twinning. Acta Genet Med Gemellol 33:61–69

Burn J, Corney G 1988 Types of twinning and determination of zygosity. In: McGillivray I, Campbell DM, Thompson B (eds) Twinning and twins. Wiley, Chichester p 7–26

Chen SC, Montelione PL 1977 Familial splenic anomaly syndrome. J Pediatr 91: 160–161

Corballis MC, Morgan MJ 1978 On the biological basis of human laterality. Evidence of a maturational left to right gradient. Behav Brain Sci 2:261–269

Czeizel A 1987 Letter to the Editor: familial situs inversus and congenital heart defects. Am J Med Genet 28:227–228

de la Monte SM, Hutchins GM 1985 Brief clinical report: sisters with polysplenia. Am J Med Genet 21:171–173

Deanfield JE, Leanage R, Stroobant J. Chrispin AR, Taylor J FN, Macartney FJ 1980 Use of high kilovoltage filtered beam radiographs for detection of bronchial situs in infants and young children. Br Heart J 44:577–583

DiStefano G, Romeo MG, Grasso S, Mazzone D, Sciacca P, Mollica F 1987 Dextrocardia with and without situs viscerum inversus in two sibs. Am J Med Genet 27: 929–934

Ehlers KH, Engle MA 1966 Familial congenital heart disease. Circulation 34: 503–516

Fuhrmann W 1968 Congenital heart disease in sibships ascertained by two affected siblings. Humangenetik 6:1–12

Gatrad AR, Read AP, Watson GH 1984 Consanguinity and complex cardiac anomalies with situs ambiguus. Arch Dis Child 59:242–245

Griffiths MI, Phillips CJ 1976 Twin research (Birmingham 1972). A series of reports to the Social Science Research Council Core Library 6: issue 3

Hurwitz RC, Caskey CT 1982 Ivemark syndrome in siblings. Clin Genet 22:7–11

Hallet JJ, Gang DL, Holmes LB 1979 Familial polysplenia and cardiovascular defects. Pediatr Res 13:344A

Jojarte G, Fekete FP 1978 Familiares asplenia-syndrom. Acta Pediatr Acad Sci Hung 19:35–40

Kartagener M, Stucki P 1962 Bronchiectasis with situs inversus. Arch Dis Child 79:193–207

Katcher AL 1980 Familial asplenia, other malformations and sudden death. Pediatrics 65:633–635

Kawagoe K, Hara K, Jimbo T, Mizuno M, Sakamoto S 1980 Occurrence of Ivemark syndrome with polysplenia in sibs of a family. Proc Jpn Acad 56: 633–637

Layton WM 1976 Random determination of a developmental process. J Hered 67:336–338

Layton WM, Manasek FJ 1980 Cardiac looping in early iv/iv mouse embryos. In van Praagh R, Takao A (eds) Etiology and morphogenesis of congenital heart disease. Futura, New York p 109–126

Moreno A, Murphy EA 1981 Inheritance of Kartagener's syndrome. Am J Med Genet 8:305–315

Mathias RS, Lacro RV, Jones KL 1987 X linked laterality sequence: situs inversus, complex cardiac defects, splenic defects. Am J Med Genet 28:111–116

Neimann N, Pernot C, Gentin G, Vert P, Worms AM 1966 Le 'syndrome d'Ivemark' Cardiomyopathie congenitale yanogene severe heterotaxie thoraco-abdominale complex et asplenie ou polysplenie. Pediatrie 21:511–532

Newman HH 1923 The physiology of twinning. University of Chicago Press, IL

Niikawa N, Kohsaka S, Mizumoto M, Hamada I, Kajii T 1983 Familial clustering of situs inversus totalis, and asplenia and polysplenia syndromes. Am J Med Genet 16:43–47

Noonan JA 1978 Twins, conjoined twins and cardiac defects. Am J Dis Child 132:17–18

Polhemus DW, Schafer WG 1952 Congenital absence of the spleen: syndrome with atrioventricularis and situs inversus. Pediatrics 9:696–708

Rose V, Izukawa T, Moes CAF 1975 Syndromes of asplenia and polysplenia: a review of cardiac and non-cardiac malformations in 60 cases with special reference to diagnosis and prognosis. Br Heart J 37:840–852

Rossi MB, Burn J, Ho SY, Thiene G, Devine W, Anderson RH 1987 Conjoined twins, right atrial isomerism and the sequential analysis. Br Heart J 58:518–24

Ruttenberg H, Nuefeld H, Lucas R et al 1964 Syndrome of congenital cardiac disease with asplenia. Am J Cardiol 13:387–406

Schidlow DV, Katz SM, Turtz MG, Donner RM, Capasso S 1982 Polysplenia and Kartagener's syndrome in a sibship: association with abnormal cilia. J Pediatr 100:401

Schonfeld E, Frischman B 1958 Syndrome of spleen agenesis, defects of the heart and vessels and situs inversus. Helv Paediatr Acta 13:636–640

Seo JW, Shin SS, Chi JG 1985 Cardiovascular system in conjoined twins: an analysis of 14 Korean cases. Teratology 32:151–161

Silver W, Steier M, Chandra N 1972 Asplenic syndrome with congenital heart defect and tetralogy of fallot in siblings. Am J Cardiol 30:91–94

Simpson J, Zellweger H 1973 Familial occurrence of Ivemark's syndrome with splenic hypoplasia and asplenia in sibs. J Med Genet 10: 303–304

Smith DW, Gong BT 1974 Scalp hair patterning: its origin and significance relative to earlybrain and upper facial development. Teratology 9:17–34

Sturgess JM, Thompson MW, Czegledy-Nagy E, Turner JAP 1986 Genetic aspects of immotile cilia syndrome. Am J Med Genet 25:149–160

Teichberg S, Markowitz J, Silverberg M et al 1982 Polysplenia & Kartagener's syndrome in a sibship association with abnormal respiratory cilia. J Pediatr 100: 399–401

Torgersen J 1950 Situs inversus, asymmetry and twinning. Am J Hum Genet 2:361–370

Tsuda T, Srimizu T, Takamizawa K, Ando M, Imai Y, Hori K 1979 Muhi syokogun no rinsho. (Jap) Shonika 20:951–962

van Mierop LHS, Gessner IH, Schiebler GL 1972 Asplenia polysplenia syndrome. In: Congenital cardiac defects—recent advances. Birth Defects Orig Artic Ser VII (1):74–82

Waite D, Steele R, Ross I, Wakefield SJ, Mackay J, Wallace J 1978 Cilia and sperm tail abnormalities in Polynesian bronchiectasis. Lancet 2:132–133

Zlotgora J, Elian E 1981 Asplenia and polysplenia syndromes with abnormalities of lateralisation in a sibship. J Med Genet 18:301–302

Zlotgora J, Schimmel MS, Glaser Y 1987 Familial situs inversus and congenital heart defects. Am J Med Genet 26:181–184

DISCUSSION

Collins: Concerning the concordance of direction of hair whorls in twins, early studies found that the frequencies of counterclockwise whorls ranged from 5–40%, the average being 16% (Lauterbach 1925, Lauterbach & Knight 1927, Newman 1928, Rife 1933). However, the overall association of directionality between the twins was not significantly different from zero (Collins 1977). This is much the same pattern as we see for the association of hand preference in twins.

Morgan: Bernstein claimed that clockwise hair whorling was dominant over counterclockwise. His evidence was very weak (see Morgan 1977).

Collins: If the direction of hair whorl was inherited in a reasonable Mendelian way, we would expect the correlation of whorl direction in twins to be greater than chance. For both handedness and hair whorl, if you know the directionality for one twin, you cannot predict that of the second twin. Furthermore, there seems to be no association between the directionalities of handedness and hair whorl (Collins 1977). Thus, they appear not to be determined by a common mechanism during embryogenesis.

McManus: John, I can see your argument for isomerism and conjoined twins, but what about ordinary non-joined monozygotic twins? What is the mechanism whereby something in one twin can affect what's happening in the other?

Burn: It depends on the timing of the twinning event. Most human twinning does not happen at the two-cell stage; it happens several days into development. That is why most twins share a single placenta. So lateral gradients may have been determined before twinning occurs.

Wolpert: Spemann & Falkenberg's old experiments (1919) showed that when you cut a newt into two at the blastula stage, the right-hand one develops with random asymmetry or abnormalities: the left one develops normally.

Burn: I would like to offer an idea which attempts to unify some of the ideas and observations we have heard. Corballis & Morgan (1978) proposed that the left side of the organism acted as a leading edge, possibly by developing more quickly. Joe Frankel (this volume) has demonstrated a capacity for intracellular laterality in ciliates, and Joe Yost (this volume) has described an extracellular influence, possibly a diffusible factor that we may call YES—Yost's extracellular substance.

The F molecules of Brown & Wolpert (1990) have a rostrocaudal polarity. Their lateral polarity relates not to the midline but to the left margin, perhaps as the cell membrane reaches the point at which it can accomodate YES and fix the F molecules to give an intracellular laterality. Cells to the right are then given their lateral polarity by relating to the cell immediately to their left. Thus, a 'queue' is formed in which intercellular contact maintains the left–right organization (Fig. 1). The twinning process may separate the 'right half' of the zygote from its point of reference on the left side with the result that the 'right half' must begin again. If YES is no longer available, laterality may not be clearly established in the right half, so there is a much greater likelihood of reversal to produce situs inversus, or of confusion leading to situs ambiguus or isomerism.

Frankel: In the case where the YES molecule is on the wrong side, should the resulting embryo be a mirror image or upside down?

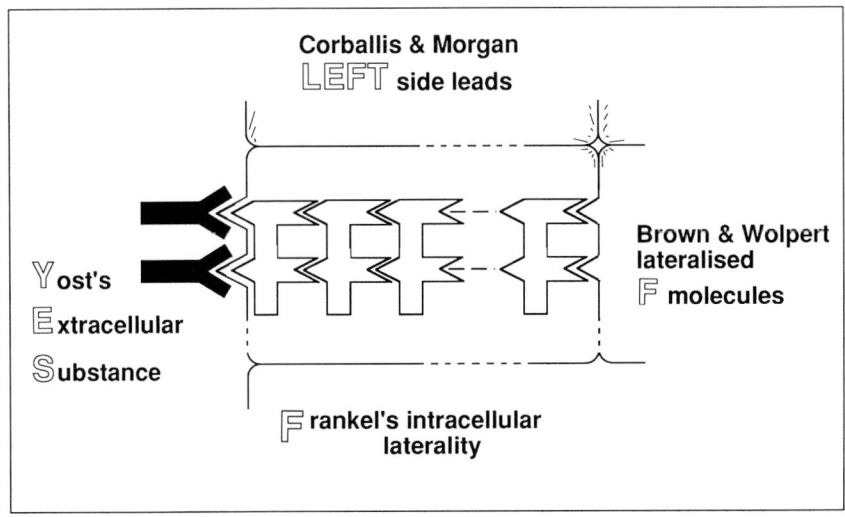

FIG. 1 (*Burn*) Model of the determination of laterality in an early embryo.

Burn: I think YES would probably be on both sides at that stage in development. YES is all over the embryo: the rule is that the back of F must meet the YES. If the F molecules all faced the wrong way, they would just relate to the YES on the opposite side. It is hard to imagine that the extracellular matrix is determinate.

Frankel: If you turn the F molecule around, are you going to tether it on its opposite side or simply have it tethered on the same side but turned around?

Burn: I was flipping it over without breaking the anteroposterior orientation.

Frankel: But you have lost a crucial element of the Brown & Wolpert hypothesis, which is that the F molecule is lined up along the anteroposterior axis *and* tethered on one side or the other.

Burn: I haven't lost it. I said that in one person in 10 000 that mechanism is destroyed or just left random, so the F is swinging like a weather vane. If the F molecules relate to one another, you don't need the extra substance in the cell, you could just have lots of F molecules.

Wolpert: Unfortunately, your model doesn't give any left–right information on its own.

Galaburda: In the twins, since YES is everywhere, after twinning there is plenty of YES on the other side of the embryo. So you need the gradient of the Brown & Wolpert model and the queue of your model.

Burn: I am not anti-gradient! The point is that the genetic literature talks about the midline as the developmental field and how everything relates to the midline. I am still attracted to the Corballis & Morgan proposal that the left margin is the lead point.

I am less concerned with that than with getting acceptance for the general possibility that late twinning may result in an organism that has lost the capacity to restore its gradient. Nigel Brown's suggestion that the right halves become fixed and are no longer flexible is equally acceptable. It comes back to the point that monozygotic twins could be morphologically qualitatively different from singletons, especially if you believe that the morphology of the brain has something to do with handedness.

McManus: Now what do we do about the fact that situs in general doesn't relate to handedness? If you are reversing the gradient as a whole, handedness and situs should both be reversed.

Burn: I know of only two studies that relate structural laterality to functional laterality. Torgersen (1950) looked at situs inversus and left-handedness and found no relationship. In an unpublished study, Chris McManus and I sent questionnaires to the families that I had traced in the isomerism family study. Of 11 mature survivors with isomerism, nine were right handed and two were left handed. The nine were very strongly right handed. So it doesn't appear that isomeric people are dyslexic and bilateral in their handedness; they seem to choose one or the other.

Galaburda: Would anybody be upset if I suggested there are different molecules setting different gradients for different systems?
Wolpert: Not at all.

References

Brown NA, Wolpert L 1990 The development of handedness in left/right asymmetry. Development 109:1–9

Collins RL 1977 Origins of the sense of asymmetry: Mendelian and non Mendelian models of inheritance. Ann NY Acad Sci 299:283–305

Corballis MC, Morgan MJ 1978 On the biological basis of human laterality. Evidence of a maturational left to right gradient. Behav Brain Sci 2:261–269

Frankel J 1991 Intracellular handedness in ciliates. In: Biological asymmetry and handedness. Wiley, Chichester (Ciba Found Symp 162) p 73–93

Lauterbach CE 1925 Studies in twin resemblance. Genetics 10:525–568

Lauterbach CE, Knight JB 1927 Variation in whorl of the head hair. J Hered 18:107–115

Morgan 1977 Embryology and inheritance of asymmetry. In: Harnad SR, Doty RW, Goldstein L, Jaynes J, Krauthamer G (eds) Lateralization in the nervous system. Academic Press, New York p 173–194

Newman HH 1928 Studies of human twins. II. Asymmetry reversal, or mirror imaging in identical twins. Biol Bull 55:298–315

Rife DC 1933 Genetic studies of monozygotic twins. III. Mirror-imaging. J Hered 24:443–446

Spemann H, Falkenberg H 1919 Uber asymmetrische Entwicklung und Situs Inversus viscerum bei Zwillingen und Doppelbildungen. Arch Entwicklungsmech Org (Wilhelm Roux) 45:371–423

Torgersen J 1950 Situs inversus, asymmetry and twinning. Am J Hum Genet 2:361–370

Yost HJ 1991 Development of the left–right axis in amphibians. In: Biological asymmetry and handedness. Wiley, Chichester (Ciba Found Symp 162) p 165–181

Laterality and motor control

Michael Peters

Department of Psychology, University of Guelph, Guelph, Ontartio, Canada N1G 2W1

Abstract. Lateralization in motor control, exemplified by handedness, is a publically observable lateral asymmetry. The understanding of causation of lateral asymmetry in movement is poor for two major reasons. First, structural limitations underlying lateral bias cannot be compelling in the sense that laterality is completely specified. Second, interacting sources of lateral bias are probably represented at several levels of the central nervous system, with some sources being many steps removed from the executing motor structures. Both factors permit flexibility in terms of which side does what. I suggest two approaches to improve understanding of mechanism. First, use of bimanual tasks which reduce flexibility in lateral expression and therefore provide more information than unimanual tasks. Second, analysis of task performances in terms of what levels of lateral specialization are likely to be involved. Such analyses will be especially useful in the understanding of hand use in left-handers who perform different skilled tasks with different hands.

1991 Biological asymmetry and handedness. Wiley, Chichester (Ciba Foundation Symposium 162) p 300–311

In the physical sciences the relation between a perceived asymmetry and underlying causes can be very clear, as Pasteur's work on the deflection of polarized light by tartaric and racemic acid first showed. At the other extreme, in population studies on lateral asymmetries stable predictions about the lateral dispositions of populations can probably be made and genetic models constructed (Annett 1985, McManus 1985). At the intermediate level, when laterality in motor control of individuals is examined, uncertainty enters the picture, and there are good reasons for this. Biochemical and biological factors that underlie asymmetries in animals are transformed during ontogeny so that a point to point mapping across developmental stages becomes tenuous. Piaget (1971), in his book on the biology of knowledge, used the term 'partial isomorphism' to acknowledge the way in which things are and are not the same at different developmental stages.

There is also a progressive loosening of the relation between asymmetry in underlying structure and behaviour (e.g. Kertesz & Geschwind 1971) that is observed as organisms become increasingly complex; this has an important function. The argument goes as follows. We presume that behavioural asymmetries (here: motor asymmetries) have an adaptive function; but, in order not to be disabling under some circumstances, such motor asymmetries *must never be entirely* controlled by the underlying structural factors. If it were so, an organism would be limited in responding to the external world (which would be completely unsympathetic with regard to any lateral fixations). As an example, bush babies—a prosimian species—prefer to reach out for insects with their left paw, but it would be quite maladaptive if they could do this with *only* their left paw.

It is in this light that the anatomical asymmetries discussed by Galaburda (this volume) must be viewed. A larger lateral corticospinal tract on the right side supports a right-sided preference but does not preclude a high degree of motor control on the left side. The relation between hand preference and skill and the larger pyramidal tract size on the right is not at all understood. Do the larger numbers of fibres in the lateral and anterior motor tracts (Nathan et al 1990) of the right side stand for a greater number of originating cortical cells in the left hemisphere or is a greater number of ipsilateral projections on the right sight involved? Does the higher number of fibres in the lateral corticospinal tract (Nathan et al 1990) stand for a higher number of originating cortical cells in the left hemisphere? To confuse the matter further, Kertesz & Geschwind (1971) observed the pattern of crossover of the pyramidal tracts that was characteristic of right-handers (left-sided tract crossing over before the right-sided tract) in all their left-handers as well. The minority of cases with an opposite crossover pattern were right-handers.

Whatever the relation between handedness and anatomy of the motor tracts, there is ample evidence that the structural givens are not as limiting in terms of behaviour as one might think. Only two sources of behavioural evidence will be cited here. First, extremely high levels of motor skill can be attained with systems that are not predestined to play a role in volitional fine motor activities. For instance, a certain Madame Hanakawa from Japan, born without arms, was able to produce wonderful origami paper folding creations with her feet. No less impressive, the Frenchman Jean de Hanau played the mandolin with his feet, by all accounts not in a pedestrian fashion; and Miss Sarah Biffin in 19th century Britain, lacking arms and feet, produced outstanding lithographs by holding the implements in her mouth (Jay 1986). Numerous similar cases are known. These show that movement intent can be expressed flexibly, and at the highest level of accomplishment, through different executing outlets. If the feet can perform delicate manipulative acts, there is no question that *both* hands will be eminently suitable for the performance of such acts. The differences between the hands that lead to the distinction between a preferred and non-preferred hand, relative to differences between feet and hands, must be quite small.

The other line of evidence concerns the prevalence of left-handedness. Handedness is very sensitive to cultural pressures: depending on the given culture and the date of the study, prevalence figures from less than 1% (Komai &

Fukuoka 1934) to more than 13% (Spiegler & Yeni-Komshian 1983) have been reported when the hand preferred for writing is used as the criterion. Sex differences in prevalence range from samples where no female left-handed writers are observed (grade 8 sample of Komai & Fukuoka 1934) to those where female outnumber male left-handed writers (Levander & Schalling 1988). The factors underlying handedness do not appear compelling because hand usage against natural preferences has no obvious detrimental effects on motor performance. The converse question, 'What would right-handers do in a left-handed world?' remains unanswered.

All this justifies the suspicion that detailed study of asymmetries in motor performance, far from revealing underlying sources that compel asymmetries, will show only to what extent habit has left its imprint on the executing motor systems. But how is that habit established? Developmental studies on human infants show that asymmetries in movement occur before the motor systems for which an anatomical asymmetry has been documented are fully functional. The most parsimonious explanation invokes the asymmetries seen in simple orienting responses which, in turn, have been related to asymmetries in dopaminergic systems in the basal ganglia, in rats (Glick & Shapiro 1984) and humans (Bracha et al 1987). I have previously stressed the importance of transformational processes. In this case it is suggested that an early lateral bias in orientation is transformed into an attentional bias that becomes important in skilled movement. The crucial point is that the attentional process, although normally concordant with structural asymmetries, remains superordinate to the level of execution, and is not therefore wholly constrained by demonstrated (Nathan et al 1990, Tan 1985) structural motor asymmetries. Thus, while we are inclined to focus attention on the preferred hand, we can readily switch attention to the other hand when switching hands in unimanual activities.

The situation becomes more interesting in bimanual activities. Here, the hands assume different roles and the allocation of attention reflects this. There are two important aspects: first, when hands become specialized in unimanual activities the switching of hand roles requires merely that the other hand also masters the skill. In bimanual tasks, *both* hands have to relearn. For this reason, there is a greater incentive for the consistency of hand roles in bimanual tasks than in unimanual tasks. Second, the attentional disposition across the two hands harmonizes with the division of motor roles. While some switching of attention is possible, fast concurrent movement of the two hands will often demand a split into primary focused attention towards the preferred hand and subsidiary attention towards the non-preferred hand (Peters 1990a). Although the preferred hand often takes the more demanding motor role, this is not always so, and the assignment of roles follows attentional preferences more than motor demands. For instance, when playing the violin, it can hardly be argued that the motor demands of the fingering hand are less than those of the bowing hand.

When the motor demands on the two hands are kept identical and the allocation of attention is manipulated, it becomes clear that performance is much better when attention is focused on the preferred hand (Peters 1985). I suggest, therefore, that lateral biases in attention take precedence over motor biases and are operative before structural biases in the executing motor machinery become functional. This does not exclude other dimensions of lateral specialization in movement. For instance, Corballis (1989) has proposed that the left hemisphere is specialized in terms of generating novel constructions from existing building blocks, using established syntactic rules. This generativity can be expressed in diverse ways, such as in skilled movement or linguistic expression. This is compatible with an attentional specialization. A case where generativity and attentional specialization coincide fully is the improvising right hand of the jazz pianist through which musical ideas are expressed.

If there were no left-handers in this world, the preceding discussion would be a reasonable contribution to our understanding of handedness rather than only to our understanding of asymmetries in hand use. But there are left-handers and these have to be accounted for. There are three principal hypotheses. First, left-handers could be flawed right-handers. 'Flawed' may refer to something as crude as outright brain damage during birth or something as mild as an unfavourable genetic make-up or minor disruption of developmental time tables. The attribute 'flawed' might apply to a few left-handers (Harris & Carlson 1988) but our own samples, of university students in particular, give no indication of 'pathology'. Our estimates of the prevalence of left-handedness among university students are as high as those for elementary school students (Peters & Pedersen 1978). If it were the case that some left-handers had been selected out before reaching the university, the relatively high proportion of left-handed university students would need an explanation. Are there more left-handers at university than expected on the basis of general population figures for adults?

Second, left-handers could be constituted like right-handers but differ in directionality. Third, left-handers might be fundamentally different from right-handers. Unfortunately, existing theories of handedness do not allow predictions that would distinguish between these alternatives. This is partly due to the incomplete character of the theories; none actually proposes a mechanism for motor asymmetry. But even if the theories made specific predictions, falsification would not be easy because of the flexibility of the motor apparatus and its interaction with cultural/environmental pressures.

It is against this decidedly gloomy background that investigations of motor asymmetries have to be carried out. My own bias, vulnerable to a variety of methodological criticisms, has been to approach motor behaviour in left-handers from a taxonomic perspective. How do left-handers behave on certain motor tasks and is it possible to distinguish meaningfully between different kinds of left-handers? An examination of motor performance of left-handers provides answers to these questions. I begin by defining, in a somewhat arbitrary way, what a right- or a left-hander is. Subjects are given a short handedness questionnaire (which hand is used for writing, hammering, throwing, unscrewing the lid of a jar, cutting bread with a knife, brushing teeth, holding a racquet). If they write with the right hand and consider themselves right-handed, they are called 'right-handed'. If they write with the left hand, and prefer the right hand for no more than one other activity, they are called 'consistently left-handed'. If they write with the left hand and prefer the right hand for two or more of the activities, they are called 'inconsistently left-handed'. The inconsistency refers to the choice of different hands for different activities, rather than the flexible use of either hand for a given activity.

Tables 1–3 show a representative selection of some of the findings, mostly based on Peters (1990b) and Peters & Servos (1989) where procedural and statistical details can be found. When hand strength is measured, we find that right-handers have a stronger right hand (Table 1). Left-handers show no significant differences between the two hands. Results of this kind have led investigators to conclude that left-handers show smaller between-hand differences than right-handers. However, when the left-handers are subdivided into consistent and inconsistent subgroups, they show strength asymmetries in opposite directions, which cancel each other in group analyses.

Table 2 shows the same kind of analysis for a throwing task, where subjects had to hit a target. As in the strength measure, left-handers as a group showed no significant hand asymmetries, but when subclassification was used the two groups of left-handers showed asymmetries in opposing directions. The left-hand superiority for consistent left-handers is expected, but the inconsistent left-handers, as a group, are stronger in the right hand and throw more accurately with the right.

When the inconsistent left-handers are tested on finger tapping speed (Table 3), they are faster with the left hand. Thus, in this group, the direction of lateral specialization is a function of the particular motor task. The rapid and selective movement of single fingers depends exclusively on those pyramidal tract

TABLE 1   Hand strength of right- and left-handers

	Male			Female		
	No.	Left hand	Right hand	No.	Left hand	Right hand
Right-handers	75	46.9	50.8	86	25.1	27.7
Left-handers	89	49.2	49.2	125	27.7	27.5
Consistent left-handers	50	49.7	46.2	69	28.1	27.1
Inconsistent left-handers	39	48.6	52.1	56	27.3	28.0

Consistent and inconsistent refer to the degree of hand preference choice expressed in reply to a questionnaire. Hand strength measured in kg using a dynamometer. Procedural and statistical details available in Peters (1990b) and Peters & Servos (1989).

TABLE 2   Throwing accuracy of right- and left-handers

	Male			Female		
	No.	Left hand	Right hand	No.	Left hand	Right hand
Right-handers	47	2.6	3.9	57	1.9	2.9
Left-handers	47	3.3	3.3	49	2.4	2.4
Consistent left-handers	32	3.6	2.7	34	2.6	2.2
Inconsistent left-handers	15	3.0	3.9	15	2.1	2.6

Consistent and inconsistent refer to the degree of hand preference choice expressed in reply to a questionnaire. Accuracy is the average throwing accuracy score out of a maximum six points. Procedural and statistical details available in Peters (1990b).

fibres that terminate directly on motoneurons in the anterior horn of the spinal cord. These fibres represent the most specialized portion of the specialized system that subserves fine manual skill. Inconsistent left-handers also perform better with the left hand on another test of fine motor skill, the Purdue Pegboard task. Thus, inconsistent left-handers show a right-side specialization for strength and ballistic activities of the arm, and a left-side specialization for fine manual activities. In addition, while consistent left-handers prefer the left foot for kicking, inconsistent left-handers often prefer the right foot.

Where does all this leave the attentional hypothesis that was posed above? In right-handers and consistent left-handers, lateral specializations for strength and skill are congruent and attentional specialization need not be considered separately. Inconsistent left-handers focus attention to the left hand for writing and fine manual skills but to the right side for throwing and kicking. Where do they focus attention when they are forced to lateralize it in bimanual tasks? So far, we have used only a simple bimanual task, where subjects are required to produce one tap in one hand for every two taps in the other hand (Peters

TABLE 3   Finger tapping speed of right- and left-handers

	Male			Female		
	No.	Left hand	Right hand	No.	Left hand	Right hand
Right-handers	75	5.3	5.9	86	4.9	5.4
Left-handers	89	5.7	5.4	125	5.3	5.1
Consistent left-handers	50	5.6	5.3	69	5.3	5.0
Inconsistent left-handers	39	5.6	5.4	56	5.3	5.2

Consistent and inconsistent refer to the degree of hand preference choice expressed in reply to a questionnaire. Finger tapping speed is the number of taps per second. Procedural and statistical details available in Peters (1990b) and Peters & Servos (1989).

& Servos 1989, unpublished replication experiments). In this task, subjects have to integrate the movements of the two hands as precisely and quickly as they can. They are asked to do the task with a 'right two/left one' or a 'left two/right one' condition. Because the fingers of each hand perform identical tapping movements and because the tapping rates in this task are very slow compared to single finger speeded tapping, asymmetries in performance between the two different conditions (right two/left one versus left two/right one) cannot be attributed to motor factors as such. The speed of performance is measured in terms of the mean duration of intertap intervals for the hand performing the two taps. Right-handers were faster by 26 msec when the right hand performed the two taps ($P < 0.0001$) than with the converse arrangement. Left-handers showed no stable asymmetry, even when subdivided into two groups as before. The asymmetry favoured the left hand by 4.2 msec in the consistent left-handers and the right hand by 0.4 msec in the inconsistent left-handers (Peters & Servos 1989, Peters 1987, unpublished replication experiments 1990). These group differences are not significant. This does not mean that left-handed individuals do not show directional asymmetries. If the absolute difference between the two conditions is measured regardless of the direction of asymmetry, consistent left-handers differ by an average of 23 msec and inconsistent left-handers by 26 msec (right-handers by 34 msec).

Why do the two groups of left-handers differ from each other on some tasks and not on others? This question can be addressed by Annett's (1985) model. According to this model, left-handers lack inherent lateral biases with regard to hand preference. Hand preference and hand performance asymmetries are seen as a function of chance, environmental factors and task characteristics. Because of the demands posed by skilled motor tasks, left-handers will show performance asymmetries, as right-handers do; however, the direction of the bias is not constrained in left-handers. With this interpretation, inconsistent left-handers do not choose the right hand for throwing and strength activities because they differ fundamentally from consistent left-handers, they are merely those left-handers who happen to do things in this way. This is the 'strong' version of Annett's (1985) model. A weaker version would emphasize the fact that left-handers share with right-handers some lateral specializations, such as a left hemisphere specialization for speech and language and, at least in some cases, a praxis specialization in the left hemisphere (Rey et al 1988, Signoret & North 1979). Such a version would state that while lateral biases in left-handers are indirect, especially with regard to hand preference, and not as strong as in right-handers, they are nevertheless present and would produce a moderate lateral bias in complex motor tasks.

An alternative hypothesis asserts that the performance differences between inconsistent and consistent left-handers do reflect underlying structural differences. This hypothesis focuses on the development of the motor systems that control the distal and axial musculature. While lateral specializations in these two systems develop to be laterally congruent in most individuals, it is possible that such specializations are not laterally congruent in a small proportion of individuals. These are the inconsistent left-handers. If so, further work with more difficult bimanual tasks should reveal differences between consistent and inconsistent left-handers when the level of motor control (axial versus distal musculature) is systematically varied. Inconsistent left-handers should show different asymmetry patterns in bilateral tasks when these involve the distal musculature than when they involve the axial musculature. Consistent left-handers should be laterally consistent across such tasks.

For reasons pointed out in the initial part of the paper, the falsification of these hypotheses may prove difficult or impossible. Thus, while work along these lines may tell us something of importance about the motor behaviour of left-handers that is of value in rehabilitation and litigation, it is possible that nothing can be learned about the nature of the underlying mechanisms.

Acknowledgement

This work was supported by National Sciences and Engineering Research Council of Canada Grant No. 7054.

References

Annett M 1985 Left, right, hand and brain. Lawrence Erlbaum Associates, London

Bracha HI, Seitz DJ, Otemaa DJ, Glick SD 1987 Rotational movement (circling) in normal humans: sex differences and relationship to hand, foot and eye preference. Brain Res 411:231–235

Corballis MC 1989 Laterality and human evolution. Psychol Rev 96:492–505

Galaburda AM 1991 Asymmetries of cerebral neuroanatomy. In: Biological asymmetry and handedness. Wiley, Chichester (Ciba Found Symp 162) p 219–233

Glick SD, Shapiro RM 1984 Functional and neurochemical asymmetries. In: Geschwind N, Galaburda AM (eds) Cerebral dominance: the biological foundations. Harvard University Press, Cambridge, MA p 147–166

Harris LJ, Carlson DF 1988 Pathological left-handedness: an analysis of theories and evidence. In: Molfese D, Segalowitz SJ (eds) Brain lateralization in children. Guilford Press, New York p 289–372

Jay R 1986 Learned pigs and fireproof women. Warner Books, New York

Kertesz A, Gechwind N 1971 Patterns of pyramidal decussation and their relationship to handedness. Arch Neurol 24:326–332

Komai T, Fukuoka G 1934 A study on the frequency of left-handedness and left-footedness among Japanese school children. Hum Biol 6:33–42

Levander M, Schalling D 1988 Hand preference in a population of Swedish college students. Cortex 24:149–156

McManus IC 1985 Handedness, language dominance, and aphasia: a genetic model. Psychol Med, Monograph Suppl 8

Nathan PW, Smith MC, Deacon P 1990 The spinocortical tracts in man. Brain 113:303–324

Peters M 1985 Constraints in the coordination of bimanual movements and their expression in skilled and unskilled subjects. Q J Exp Psychol A Hum Exp Psychol 37:171–196

Peters M 1987 A nontrivial motor performance difference between right-handers and left-handers: attention as intervening variable in the expression of handedness. Can J Psychol 41:91–99

Peters M 1990a Interaction of vocal and manual movements. In: Hammond G (ed) Cerebral control of speech and limb movement. Advances in Psychology. North-Holland, Amsterdam p 535–574

Peters M 1990b Subclassification of lefthanders poses problems for theories of handedness. Neuropsychologia 28:279–289

Peters M, Pedersen K 1978 Incidence of left-handers with inverted writing position in a population of 5910 elementary school children. Neuropsychologia 16:743–746

Peters M, Servos P 1989 Performance of subgroups of lefthanders, and righthanders. Can J Psychol 43:341–358

Piaget J 1971 The biology of knowledge. University of Chicago Press, Chicago, IL

Rey M, Dellatolas G, Baucaud J, Talairach J 1988 Hemispheric lateralization of motor and speech functions after early brain lesions: a study of 73 epileptic patients with intracarotid amytal test. Neuropsychologia 26:167–172

Spiegler BJ, Yeni-Komshian GH 1983 Incidence of left-handed writing in a college population with reference to family patterns of hand preference. Neuropsychologia 1:651–659

Signoret JL, North 1979 Les apraxies gestuelles. Masson, Paris

Tan Ü 1985 Relationships between hand skill and the excitability of motor neurons innervating the postural soleus muscle in human subjects. Int J Neurosci 26:289–300

DISCUSSION

Wolpert: You seem to be making a major point as to why there is laterality—the brain has to be lateralized in order to function. That is a totally novel idea to me.

Peters: Whether there is anatomical asymmetry or not, there has to be functional asymmetry. Practically all higher activities in thought and movement involve complementary contributions from the two sides of the brain.

Galaburda: I have a couple of arguments against your specific model—in principle it may be OK. This notion that cerebral dominance has to do with the unilateral control of both sides is an old one.

Peters: I am saying there is unilateral control at a higher level: at lower levels of motor control, there is separate control of each hand by the opposite hemisphere.

Galaburda: I understand, the unilateral control is the command part. There is no evidence that the connectivity is different from left to right as opposed to from right to left. That was the first thing we looked for when we did our connectivity studies.

The other thing is we know that the command posts in the primary brain are in the supplementary paralimbic cortex, which is dorsal, near the midline.

This means that the distance between one side and the other is much shorter. Secondly, there is basically no asymmetry in the supplementary paralimbic cortex. Clinically, a lesion of this area on either side produces an identical syndrome and recovery is prompt where the lesion is bilateral. Thus, the notion of unilateral control is not supported.

Peters: My argument is not that one supplementary motor area communicates with the other, but that the supplementary motor area on one side, most commonly the left, in conjunction with whatever other areas are involved, sends commands to initiate and terminate movement to both the right and left hand control areas in the respective motor cortices. Whether or not lesions of the implicated area on either side produce identical syndromes is a moot point. The fact is we can see strong evidence of functional asymmetries in the coordination of concurrent bimanual movements and the asymmetries are directionally consistent in right-handers. The lesions Al Galaburda talks about cannot be expected to affect the entire constellation of brain regions that together initiate and terminate the movements of concurrently moving hands.

Berg: Has this question been studied by PET (positron emission tomography), or is the time-resolution of that technique too poor?

Peters: There are some electrophysiological data, but PET is still too crude. EEG (electroencephalography) shows unilateral activity for a very short moment, then it's bilateral.

Berg: If one were finger tapping synchronously, there might be some strong asymmetry in the PET scan.

Peters: That is possible, if one has fine enough resolution and knows where to look.

Frankel: Are these command centres functionally unilateral or replaceable? If you destroyed the left, would there be a subsidiary command centre in the right side ready to go into operation?

Peters: I did a terrible thing by allowing Albert to give the impression that command centres are clearly localized entities that do clearly defined things. Whatever cortical and subcortical machinery is involved in the planning and implementation of movement must be distributed widely. In principle, each brain half should be able to function in the planning and implementation of movement: their activities may be quite symmetrical to begin with. However, as Dilip Kondepudi described in physical systems (p 10–11), very tiny asymmetries can accumulate and progressively reinforce each other so that large functional asymmetries result. Planning and initiation of movement represents a very high level of motor function and the effect of cumulative minor asymmetries will, at this level, produce powerful functional asymmetries.

Jefferies: Most primates have stereoscopic vision, which implies that one eye is dominant. When does the right eye begin to become dominant among the primates with stereoscopic vision?

Peters: I am not sure how eye dominance relates to stereoscopic vision. Eye movement control relates to lateral asymmetry just as the problems of activating a bilateral speech musculature relate to lateralized control. In eye movement, the medial rectus of one eye has to be coordinated with the lateral rectus muscle of the other eye. I find it difficult to imagine how smooth lateral eye movement could be managed by bilateral synchronized activation. Most likely, one brain half would have to initiate action in both recti at the same time.

Jefferies: There's some dubious evidence that australopithecines were right handed—they clobbered baboons consistently on the left side of the skull (Dart 1949). They probably couldn't speak because they had rather small brains. So if anything the sequence is from 'right-eyedness' to right-handedness to speech.

Peters: Humans with a microcephalic condition have been described who, despite having a very small brain, were capable of speech. Perhaps the emergence of rudimentary speech as a species-specific trait preceded the evolution of large brains.

Annett: In hominid brains there is evidence for the torque that Al Galaburda was talking about. How many of those hominids were australopithecines, I don't know.

Corballis: It is pretty much in dispute, whether the anatomical asymmetries observed in the human brain were present in the australopithecines. They had probably evolved by *Homo habilis*, however.

Fujinaga: I understand why laterality is necessary at the level of the cortex because language and motor functions need a very fine network of synapses, which is more efficient if it is located on one side of the brain rather than on both sides. Is there any asymmetry at the lower level of the brain?

Peters: Albert Galaburda alluded to at least one asymmetry in the pyramidal tracts. Where these cross over in the lower brain stem, the fibres from the left cortex begin to cross over to the right more rostrally than fibres from the right cortex cross to the left. Unfortunately, there was no correlation between the crossing patterns and handedness. There is also an asymmetry in the size of the pyramidal tracts. Both the crossed dorsolateral corticospinal tract that goes to the motor neurons supplying the distal musculature and the uncrossed anterolateral corticospinal tract that supplies muscles of the axial musculature are larger on the right side in three-quarters of cases. Again, there is no clear correlation with handedness.

The cerebellum must also be involved, but in a complex way. The right cerebellum is involved with right body half musculature, and the left cerebellum with the left body musculature. There is a complicated system of double crossing of fibres from the cerebellum, which brings information from one side of the cerebellum together with that from the opposite cerebral hemisphere. This must be involved in skill and lateralization of motor control, but it has not been studied.

Collins: Mice that show a paw preference in skill are also stronger with the preferred paw on a test of grip strength (Collins 1970).

Secondly, the dissociation of left-handers into consistent and inconsistent groups is what you expect if you take a U-shaped distribution, impress it against a dextral gradient then divide the resulting J-shaped distribution into an extreme right portion, in which individuals are dextral, and a left portion, in which individuals are sinistral.

Peters: Unfortunately, I can't say what the frequencies of consistent and inconsistent left-handers are because I am not sampling randomly.

McManus: I think that the coin which is tossed to decide whether you are right or left handed in the *CC* genotype according to my model (or the RS-genotype according to the Annett model), is also being tossed to decide whether you are right or left cerebral dominant for language. That model can readily be extended by saying that since Michael is suggesting there are two separate motor control systems, the coins that determine their laterality should be tossed independently. Such a model predicts that some people have the two motor control systems in different hemispheres. Specifically, my genetic model predicts that about 70% of left-handers should be of the inconsistent type.

Annett: The simplest solution is to say that there is no bias and all these things that can lateralize do so independently in the neutral or null case. Some of these chance lateralizations turn out to be better for us than others.

References

Collins RL 1970 The sound of one paw clapping: an inquiry into the origin of left-handedness. In: Lindzey G, Thiessen D (eds) Contributions to behavior-genetic analysis. Appleton-Century-Crofts, New York p 115–136

Dart RA 1949 The bone-bludgeon hunting technique of *Australopithecus*. S Afr J Sci 2:150–152

Final general discussion

Is bilateral symmetry primitive or derived?

Frankel: I am intrigued by isomerisms: right–right individuals who have two right halves and some left structures missing, and left–left individuals with some right structures missing. I found out from John Burn's talk that such isomerisms exist in humans. In my paper I described isomerisms in ciliates. On the other hand, in the two best studied multicellular genetic systems, *Caenorhabditis* and *Drosophila*, there are no known left–right isomerisms. *Drosophila* has well known posterior–posterior isomerisms called *bicaudal*. There are also rare anterior–anterior isomerisms.

Wolpert: You can't tell the difference between left and right in *Drosophila* anyway.

Frankel: But in *Caenorhabditis* you could tell if there were a left–right isomerism. It would be lethal, but you would be able to tell. Do they exist?

Wood: They haven't been seen, and I doubt they could occur. The asymmetry normally arises in the placement of cells at the 6-cell stage, and you can't have two left sides at that point! You asked earlier how an embryo with no asymmetry at this stage would develop: again, such embryos haven't been seen and I'm afraid it would be very difficult to produce them experimentally.

Frankel: Before this symposium I thought the question of asymmetry in a crystal or a ciliate was different from the question of bilateral asymmetry. It now seems to me that they are similar questions, for two reasons. First, Dr Jefferies described how the organisms from which we evolved were predominantly asymmetrical and had a small region of symmetry that slowly spread forwards during evolution. Bilateral symmetry was secondary. Second, in *Caenorhabditis* the developmental mechanism that distinguishes left from right is essentially the same as the one that distinguishes dorsal from ventral but it occurs one round of cell division later. So *Caenorhabditis* produces an anteroposterior axis, then a dorsoventral axis, and then a left–right axis which is just as polar as the first two axes. Therefore, rather than asymmetry being a secondary superimposition on bilateral symmetry, it could be the reverse. That fits very nicely with the Corballis & Morgan hypothesis that there is a basic left–right asymmetry.

Wood: The question is whether there is anything special about the left–right axis. Is it fundamentally different from the anteroposterior or dorsoventral axes?

Wolpert: Yes, because it only has meaning with respect to those other axes. We are back to where we started.

UNIVERSITY OF OXFORD · DEPARTMENT OF PHYSICS

Clarendon Laboratory · Parks Road · Oxford · OX1 3PU

From the Head of
ATOMIC & LASER PHYSICS
Professor P G H Sandars

Direct Line: (0865) 272234
EMail: SANDARS@PHYSICS.OXFORD.AC.UK

Secretary: (0865) 272254
Switchboard: (0865) 272200
Fax: (0865) 272375

29/May/1995

Richard,

Many thanks indeed.
It is much more than I deserve.

Peter

With Compliments

Wood: I mean is it different in the sense of having a fundamental asymmetry, such that bilateral symmetry must be superimposed on it.

Jefferies: I believe that the dorsoventral axis of vertebrates is homologous to the left–right axis of an insect.

Wolpert: I personally believe that there is a symmetrical organism on which the left–right axis is imposed, and there is embryological evidence to support that. The evolutionary evidence may disagree, but I doubt it.

McManus: You have to account for why *Drosophila* is not asymmetrical.

Wood: You could imagine superimposing perfect bilateral symmetry on a basically asymmetrical developmental process.

Wolpert: The primitive condition is symmetry. Before there were left- and right-sided organisms there were symmetrical organisms with a front and a back.

Jefferies: Primitive hemichordates, for example.

Wolpert: No, I am talking about primitive multicellular animals; bilaterians or coelenterates or whatever.

Frankel: I only brought up this point because initially I thought that ciliates were unique in their asymmetry. But Dr Jefferies and Dr Wood have described other cases where bilateral symmetry is secondary in two totally different contexts.

Wood: Nematodes clearly start as fundamentally asymmetrical embryos that acquire bilateral symmetry during development.

The evolution of human laterality

Corballis: I wanted to speculate about human laterality and why it might be important in human evolution, although I should say at once that *any* speculation about human evolution is dangerous. We have heard about several examples of asymmetry turning into symmetry, but cerebral asymmetry in humans has gone the other way; that is, handedness and cerebral dominance for language are superimposed on a structure that was largely symmetrical through evolution until humans. You don't see handedness in the primates—there is no population bias as there is in humans. This is something that seems to distinguish humans from other species.

Language itself also distinguishes humans from other species. The human language has distinctive properties, and special and flexible rules, which differentiate it from any other form of animal communication. Language was one of the properties Descartes referred to in separating the human mind from the animal mind.

So human handedness and cerebral dominance for language are, as it were, the stuff of humanity and I think that's one of the reasons they are so fascinating. Also, one doesn't see these asymmetries clearly in a structural sense but they are very clearly there in a functional sense, which adds to the sense of mystery.

An important question about human laterality is why is it directional? Why is it the right hand and the left brain? I just want to speculate a little. Marian Annett has some speculations about this too. I think there may have been a reason to have control of the hands in praxis and in sequential and planned movements in the same hemisphere as control of the vocal apparatus. The reason is that language may have evolved from a gestural system into a vocal system, although for much of the time it was probably a mixture of both. When people tried to teach language to the chimpanzees it soon became clear that their vocal apparatus simply could not do it. It took a long time in evolution for the vocal apparatus to develop the characteristics that enable us to speak. It was probably not until *Homo erectus*, some people say not until *Homo sapiens sapiens*, that the shape of the vocal apparatus permitted us to speak in a flexible manner. It is difficult to believe that language has evolved entirely since then. Therefore there must have been a mechanism for language much earlier.

The distinctive characteristic of hominids was that they stood upright, thereby freeing the hands and the mouth for communication and tool use and so forth. So there are good reasons to suppose that we started to communicate and compute messages that were open-ended by using manual gestures rather than the voice. This idea has been around for centuries. It is supported by the observation that sign language even now is as natural as vocal language. Deaf children who learn sign language learn it very easily; they go through the same critical periods, the same stages, they even babble in sign (Petitto & Marentette 1991).

So as language evolved there was a gradual transition from manual gesture to communication by voice, and it would have been an advantage to have the control of the voice in the same hemisphere as control of the hands. I say hands in the plural because the praxic skills that Michael Peters was talking about seem to be organized within a single hemisphere, regardless of which hand is being used.

Linguists claim that they can trace vocal languages back to the root language known as protoWorld. It is thought that protoWorld is probably no older than 100 000 years. That is very recent. So maybe the switch from a gestural language to a vocal one occurred at about this time, and that is what distinguishes *Homo sapiens* from the other hominids. So right-handedness may be a relic of a past that was involved in the development of language. And it may have been more important then than now to have these control systems in the same hemisphere.

Wolpert: That explains lateralization, but no one at this meeting has explained why lateralization is handed, not random.

Corballis: Wouldn't the directionality guarantee that the two control systems were on the same side? One consequence of randomness would be that the two control systems could be on different sides.

McManus: Once one control system was randomly set up on one side, the rest would have to follow.

Wolpert: That doesn't tell you that you always have to do it on the left or on the right.

Corballis: You have to do it always on the same side. The simplest way to link may be to have a consistent bias.

Peters: The difficulty with the evolutionary argument is that one can easily take the converse position without any loss of logical continuity. Primates and human infants use the mouth as the instrument of precision manipulation. Goethe, in his morphological writings, called the mouth 'the hand of the head'. Human infants use the hands primarily to transport things to the mouth to be explored.

All primates manipulate food with the mouth: this manipulation requires fine control of pressure with the tongue. All the general control properties that later characterize handedness are already used by oral manipulative movements. So I would argue that because delicate and precise control systems were available for oral manipulation, these were adapted for development of fine motor control in the hands.

Berg: In sign language, is one hand used a lot more than the other? If I were using sign language, could someone follow me in a mirror or would they get confused?

Peters: The right hand is more active. In formal sign language, there is a relation between the left hemisphere and signing. The literature reports that aphasia in signing is common after left but not after right hemisphere damage. In terms of hand gestures during speech, it is the right hand that more frequently accompanies the speech with supportive gestures in right-handers. In left-handers, the situation is less clear, as usual (Kimura 1976).

As to whether a person could read sign language if only a mirror image were visible, I imagine there would be an initial problem. From the perspective of the viewer, your hands would have undergone a left–right reversal. Because the two hands often perform complementary activities in signing, and each hand is assigned a specific role, the viewer would have to attend to the specific movement rather than to the specific hand.

Jefferies: Has anybody looked at the neural structure of the brains of people who have never learnt to speak, such as deaf mutes?

Galaburda: There is amazingly little known about that. I have tried to get brain specimens from Ursula Bellugi who has done much interesting work on sign language. She has followed a huge population of congenitally deaf people who have only ever spoken with sign language, but has been unable to provide me with any brain specimens. There were a couple of reports in the early 1920s of curtailment of the development of the speech areas, but I have no way of ascertaining the validity of those.

References

Kimura D 1976 The neural basis of language qua gesture. In: Whitaker H, Whitaker HA (eds) Studies in neurolinguistics. Academic Press, New York p 145–156

Petitto LA, Marentette PF 1991 Babbling in the manual mode: evidence for the ontogeny of language. Science (Wash DC) 251:1493–1496

Summing-up

Lewis Wolpert

I started with a question: if we were made of D-amino acids would everything we have discussed have been a mirror image? I suppose the answer is still 'both yes and no'. The important thing is what I think of as 'Chothia's gap'. We've come a long way: from a 1 in 10^{17} advantage in molecular stability right up to the difference between tapping with fingers of the right and left hands. Certainly, there are gaps on the way, but possibly the most important gap is where one begins to lose chemical information about asymmetry—'Chothia's gap'. This gap, however, may merely reflect our ignorance.

I am nevertheless impressed, not by the degree of agreement between us, which certainly isn't the case, but with the sense that we are dealing with a unitary phenomenon. Whatever is underlying all this leftness and rightness, whether it's in the brain or in *Xenopus* or in *Caenorhabditis*, it is going to turn out to be universal. That's a belief. I believe the phenomenology is the same, that it is under genetic control and that when we understand it, it will all relate to the same basic mechanism. I am afraid the noble efforts of the geneticists in identifying the genes are the only way we are ever going to find out. Then we have to study their effects all the way through development if we are ever going to understand how it works.

Handedness is part of our ancient past. Our handedness was determined, probably by chance, when Dr Jefferies' primitive ancestor developed asymmetry and fell over onto one side. We still carry within us the consequences of that event, which occurred hundreds of millions of years ago.

Thank you all. It's been a lovely meeting.

Index of contributors

Non-participating author and co-authors are indicated by asterisks. Entries in bold type indicate papers; other entries refer to discussion contributions

Indexes compiled by John Rivers

317

Subject index